MODELING ANOMALOUS DIFFUSION

From Statistics to Mathematics

MODELING ANOMALOUS DIFFUSION

From Statistics to Mathematics

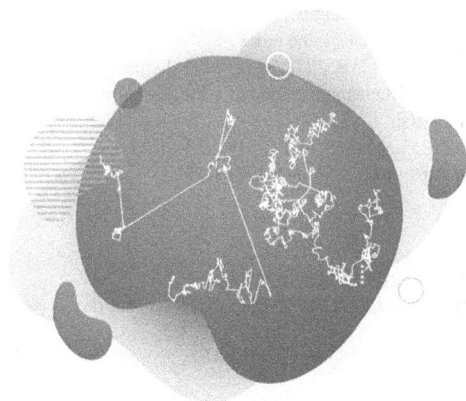

Weihua Deng
Ru Hou
Wanli Wang
Pengbo Xu

Lanzhou University, China

World Scientific

NEW JERSEY · LONDON · SINGAPORE · BEIJING · SHANGHAI · HONG KONG · TAIPEI · CHENNAI · TOKYO

Published by

World Scientific Publishing Co. Pte. Ltd.

5 Toh Tuck Link, Singapore 596224

USA office: 27 Warren Street, Suite 401-402, Hackensack, NJ 07601

UK office: 57 Shelton Street, Covent Garden, London WC2H 9HE

Library of Congress Cataloging-in-Publication Data

Names: Deng, Weihua, author. | Hou, Ru, 1990– author. |
   Wang, Wanli, 1987– author. | Xu, Pengbo, author.
Title: Modeling anomalous diffusion : from statistics to mathematics /
   Weihua Deng, Ru Hou, Wanli Wang, Pengbo Xu.
Description: New Jersey : World Scientific, [2020] |
   Includes bibliographical references and index.
Identifiers: LCCN 2019046225 | ISBN 9789811212994 (hardcover)
Subjects: LCSH: Diffusion--Mathematical models. | Stochastic processes. |
Differential equations, Partial.
Classification: LCC QC185 .D47 2020 | DDC 519.2/33--dc23
LC record available at https://lccn.loc.gov/2019046225

British Library Cataloguing-in-Publication Data

A catalogue record for this book is available from the British Library.

For any available supplementary material, please visit
https://www.worldscientific.com/worldscibooks/10.1142/11630#t=suppl

Contents

This book introduces the diffusion from a physical and atomistic way, by considering the random walk of the diffusing particles. In other words, diffusion is firstly analyzed from its microscopic models—stochastic processes, including continuous time random walk (CTRW), CTRW with multiple internal states, Lévy process, subordinated Lévy process, generalized Langevin equation, subordinated Langevin equation, fractional Brownian motion, etc. Anomalous and nonergodic diffusions are typical multiscale phenomena, which are embodied by the broad distribution of jump lengths with divergent first or second moment and/or long range correlation of the process in time or waiting times with divergent first moment. Most of the time, a specific anomalous and nonergodic diffusion process can be modeled by several different microscopic models, e.g., the CTRW model and subordinated Langevin equation. But there are still a number of instances, in which a particular type of microscopic model has obvious advantages or has to be used. For example, the external potential can be easily described by a generalized Langevin equation, while it is not easy to do in the CTRW framework. This book also discusses the anomalous diffusion processes with reactions.

Once the microscopic models of the anomalous and nonergodic diffusions are established, it is natural to analyze the models for uncovering the potential mechanism, explaining the observed natural phenomena, and extending their applications. One of the basic strategies is to get the probability density functions (PDF) of interesting/valuable statistical observables, such as, positions of the particles, functional of the particles' trajectories, first passage time, escape probability, etc. This book derives the partial differential equations with integral-differential operators governing the PDFs of the various statistical observables, including the Fokker-Planck equations, Feynman-Kac equations, aging Fokker-Planck and Feynman-Kac equations, Fokker-Planck and Feynman-Kac equations with multiple internal states, reaction diffusion equations and the corresponding Feynman-Kac equations. Besides deriving these equations, this book also gives their applications; in particular, the PDFs of the rare events for some statistical observables are analyzed.

Finally, we would like to thank our group members for their invaluable scientific contributions to the contents of this book, and their help to the improvement of the presentation.

Preface

In 1827, while examining grains of pollen of the plant Clarkia pulchella suspended in water under a microscope, Botanist Robert Brown observed minute particles, now known to be amyloplasts (starch organelles) and spherosomes (lipid organelles), ejected from the pollen grains, executing a continuous jittery motion. This phenomenon is now known as Brownian motion. In 1905, Albert Einstein made an investigation on the theory of the Brownian motion, relating macroscopic kinetic parameters such as the diffusion constant and friction coefficient to the correlation functions characterizing fluctuations of microscopic variables—known as a fluctuation-dissipation relation, and providing a derivation of the diffusion equation starting from the microscopic irregular motion of a particle. For an extremely long period of time, people thought almost all the diffusion phenomena are Brownian motion, and even it is called normal diffusion.

In recent decades, anomalous and nonergodic diffusions are topical issues in almost all disciplines. In 2004, the phrase 'anomalous is normal' was used in a title of a PRL paper, which reveals that the diffusion of classical particles on a solid surface has rich anomalous behavior controlled by the friction coefficient, meaning that anomalous diffusion phenomena are ubiquitous in the natural world. From the phenomenological point of view, diffusion is the movement of a substance from a region of high concentration to a region of low concentration without bulk motion. In some sense, the decided difference between normal and anomalous diffusions is their speed. Roughly speaking, the square of the diffusion radius of normal diffusion is a linear function of the time t, while it is not for anomalous diffusion; for example, if it is a power function of t with the power bigger than one, the phenomenon is called superdiffusion, and similarly termed as subdiffusion if the power is less than one.

Chapter 1

Stochastic Models

This chapter first briefly presents the basic knowledge in statistics and probability, then introduces the stochastic models for the anomalous and nonergodic diffusions, which have the multiscale features.

1.1 Background Knowledge in Statistics and Probability

As is well known, the basic concept in statistics and probability is the random variable. Thus, the mathematical definition of a random variable is presented in the first place. Then the distributions and moments are also described.

1.1.1 *Random Variables and Distributions*

To start with this section, the definition of the random variable is presented as follows. A random variable $X \colon \Omega \to P$ is a measurable function from a set of possible outcomes Ω to a measurable space P. Specially, let $(\Omega, \mathcal{A}, P)$ be a probability space. A single real-valued function $X = X(\omega)$ defined on Ω is called a random variable if for any real x the set $\{\omega \colon X(\omega) < x\}$ belongs to the class $\mathcal{A}$, which is called σ-algebra or event field defined as:

(1) $\Omega \in \mathcal{A}$;
(2) If $A \in \mathcal{A}$, then the complement of A also belongs to $\mathcal{A}$;
(3) If $A_n \in \mathcal{A}$, then $\cup_{n=1}^{\infty} A_n \in \mathcal{A}$.

The measure P_X defined on $\mathcal{B}_1$ (all Borel subsets of $\mathbb{R}^1$) by the equation $P_X(B) = P\{X(\omega) \in B\}, B \in \mathcal{B}_1$, is called the probability distribution of X. This measure is uniquely determined by the distribution function of X: $F_X(x) = P_X\{-\infty, x\} = P\{\omega : X(\omega) < x\}$.

The basic properties of distribution function include

(1) (Monotonicity) If $a < b$, then $F_X(a) \leq F_X(b)$;
(2) $\lim\limits_{x \to -\infty} F_X(x) = 0$, $\lim\limits_{x \to +\infty} F_X(x) = 1$;
(3) (Left continuity) $F_X(x - 0) = F_X(x)$.

Generally, there are several important distributions of discrete random variables, for example,

(1) Binomial distribution

$$b(k; n, p) = \binom{n}{k} p^k (1 - p)^{n-k}, k = 0, 1, 2, \cdots, n, \tag{1.1}$$

which indicates the discrete probability distribution of p successes in a sequence of n independent experiments. Each experiment has two outcomes, namely success with the probability p and failure with the probability $1 - p$.

(2) Poisson distribution

$$p(k; \lambda) = \frac{\lambda^k}{k!} e^{-\lambda}, k = 0, 1, 2, \cdots, \tag{1.2}$$

which can be used to approximate binomial distribution and describe the number of events occurring in a specified interval of time or space.

(3) Geometric distribution

$$g(k, p) = (1 - p)^{k-1} p, k = 1, 2, \cdots, \tag{1.3}$$

which is the distribution of first success in Bernoulli trails or the number of failures before the first success. Like its continuous analogue (the exponential distribution), the geometric distribution is memoryless.

As for the continuous random variable X, the probability density function (PDF) is defined as

$$p(x) = F'_X(x). \tag{1.4}$$

Thus, we have

$$P\{a \leqslant X < b\} = \int_a^b p(x) dx. \tag{1.5}$$

Typical distributions of continuous random variables include

(1) Uniform distribution in the interval $[a, b]$

$$p(x) = \frac{1}{b - a} \tag{1.6}$$

for $a \leqslant x \leqslant b$.

(2) Normal distribution $N(\mu, \sigma^2)$

$$p(x) = \frac{1}{\sqrt{2\pi}\sigma} e^{-\frac{(x-\mu)^2}{2\sigma^2}} \tag{1.7}$$

for $-\infty < x < +\infty$, where $\sigma > 0$, μ and σ are constants. Especially, when $\mu = 0$ and $\sigma = 1$, we have the standard normal distribution, denoted by $N(0,1)$. If a random variable X obeys $N(\mu, \sigma^2)$, then $\frac{X-\mu}{\sigma}$ obeys $N(0,1)$.

(3) Exponential distribution

$$p(x) = \lambda e^{-\lambda x} \tag{1.8}$$

for $x \geqslant 0$ and the parameter λ is positive. Similar to its discrete analogue (the geometric distribution), the exponential distribution is also memoryless.

(4) Gamma distribution

$$p(x) = \frac{\lambda^r}{\Gamma(r)} x^{r-1} e^{-\lambda x} \tag{1.9}$$

for $x > 0$. Both the shape parameter r and the scale parameter λ are positive real numbers. One special case of the Gamma distribution is the Erlang distribution, which is defined as

$$p(x) = \begin{cases} \frac{\lambda^n}{(n-1)!} x^{n-1} e^{-\lambda x}, & \text{if } x > 0, \\ 0, & \text{if } x \leq 0 \end{cases}$$

with n being a natural number.

If the random variables $X_1(\omega), X_2(\omega), \cdots, X_n(\omega)$ are defined on the same probability space $(\Omega, \mathcal{A}, P)$, then $\vec{X}(\omega) = (X_1(\omega), X_2(\omega), \cdots, X_n(\omega))$ is n-dimensional random vector. The function $F(x_1, x_2, \cdots, x_n) = P\{X_1(\omega) < x_1, X_2(\omega) < x_2, \cdots, X_n(\omega) < x_n\}$ is the joint distribution function of $\vec{X}(\omega)$. We say the random variables $X_1, \cdots, X_n$ are independent if $P\{X_1 < x_1, \cdots, X_n < x_n\} = P\{X_1 < x_1\} \cdots P\{X_n < x_n\}$ for arbitrary $x_1, \cdots, x_n$.

For the continuous case, there exists some function satisfying

$$F(x_1, x_2, \cdots, x_n) = \int_{-\infty}^{x_1} \cdots \int_{-\infty}^{x_n} p(y_1, \ldots, y_n) dy_1 \cdots dy_n \tag{1.10}$$

and $p(x_1, x_2, \cdots, x_n)$ is PDF of a random vector with $p(x_1, x_2, \ldots, x_n) \geqslant 0$ and $\int_{-\infty}^{\infty} \cdots \int_{-\infty}^{\infty} p(x_1, \cdots, x_n) dx_1 \cdots dx_n = 1$.

Typical multivariate probability distributions are listed as follows:

(1) Multinomial distribution

$$P\{X_1 = k_1, X_2 = k_2, \cdots, X_r = k_r\} = \frac{n!}{k_1! k_2! \cdots k_r!} p_1^{k_1} p_2^{k_2} \cdots p_r^{k_r}$$

$$(1.11)$$

with $k_1 + k_2 + \cdots + k_r = n$, which indicates the numbers of different possible outcomes in n independent experiments. For each experiment, the possible outcomes are $A_1, A_2, \cdots, A_r$ with the possibilities $p_1, p_2, \cdots, p_r$, respectively, and $p_1 + p_2 + \cdots p_r = 1$.

(2) Multivariate normal distribution

$$p(x_1, \cdots, x_n)$$

$$= \frac{1}{(2\pi)^{n/2}(\det \Sigma)^{1/2}} \exp\left[-\frac{1}{2} \sum_{j,k=1}^{n} \gamma_{jk}(x_j - \mu_j)(x_k - \mu_k)\right], \quad (1.12)$$

where $\Sigma = (\sigma_{ij})$ is an $n \times n$ positive definite symmetric matrix and its inverse matrix is denoted as $\Sigma^{-1} = (\gamma_{ij})$; $\det \Sigma$ is its determinant and $(\mu_1, \cdots, \mu_n)$ is an arbitrary real-valued row vector. As defined later, $\Sigma = (\sigma_{ij})$ is the covariance matrix and $(\mu_1, \cdots, \mu_n)$ is the vector of expectations.

Suppose that (X, Y) is a two-dimensional random vector with the joint probability distribution $F(x, y)$. Then we can calculate the marginal distribution functions via

$$F_1(x) = P\{X < x\} = F(x, +\infty), \quad (1.13)$$

and

$$F_2(y) = P\{Y < y\} = F(+\infty, y). \quad (1.14)$$

Especially, for the continuous random vector with joint density function $p(x, y)$, the marginal density functions are

$$p_1(x) = \int_{-\infty}^{+\infty} p(x, y) dy, \quad (1.15)$$

and

$$p_2(y) = \int_{-\infty}^{+\infty} p(x, y) dx. \quad (1.16)$$

1.1.2 Moments

Moments are important statistical quantities for random variables. Let X be a continuous random variable with the density function $p(x)$. If the integral $\int_{-\infty}^{+\infty} xp(x)dx$ is absolutely convergent, it is called the expectation of X, usually denoted as EX. The basic property of expectation is linearity, namely

$$E\left(\sum_{i=1}^{n} c_i X_i + b\right) = \sum_{i=1}^{n} c_i EX_i + b \tag{1.17}$$

for arbitrary constants $c_i, i = 1, \cdots, n$ and b.

If $E(X - EX)^2$ exists, then it is called the variance of the random variable X, usually denoted as DX, while $\sqrt{DX}$ is the standard deviation. Variance describes the dispersion to expectation and it has the well-known formula

$$DX = EX^2 - (EX)^2. \tag{1.18}$$

When c is a constant, we have $D(X + c) = D(X)$ and $D(cX) = c^2 DX$.

For the random vector $\vec{X}(\omega) = (X_1(\omega), X_2(\omega), \cdots, X_n(\omega))$, define its variance as $(DX_1, DX_2, \cdots, DX_n)$. The covariance of X_i and X_j is defined as

$$\text{cov}(X_i, X_j) = E[(X_i - EX_i)(X_j - EX_j)] \tag{1.19}$$

with $i, j = 1, 2, \cdots, n$. Obviously, according to the definition of the covariance and variance,

$$\text{cov}(X_i, X_j) = EX_i X_j - EX_i \cdot EX_j \tag{1.20}$$

and

$$D(X_i + X_j) = D(X_i) + D(X_j) + 2\text{cov}(X_i, X_j). \tag{1.21}$$

The correlation coefficient of X_i and X_j is defined as

$$\rho_{ij} = \frac{\text{cov}(X_i, X_j)}{\sqrt{DX_i}\sqrt{DX_j}}. \tag{1.22}$$

Particularly, when $\rho_{ij} = 0$, X_i and X_j are uncorrelated.

1.2   Algorithm for the Generation of Random Variables

Generally speaking, to generate random variables one usually needs to first generate a random variable with uniform distribution in $[0, 1]$. Assume a random variable X obeys the distribution function $F(x)$. Define

$$F^{-1}(y) = \inf\{x : F(x) > y\} \tag{1.23}$$

for arbitrary $0 \leqslant y \leqslant 1$ as the inverse function of $F(x)$. Thus, the random variable $Y = F(X)$ obeys the uniform distribution in the interval $[0, 1]$, since $P\{Y < x\} = P\{X < F^{-1}(x)\} = x$.

Inversely, if Y obeys the uniform distribution in the interval $[0, 1]$, $X = F^{-1}(Y)$ has the distribution function $F(x)$, because $P\{X < x\} = P\{Y < F(x)\} = F(x)$. Consequently, in practice, we can first generate random numbers following the uniform distribution in the interval $[0, 1]$ and random numbers of other distributions can be further generated according to the above relation.

There are some other ways to generate the random numbers of normal distribution. For example, first we can generate two independent random numbers U_1 and U_2 of the uniform distribution in the interval $[0, 1]$. Let

$$X = (-2 \ln U_1)^{1/2} \cos(2\pi U_2), \tag{1.24}$$

and

$$Y = (-2 \ln U_1)^{1/2} \sin(2\pi U_2). \tag{1.25}$$

It can be proved that X and Y are independent random numbers obeying the standard normal distribution.

1.3   Continuous Time Random Walk and Lévy Process

In mathematics, continuous time random walk (CTRW) is a stochastic process with arbitrary distributions of jump lengths and waiting times, originally discussed by E. Montroll and G. H. Weiss [Montroll and Weiss (1965); Kenkre *et al.* (1973); Gorenflo *et al.* (2007)] and first applied to physical systems by Scher and Lax [Scher and Lax (1972)], where the random variables of waiting times and displacements are independent and identically distributed (IID), respectively. For the past several decades, CTRW model is used to describe different kinds of anomalous systems, ranging from amorphous semiconductors to DNA molecules.

1.3.1 *Continuous Time Random Walk*

The CTRW model is a stochastic process, characterizing the motion of the particle. It can be described as follows. Events occur at the random epochs of time t_1, t_2, ..., t_N, ..., from $t = 0$. A walker is trapped at the initial position for the time $\tau_1 = t_1$, then makes a jump and the displacement is $\mathbf{x_1}$; the walker is further trapped at $\mathbf{x_1}$ for time $\tau_2 = t_2 - t_1$, and then jumps to a new position; this process is then renewed. These two kinds of random variables are characterized by a set of waiting times $\{\tau_1, \tau_2, \cdots, \tau_N, \cdots\}$ and the displacements $\{\mathbf{x_1}, \mathbf{x_2}, \cdots, \mathbf{x_N}, \cdots\}$, respectively. We assume that all $\mathbf{x_i}$ are IID with respect to (w.r.t.) a common PDF $w(\mathbf{x})$ and all τ_i are also IID random variables with a common PDF $\phi(\tau)$.

Motivated by previous studies of complex systems, here we consider the waiting time PDFs with power law tails, i.e., for large τ

$$\phi(\tau) \sim \frac{1}{\tau^{1+\alpha}}, \alpha > 0.$$

In this case, the first moment of waiting time is divergent for $0 < \alpha < 1$.

An example is the heavy-tailed PDF

$$\phi(\tau) = \begin{cases} 0, & \tau < \tau_0; \\ \alpha \frac{\tau_0^\alpha}{\tau^{1+\alpha}}, & \tau > \tau_0. \end{cases} \tag{1.26}$$

Here τ_0 is a time scale. Using the Tauberian theorem [Feller (1971)], in Laplace space

$$\widehat{\phi}(s) \sim 1 - b_\alpha s^\alpha \tag{1.27}$$

for small s, where s is conjugate to τ, $b_\alpha = \tau_0^\alpha |\Gamma(1 - \alpha)|$, $0 < \alpha < 1$. In order to simplify the expression, we denote $\widehat{\phi}(s)$ as the Laplace transform of $\phi(\tau)$. When $1 < \alpha < 2$, the first moment $\langle \tau \rangle = \int_0^\infty \tau \phi(\tau) d\tau$ is finite and the corresponding Laplace form is

$$\widehat{\phi}(s) \sim 1 - \langle \tau \rangle s + b_\alpha s^\alpha \tag{1.28}$$

for small s. Notice that $\widehat{\phi}(0) = 1$, which means that the PDF is normalized. We would like to further introduce the one sided Lévy distribution $\phi(\tau) = \ell_\alpha(\tau)$ with index α. In Laplace space, one sided stable Lévy distribution $\phi(\tau)$ is

$$\int_0^\infty \exp(-s\tau)\phi(\tau)d\tau = \exp(-s^\alpha) \tag{1.29}$$

and the small s expansion is given by $\widehat{\phi}(s) \sim 1 - s^\alpha$ and here $0 < \alpha < 1$. For specific choices of α, the closed form of the $\ell_\alpha(\tau)$ is tabulated for example in Mathematica [Burov and Barkai (2012)]. In particular, a useful special case is for $\alpha = 1/2$

$$\ell_{1/2}(\tau) = \frac{1}{2\sqrt{\pi}}\tau^{-\frac{3}{2}}\exp\left(-\frac{1}{4\tau}\right). \tag{1.30}$$

It implies that for large τ, $\ell_{1/2}(\tau) \sim 1/\sqrt{4\pi}\tau^{-3/2}$. So the first moment of the sojourn time diverges.

Now we would like to discuss another heavy-tailed density,

$$w(x) \sim \frac{A}{|x|^{1+\beta}} \tag{1.31}$$

with $0 < \beta < 2$. It has the characteristic function

$$\widetilde{w}(k) \sim 1 - A_\beta |k|^\beta, \tag{1.32}$$

where $A_\beta = A\pi/(\Gamma(1 + \beta)\sin(\pi\beta/2))$. If $\beta > 2$ in Eq. (1.31), we can see that the second moment of random variable is finite. For this case, $\widetilde{w}(k) \sim 1 - A_2 k^2$ with A_2 being a constant.

Let us consider a process starting at $t = 0$. It's natural to ask how often the jumping happens, which leads to the definition of $p_N(t)$ as the probability of taking N steps up to time t,

$$p_N(t) = \int_0^t Q_N(\tau)\Phi(t - \tau)d\tau, \tag{1.33}$$

where $Q_N(t_N)$ is the probability density of the occurrence of the N-th step at time $t_N = \tau_1 + \tau_2 + \cdots \tau_N$, defined by

$$Q_N(t_N) = \int_0^{t_N} \phi(\tau)Q_{N-1}(t_N - \tau)d\tau \tag{1.34}$$

and $\Phi(t)$ is the survival probability

$$\Phi(t) = \int_t^\infty \phi(y)dy, \tag{1.35}$$

this is, the waiting time exceeds the observation time t. Taking Laplace transform of Eq. (1.34), and utilizing the convolution property of Laplace transform lead to

$$\widehat{Q}_N(s) = \widehat{Q}_{N-1}(s)\widehat{\phi}(s). \tag{1.36}$$

Noticing $Q_1(\tau) = \phi(\tau)$, Eq. (1.36) becomes

$$\widehat{Q}_N(s) = \widehat{\phi}^N(s). \tag{1.37}$$

Thus we have

$$\widehat{p}_N(s) = \widehat{\phi}^N(s)\frac{1 - \widehat{\phi}(s)}{s}. \tag{1.38}$$

The distribution $p_N(t)$ is normalized since $\sum_{N=0}^{\infty} \widehat{p}_N(s) = 1/s$. If considering N in Eq. (1.38) as a continuous variable, the Laplace transform of Eq. (1.38) has the form:

$$\widehat{p}_u(s) = \frac{1 - \widehat{\phi}(s)}{s} \int_0^{\infty} \exp(-uN) \exp(N \log(\widehat{\phi}(s))) dN. \qquad (1.39)$$

The detailed discussions of the inversion of Eq. (1.38) will be given in the Chap. 7.

From Eq. (1.38) we can easily obtain the moments of N in Laplace space. For instance

$$\langle \widehat{N}(s) \rangle = \sum_{N=0}^{\infty} N \widehat{p}_N(s) = \frac{\widehat{\phi}(s)}{s(1 - \widehat{\phi}(s))} \qquad (1.40)$$

and

$$\langle \widehat{N^2}(s) \rangle = \sum_{N=0}^{\infty} N^2 \widehat{p}_N(s) = \frac{\widehat{\phi}(s)(1 + \widehat{\phi}(s))}{s(1 - \widehat{\phi}(s))^2}. \qquad (1.41)$$

In fact, what we are interested in is the long time behaviors of $\langle N(t) \rangle$ and $\langle N^2(t) \rangle$. In this case, substituting an asymptotic form of $\widehat{\phi}(s)$ into Eqs. (1.40) and (1.41), and taking inverse Laplace transform, then we can obtain the corresponding moments in real space. When the first moment of waiting time is finite, we have $\langle N(t) \rangle \sim t/\langle \tau \rangle$ for large t.

1.3.2 *Propagator Function*

In CTRW model, each step is characterized by a waiting time τ_i, and a displacement $\triangle \mathbf{x_i} = (x_{i,1}, x_{i,2}, \ldots, x_{i,d})$ where $(x_{i,1}, x_{i,2}, \ldots, x_{i,d})$ is the corresponding step in a d dimensional space. Now we turn to find the PDF of $\mathbf{x}(t)$, denoted as $p(\mathbf{x}, t)$. This is equal to the summation over all N of the probability that the particle ends up at $\mathbf{x}$ in N steps, given that it has finished N steps over time t:

$$p(\mathbf{x}, t) = \sum_{N=0}^{\infty} p_N(t) \chi_N(\mathbf{x}), \qquad (1.42)$$

where the waiting time and step size are independent. Taking the Fourier-Laplace transform and using convolution property of Fourier transform yield

$$\widetilde{p}(\mathbf{k}, s) = \sum_{N=0}^{\infty} \widehat{p}_N(s) \widetilde{\chi}_N(\mathbf{k})$$

$$= \sum_{N=0}^{\infty} \frac{1 - \widehat{\phi}(s)}{s} \widehat{\phi}^N(s) \widetilde{w}^N(\mathbf{k}) \qquad (1.43)$$

$$= \frac{1 - \widehat{\phi}(s)}{s} \frac{1}{1 - \widehat{\phi}(s)\widetilde{w}(\mathbf{k})},$$

where we use the relation $\chi_N(\mathbf{x}) = \int \chi_{N-1}(\mathbf{y}) w(\mathbf{x} - \mathbf{y}) d\mathbf{y}$. Equation (1.43) is called the Montroll-Weiss equation [Montroll and Weiss (1965)]. By taking the inverse Laplace and Fourier transforms of Eq. (1.43), we can get the governing equation of $p(x, t)$ or directly obtain the exact/asymptotic expression of $p(x, t)$.

For the convenience of presentation, we focus on the case of $d = 1$. Inserting Eq. (1.43) into the equation below leads to the n-th moment of $x(t)$ in Laplace space, i.e.,

$$\langle \widehat{x}^n(s) \rangle = (-i)^n \frac{d^n}{dk^n} \widetilde{\widehat{p}}(k, s) \Big|_{k=0}. \qquad (1.44)$$

For example, let us consider the mean squared displacement (MSD) of CTRW model. When $\phi(\tau)$ is power law with finite first moment and $w(x)$ is a symmetric Gaussian distribution, there is

$$\langle \widehat{x}^2(s) \rangle = \frac{\widehat{\phi}(s)}{s(1 - \widehat{\phi}(s))} \int_{-\infty}^{\infty} x^2 w(x) dx. \qquad (1.45)$$

Using $\widehat{\phi}(s) \sim 1 - s\langle \tau \rangle$ and taking inverse Laplace transform of the above equation lead to the MSD

$$\langle x^2(t) \rangle \sim \frac{\int_{-\infty}^{\infty} x^2 w(x) dx}{\langle \tau \rangle} t. \qquad (1.46)$$

Equation (1.46) implies that $\langle x^2(t) \rangle$ grows linearly with the observation time t if the second moment of x and the first moment of τ exist. Note that the prefactor of the time in Eq. (1.46) is connected with the diffusion coefficient denoted by $(\int_{-\infty}^{\infty} x^2 w(x) dx)/(2\langle \tau \rangle)$.

While, if $\phi(\tau)$ is broad distribution of waiting time with index $0 < \alpha < 1$, then we have

$$\langle x^2(t) \rangle \sim K_\alpha t^\alpha \qquad (1.47)$$

with the generalized diffusion coefficient $K_\alpha = \int_{-\infty}^{\infty} x^2 w(x) dx / (\Gamma(1+\alpha) b_\alpha)$, where b_α is defined in Eq. (1.27). Due to the broad distribution of waiting time, the process shows subdiffusion when $\alpha < 1$, which is slower than Brownian motion. In the particular case $\alpha \to 1$, Eq. (1.47) reduces to Eq. (1.46).

Now we consider some special cases of $\phi(\tau)$ and $w(x)$ to derive the corresponding propagators with the parameters A_β and A_2, respectively given in Eq. (1.32) and the expression immediately below Eq. (1.32), and b_α defined in Eq. (1.27):

(1) The random variables of waiting time and step length have the finite first and the second moments (see Eqs. (1.26) and (1.31)), respectively. Then the PDF in Fourier-Laplace space asymptotically becomes

$$\widetilde{p}(k,s) \sim \frac{1}{s + \frac{A_2}{\langle \tau \rangle}k^2} \tag{1.48}$$

in the $(k,s) \to (0,0)$ diffusion limit. For long time scale, $p(x,t)$ has Gaussian shape

$$p(x,t) \sim \frac{1}{2\sqrt{\frac{A_2}{\langle \tau \rangle}\pi t}} \exp\left(-\frac{x^2}{4\frac{A_2}{\langle \tau \rangle}t}\right). \tag{1.49}$$

(2) The first moment of the waiting time is finite but the second moment of step length is divergent. As discussed above, we have

$$\widetilde{p}(k,s) \sim \frac{1}{s + \frac{A_\beta}{\langle \tau \rangle}|k|^\beta}, \tag{1.50}$$

the inversion of which is

$$p(x,t) \sim \left(\frac{\langle \tau \rangle}{A_\beta}\right)^{1/\beta} l_\beta\left(\left(\frac{\langle \tau \rangle}{A_\beta}\right)^{1/\beta}\frac{|x|}{t^{1/\beta}}\right) \tag{1.51}$$

with $l_\beta(z)$ denoting the symmetric β-stable Lévy distribution. We usually call such process as Lévy flight and it will be further introduced in Sec. 1.4.

(3) The first moment of waiting time and the second moment of step length are divergent and finite, respectively. For $(k,s) \to (0,0)$,

$$\widetilde{p}(k,s) \sim \frac{1}{s\left(1 + \frac{A_2}{b_\alpha}|k|^2 s^{-\alpha}\right)}; \tag{1.52}$$

taking the inverse Fourier-Laplace transform of the above equation yields the closed-form solution

$$p(x,t) \sim \frac{1}{2\sqrt{\pi\frac{A_2}{b_\alpha}t^\alpha}}H_{1,2}^{2,0}\left[\frac{x^2}{4\frac{A_2}{b_\alpha}t^\alpha}\,\middle|\,{}^{(1-\frac{\alpha}{2},\ \alpha)}_{(0,1),(\frac{1}{2},1)}\right], \tag{1.53}$$

where $H_{p,q}^{m,n}(z)$ is the Fox function, defined by

$$
H_{p,q}^{m,n}(z) = H_{p,q}^{m,n} \left[z \left| \begin{matrix} (r_1,R_1),(r_2,R_2),\ldots,(r_p,R_p) \\ (b_1,B_1),(b_2,B_2),\ldots,(b_q,B_q) \end{matrix} \right. \right] \tag{1.54}
$$
$$
= \frac{1}{2\pi i} \int \chi(\tau) z^\tau d\tau
$$

with

$$
\chi(\tau) = \frac{\prod_{j=1}^m \Gamma(b_j - B_j\tau) \prod_{j=1}^n \Gamma(1 - r_j + R_j\tau)}{\prod_{j=m+1}^q \Gamma(1 - b_j + B_j\tau) \prod_{j=n+1}^p \Gamma(r_j - R_j\tau)}.
$$

1.3.3 *Lévy Process*

A continuous time process $X_t = X(t)$ with values in $\mathbb{R}^d$ is named a Lévy process if

(1) $X(0) = 0$;
(2) It has independent increments, i.e., for all $0 = t_0 < t_1 < \cdots < t_k$, the increments $X(t_i) - X(t_{i-1})$ are independent;
(3) It has stationary increments, meaning for all $0 \le s \le t$ the random variables $X(t) - X(s)$ and $X(t-s) - X(0)$ have the same distribution;
(4) It is stochastically continuous, i.e., for every $t \ge 0$ and $\epsilon > 0$, there exists

$$
\lim_{s \to t} P\{|X_s - X_t| > \epsilon\} = 0.
$$

The default initial condition is $X(0) = 0$. Furthermore, note that every process with stationary and independent increments has a version with paths, which are right-continuous and have the left limits. The most common example of a Lévy process is Brownian motion, where $X(t) - X(s)$ is normally distributed with zero mean and variance $t - s$. Other examples are Poisson process, Cauchy process, compound Poisson process, gamma process, and the variance gamma process, etc.

The Lévy-Khinchin formula [Applebaum (2009)] says that every Lévy process has a specific form for its characteristic function, i.e., for all $t \ge 0$, $\mathbf{v} \in \mathbb{R}^n$,

$$
E\left[\exp(i(\mathbf{v}, X(t))) \right] = \exp(t\eta(\mathbf{v})), \tag{1.55}
$$

where

$$\eta(\mathbf{v}) = i(\mathbf{a}, \mathbf{v}) - \frac{1}{2}(\mathbf{v}, \mathbf{bv})$$

$$+ \int_{\mathbb{R}^n \setminus \{0\}} \left[e^{i(\mathbf{v}, \mathbf{y})} - 1 - i(\mathbf{v}, \mathbf{y}) \chi_{\{|\mathbf{y}| < 1\}} \right] \nu(d\mathbf{y}),$$

and χ_I is the indicator function of the set I, $\mathbf{a} \in \mathbb{R}^n$, $\mathbf{b}$ is a positive definite symmetric $n \times n$ matrix and ν is a σ-finite Lévy measure on $\mathbb{R}^n \setminus \{0\}$, i.e.,

$$\int_{\mathbb{R}^n \setminus \{0\}} (\mathbf{y}^T \mathbf{y} \wedge 1) \nu(d\mathbf{y}) < \infty. \tag{1.56}$$

For example, we discuss the one-dimensional purely discontinuous real-valued Lévy process with $\mathbf{a}, \mathbf{b} = 0$, and $d\nu(y) = dy/(\pi y^2)$ satisfying Eq. (1.56). Using the Lévy-Khintchine formula, the corresponding characteristic function is

$$\eta(v) = -|v|, \tag{1.57}$$

which is the Lévy symbol of Cauchy process.

1.4 Lévy Flight, Lévy Walk, and Subordinated Processes

In Sec. 1.3, the CTRW model is introduced and it is also used to depict subdiffusion. Further under the CTRW framework, this section discusses the Lévy flight and Lévy walk, and the Lévy flight is also characterized by the Brownian motion coupled with subordinator. Both Lévy flight and Lévy walk reflect superdiffusion. For Lévy flight, the CTRW model is with finite first moment of waiting time and divergent second moment of jump length, while the jump length of Lévy walk is a function of running time, being a random variable generally with a power law distribution.

1.4.1 *Lévy Flight*

Lévy flight is a stochastic process characterized by the occurrence of the extremely long jumps, leading to the discontinuous trajectories. Lévy statistics provides a new framework for the description of many natural phenomena, such as wanders of albatrosses, spider monkeys, marine predators, bees, etc. As we have seen, the symmetric random walks with the power law step length distribution $w(x) \sim |x|^{-\beta-1}$ with $0 < \beta < 2$ possessing the characteristic function $\widetilde{w}(k) = \exp(-a|k|^\beta)$, where a is a constant. Because of $0 < \beta < 2$ the second moment of jump length is divergent, which is different from the ordinary Brownian motion, for which all the moments of displacements are finite.

Mathematically, it is well known that the boundary conditions of the classical diffusion equation just rely on the given information of the solution along the boundary of a domain; however, for the Lévy flights in a bounded domain, the boundary conditions of the corresponding macroscopic governing equation involve the information of a solution in the complementary set of Ω, i.e., $\mathbb{R}^n \backslash \Omega$, with the potential reason that paths of the corresponding stochastic process are discontinuous [Deng *et al.* (2018)].

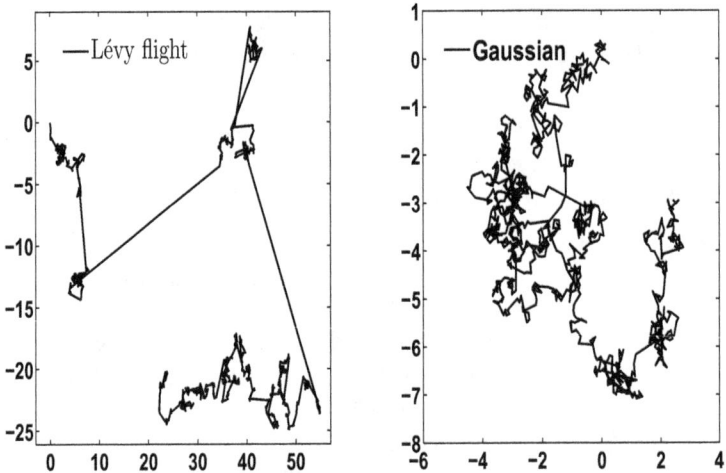

Fig. 1.1 Two examples of 1000 steps of a Lévy flight and Brownian motion in two dimensions, the former with index $\beta = 1.5$. Here the particles start at the position $(0,0)$ and the angle direction is uniformly distributed. Compared with the classical Brownian motion (right), the Lévy flight trajectory (left) is characterized by the big jump.

The position of a random walk can be defined by the sum of N IID displacement Δx_i with a common PDF $w(x)$, i.e.

$$x_N = \sum_{i=1}^{N} \Delta x_i. \tag{1.58}$$

Here we consider the particle moving symmetrically in one dimension. Using the example of a Gaussian PDF $w(x)$ and the central limit theorem, then the PDF of scaled position $y_N = x_N/\sqrt{N}$ is

$$\lim_{N \to \infty} p_N(y) = p(y) = \frac{1}{\sqrt{2\pi\delta^2}} \exp\left(-\frac{y^2}{2\delta^2}\right), \qquad (1.59)$$

where δ^2 is the variance of the single step $\triangle x_N$. It means that $x_N \sim N^{0.5}$ for the normal diffusion. Utilizing Eq. (1.59) and $y_N = x_N/\sqrt{N}$, yields the PDF of $p_N(x)$ for large N, i.e.

$$p_N(x) \sim \frac{1}{\sqrt{2\pi N\delta^2}} \exp\left(-\frac{x^2}{2N\delta^2}\right), \qquad (1.60)$$

which depends on N. We can see that on large scales the trajectories of the particle with finite second moment of jump length resemble ordinary Brownian motion.

Now we consider the case of Lévy fight, belonging to a class of the random walks for which the central limit theorem does not work. Similar to the discussion of the ordinary random walks, we consider that the second moment of jump length is divergent. Motivated by the earlier studies of complex systems, we choose the power law jump length distribution, i.e., for large x

$$w(x) \sim |x|^{-\beta-1} \qquad (1.61)$$

with $0 < \beta < 2$. Applying the generalized central limit theorem, i.e., the Lévy-Khinchin theorem, the position of a Lévy flight is scaled according to

$$y_N = \frac{x_N}{N^{1/\beta}}, \qquad (1.62)$$

and the scaled variable y_N possesses a PDF having no relation with the number of events N in the limit $N \to \infty$, namely

$$\lim_{N \to \infty} p_N(y) = l_\beta(y). \qquad (1.63)$$

Here the PDF $l_\beta(y)$ is two sided Lévy distribution with the index β, which reads

$$l_\beta(y) = \frac{1}{2\pi} \int_{-\infty}^{\infty} \exp(-iky) \exp\left(-a|k|^\beta\right) dk \qquad (1.64)$$

with a being a constant. Note that the asymptotic behavior of a Lévy distribution follows the power law

$$l_\beta(x) \sim \frac{1}{|x|^{\beta+1}}. \qquad (1.65)$$

From Eq. (1.65), it can be noted that the variance of $l_\beta(x)$ with $0 < \beta < 2$ diverges.

Combining Eqs. (1.62) and (1.63), we get an explicit expression for the PDF of x_N for sufficiently large N

$$p_N(x) \sim \frac{1}{N^{1/\beta}} l_\beta \left(\frac{x}{N^{1/\beta}} \right). \tag{1.66}$$

More importantly, note that the width of the distribution $|x| \sim N^{1/\beta}$. In this case, for $\beta < 2$, we have $N^{1/\beta} > \sqrt{N}$, illustrating a superdiffusive process.

1.4.2   *Lévy Walk*

Lévy distribution introduced in the above section describes the enhanced diffusion on the basis of the broad distributions of single motion events. Note that Lévy distribution $l_\beta(x)$ exhibits diverging moments for $\beta < 2$, which makes the description of anomalous diffusion problematic in some sense in terms of stable laws. One of the most straightforward ways to resolve this inconsistency is to regularize the power law distributions by truncating them at large values [Mantegna and Stanley (1994)]. Even though this makes the moments of the distribution finite, the truncation introduces a certain arbitrariness, which is difficult to be justified for different kinds of experiments. Thus, space-time correlation is introduced, for instance, assuming motion at a constant velocity. Then the Lévy walk model appears.

Now we briefly outline the main ingredients of the Lévy walk process [Zaburdaev *et al.* (2015)], which supposes that the particle moves continuously at a constant velocity and changes directions randomly. The motion takes place at a constant velocity for some time after which the direction and length of the next motion event are chosen randomly but at the same velocity. Each motion of the process is uncorrelated. Consider a sequence of IID random variables $\tau_1, \tau_2, \cdots$, with a common PDF $\phi(\tau)$. The number of renewal events in the time interval between 0 and t is $N(t) = \max[N, t_N \leq t]$. So the displacement of jump is

$$\mathbf{x}_i = v\tau_i \mathbf{V}_i, i = 1, 2, \cdots,$$

where $\mathbf{V}_i$, deciding the direction of i-th jump, is a sequence of IID random units vectors distributed uniformly on the n-dimensional sphere. Here v represents the speed of the particles, which is a constant. Obviously, the length of each jump $\mathbf{x}_i$ is equal to $v \times \tau_i$. Then the position of the walker at time t is given by

$$\mathbf{x}(t) = \sum_{i=1}^{N} v\tau_i \mathbf{V}_i + vB_t \mathbf{V}_{N+1}, \tag{1.67}$$

where the backward recurrence time B_t is equal to $t - t_N = t - \sum_{i=1}^{N} \tau_i$; see Fig. 1.2.

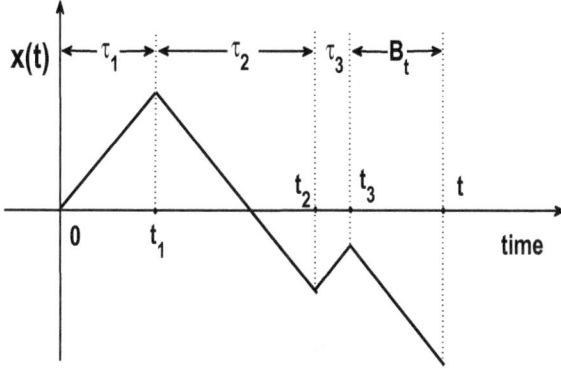

Fig. 1.2   Illustration of a Lévy walk in one dimension. Denote t_i as the time when the i-th event occurs. B_t represents the backward recurrence time. Just as the figure shows, three events have finished before observation time t.

For the convenience of representation, in the following we mainly focus on one-dimensional symmetric process. Equation (1.67) can help us to understand the model in a simple way but it is not easy to study this model analytically, so we turn to another way. To account for the space-time correlation discussed above, the probability to move a distance r in time t in a single motion event and to stop at r for initiating a new motion randomly is

$$\phi(r,t) = \frac{1}{2}\delta(|r| - vt)\phi(t), \tag{1.68}$$

where the time and the length are given in dimensionless units, and in this section we still consider $\phi(t)$ as the power law distribution defined in Eq. (1.26). Equation (1.68) is a normalized PDF.

We further introduce the probability of moving a distance r in time t in a single motion event and not necessarily stopping at r, defined by

$$\Phi(r,t) = \frac{1}{2}\delta(|r| - vt)\Phi(t), \tag{1.69}$$

where $\Phi(t)$ is the survival probability. The next step is to calculate the propagator $p(r, t)$, which gives

$$p(r,t) = \Phi(r,t) + \int_{-\infty}^{\infty} dz \int_0^t \phi(z,\tau)\Phi(r-z,t-\tau)d\tau + \cdots ; \qquad (1.70)$$

the density is a sum of outgoing particles performing different steps to reach r at time t. Taking Fourier-Laplace transform, and using the convolution property of the Fourier-Laplace transform, Eq. (1.70) becomes

$$\widetilde{\widehat{p}}(k, s) = \frac{\widetilde{\widehat{\Phi}}(k, s)}{1 - \widetilde{\widehat{\phi}}(k, s)}, \qquad (1.71)$$

where k is the Fourier pair of x, and s is the Laplace pair of t.

According to Eq. (1.27) and Eq. (1.28), substituting the forms of $\widetilde{\widehat{\phi}}(k, s)$ (for other cases of the Laplace transform, one can refer to [Zaburdaev et al. (2015)]) and $\widetilde{\widehat{\Phi}}(k, s)$ into Eq. (1.71), then taking the second order derivative with respect to k and setting $k = 0$, the dominant term of the MSD after inverse Laplace transform is

$$\langle x^2(t)\rangle \sim \begin{cases} t^2, & 0 < \alpha < 1; \\ t^2/\ln(t), & \alpha = 1; \\ t^{3-\alpha}, & 1 < \alpha < 2; \\ t\ln(t), & \alpha = 2; \\ t, & \alpha > 2. \end{cases} \qquad (1.72)$$

Next we consider deriving the propagator and focus on the case of $1 < \alpha < 2$. Here we suppose that both s and k are small and $|k|^\alpha/s$ is fixed. The propagator is well presented by the Lévy stable distribution

$$p(r,t) \sim t^{-1/\alpha} l_\alpha\left(-c\frac{|r|}{t^{1/\alpha}}\right), \qquad (1.73)$$

where c is a constant. Note that Eq. (1.73) describes the scaling behaviors of $p(r, t)$ at the center region only, which implies the MSD $\langle x^2(t)\rangle$ calculated by using (1.73) is not always correct. When s is order of k, the corresponding scaling describes the ballistic behavior giving the information of the rare fluctuations [Rebenshtok et al. (2014b)]. Besides, for the case of $0 < \alpha < 1$, the analytical solution of $p(r, t)$ can be obtained by the method discussed in [Rebenshtok and Barkai (2008)].

We now extend this derivation to the case where the velocity of the particle is not fixed but is a random variable itself. We suppose that the PDF of velocity is $h(v)$. The corresponding transport equation of the random walk is

$$\nu(x,t) = \int_{-\infty}^{\infty} \int_0^t \nu(x - v\tau, t - \tau)h(v)\phi(\tau)d\tau dv + \delta(t)p(x, t = 0). \quad (1.74)$$

It shows that a particle changes its velocity at the point (x,t), under the condition that it has already changed its velocity to the value v at time $t-\tau$ and position $x - v\tau$. The first term on the right-hand side of Eq. (1.74) integrates over all these events, taking into account that $h(v)\phi(\tau)$ is the probability for a certain velocity v and a flight time τ to occur. Besides, the last term of Eq. (1.74) assumes that there is an initial distribution of particles $p(x, t = 0)$, and that the particles immediately change their velocities at $t = 0$, thus starting the whole evolution.

Based on the above equation, we get the density of particles $p(x,t)$

$$p(x,t) = \int_{-\infty}^{\infty} \int_0^t \nu(x - v\tau, t - \tau)\Phi(\tau)h(v)d\tau dv, \quad (1.75)$$

where $\Phi(\tau)$ is the survival probability, that is, the probability for a particle not to renew until time τ. Fourier-Laplace transforming Eq. (1.75) with respect to x and t, respectively, yields

$$\widetilde{p}(k, s) = \frac{\int_{-\infty}^{\infty} \widehat{\Phi}(s + ikv)h(v)dv}{1 - \int_{-\infty}^{\infty} \widehat{\phi}(s + ikv)h(v)dv}. \quad (1.76)$$

Here we suppose $p(x, t = 0) = \delta(x)$. In the particular case $h(v) = \delta(v - 1)/2 + \delta(v + 1)/2$, Eq. (1.76) reduces to the standard Lévy walk model Eq. (1.71).

An interesting case for Eq. (1.76) is when

$$h(v) = \frac{1}{\pi(1 + v^2)}, \quad (1.77)$$

the density $p(x,t)$ can be explicitly obtained as

$$p(x,t) = \frac{t}{\pi(x^2 + t^2)}. \quad (1.78)$$

1.4.3   *Subordinator*

In the last part of this section, we mainly discuss the subordinated processes. Here we first introduce the Brownian motion denoted by $B(t)$ and the PDF denoted as $p_B(x,t)$ whose Fourier transform w.r.t. x is $\widetilde{p}_B(k,t) = \exp(-tk^2)$. Then we construct a Lévy process $S(t)$ with the characteristic function $E(e^{-uS(t)}) = e^{-u^{\beta/2}t}, \beta \in (0, 2)$. And we use the notation $p_S(\tau, t)$ to represent the PDF of the process $S(t)$. Next we will focus on the process $X(t) = B(S(t))$ and denote its PDF as $p_X(x,t)$. Thus there exists

$$\tilde{p}_X(k,t) = \int_{-\infty}^{\infty} e^{ikx} p_X(x,t) dx$$

$$= \int_0^{\infty} \int_{-\infty}^{\infty} e^{ikx} p_B(x,\tau) dx p_S(\tau,t) d\tau \qquad (1.79)$$

$$= \int_0^{\infty} e^{-\tau k^2} p_S(\tau,t) d\tau$$

$$= e^{-t|k|^\beta}.$$

It is obvious that the result above is the same as the case of Lévy flight. And that indicates the process $X(t) = B(S(t))$ is superdiffusion. Besides if we take $\tau = E(t)$ to be the inverse subordinator, that is $E(t) = \inf\{\tau \geq 0, S(\tau) \geq t\}$, we can obtain a subdiffusion process by constructing the process $X(t) = B(E(t))$. Specifically, we first consider the PDF of process $E(t)$ denoted as $p_E(\tau,t)$. From the correlation between $S(\tau)$ and $E(t)$, we have $P\{E(t) \leq \tau\} = P\{S(\tau) \geq t\}$. Thus there exists

$$p_E(\tau,t) = \frac{\partial}{\partial \tau} P\{E(t) \leq \tau\} = \frac{\partial}{\partial \tau} P\{S(\tau) \geq t\}$$

$$= \frac{\partial}{\partial \tau}[1 - P\{S(\tau) < t\}] = -\frac{\partial}{\partial \tau} \int_0^t p_S(x,\tau) d\tau. \qquad (1.80)$$

On the other hand, we have already known $E(e^{-sS(t)}) = e^{-s^\alpha t}$, where $\alpha \in (0,1)$. Taking Laplace transform w.r.t. t on Eq. (1.80) leads to

$$\hat{p}_E(\tau,s) = -\frac{\partial}{\partial \tau} s^{-1} e^{-\tau s^\alpha} = s^{\alpha-1} e^{-\tau s^\alpha}. \qquad (1.81)$$

Besides for the process $X(t) = B(E(t))$, we have

$$p_X(x,t) = \int_0^{\infty} p_B(x,\tau) p_E(\tau,t) d\tau. \qquad (1.82)$$

After taking Fourier-Laplace transform w.r.t. x and t on both sides of Eq. (1.82), we have

$$\tilde{\hat{p}}_X(k,s) = \int_0^{\infty} e^{-\tau k^2} s^{\alpha-1} e^{-\tau s^\alpha} d\tau = \frac{s^{\alpha-1}}{s^\alpha + k^2}. \qquad (1.83)$$

The inversion of Eq. (1.83) $p_X(x,t)$ solves

$$\frac{\partial}{\partial t} p(x,t) = {}_0D_t^{1-\alpha} \frac{\partial^2}{\partial x^2} p(x,t) \qquad (1.84)$$

with the initial condition $p(x,0) = \delta(x)$. We can also construct some other kinds of processes and subordinators as stochastic representations of the equations, such as time tempered fractional Fokker-Planck equation [Gajda and Magdziarz (2010)], fractional Feynman-Kac equation [Cairoli and Baule (2015)], Klein-Kramers equation [Magdziarz and Weron (2007)], etc.

1.5 Langevin Pictures for Lévy Flights

Here we discuss the Lévy flight described by Langevin equation. For the Langevin picture, the external force can be naturally added.

As mentioned, let us define $w(\eta)$ as the distribution of step length. Here η is the step length of each step. The time intervals between two consecutive steps are random variables with a common PDF $\phi(\tau)$. Thus, the position of the particle in the continuous-time limit by means of the parameter gives

$$x(n) = \int_0^n \eta(n')dn'. \tag{1.85}$$

Taking derivative w.r.t. n, Eq. (1.85) gives the classical Langevin equation

$$\frac{d}{dn}x(n) = \eta(n). \tag{1.86}$$

If introducing external force F, there exists

$$\frac{d}{dn}x(n) = F(x,n) + \eta(n). \tag{1.87}$$

For simplification, here we focus on the space-time independent case, i.e., the waiting time distribution $\phi(\tau)$ and the step length distribution $w(\eta)$ are decoupled. We further introduce the total elapsed time after n 'steps' in the continuous limit, defined by

$$t(n) = \int_0^n \tau(n')dn', \tag{1.88}$$

which results in

$$\frac{d}{dn}t(n) = \tau(n). \tag{1.89}$$

It implies that the CTRW model is shown by two types of Langevin equations, namely Eqs. (1.86) or (1.87), and (1.89). These two equations, describing the dynamic of position and total elapse time, respectively, give the information of the dynamic of anomalous diffusion.

Notice that Eq. (1.89) can be further simplified for a specific $\phi(\tau)$, such as delta function, i.e., $\phi(\tau) = \delta(\tau - a)$. Utilizing Eq. (1.89), we can see that $t = an$. Then the Langevin equation is obtained

$$\frac{d}{dt}x = \frac{1}{a}F(x,t) + \frac{1}{a}\eta(t). \tag{1.90}$$

In the following, utilizing Eqs. (1.86) and (1.89), we derive some useful relations. In the absence of the force F, the probability distributions of the stochastic processes described by Eqs. (1.86) and (1.89) are denoted by $p_x(x, n)$ and $p_t(t, n)$, respectively. Now we focus on power law distributions for the waiting times and step length, i.e., $w(\eta) \sim \eta^{-\beta-1}$ and $\phi(\tau) \sim \tau^{-\alpha-1}$. Using the relations $p_x(x, n) = \langle \delta(x - x(n)) \rangle$ and $p_t(t, n) = \langle \delta(t - t(n)) \rangle$, there exist [Jespersen *et al.* (1999); Fogedby (1994)]

$$p_x(x, n) = \frac{1}{2\pi} \int_{-\infty}^{\infty} \exp(ikx - |k|^{\mu} n) dk$$
$$= n^{1/\mu} l_{\mu} \left(\frac{x}{n^{1/\mu}} \right)$$

(1.91)

and

$$p_t(t, n) = \frac{1}{2\pi} \int_{-\infty}^{\infty} \exp(i\omega t - (-i\omega)^{\nu} n) d\omega$$
$$= n^{-1/\nu} l_{\nu} \left(\frac{t}{n^{1/\nu}} \right).$$

(1.92)

Here we choose a special scale for t and n so that the coefficients of $|k|^{\mu}$ and $(-i\omega)^{\nu}$ are unities. Here the indexes μ and ν are determined by β and α, respectively. In particular case $\beta > 2$, we have $\mu = 2$. It implies that $l_{\mu}(x) = \exp(-x^2)/\sqrt{\pi}$, showing the characteristic of ordinary Brownian motion. While if $\beta < 2$ and $\nu < 1$, we have $\mu = \beta$ and $\nu = \alpha$.

The probability distribution of the walker w.r.t. x is defined by

$$p(x, t) = \int_0^{\infty} p_x(x, n) p_n(t, n) dn.$$

(1.93)

It is different with $p_t(t, n)$ that the roles of n and t in $p_n(t, n)$ are random variable and parameter, respectively. In fact, $p(x, t)$ can be obtained by the Montroll-Weiss formula shown in Eq. (1.43). Thus, with the help of Eq. (1.91), we again obtain $p_n(t, n)$ and its scaling form.

1.6 Continuous Time Random Walk and Lévy Walk with Multiple Internal States

In Sec. 1.3 and Sec. 1.4, we have introduced the CTRW model to characterize subdiffusion and superdiffusion (Lévy flight), and the Lévy walk to describe superdiffusion. All the models have just one internal state; here we extend them to have multiple internal states. In fact in the natural world the multiple internal states are indeed needed, for example the particles move in the multiphase viscous liquid composed of materials with different chemical properties, in which each phase has different kinds of waiting time and jump length distributions; the animals' searching food strategies can also be considered as a process with multiple internal states, since the next step of the animal's waiting and moving are affected by the current state (such as the energy, environment, etc). For the CTRW with multiple internal states, each internal state is specified by the particular waiting time and jump length distributions (for Lévy walk, each internal state represents the particular pair of velocity and walking time distributions).

Here we only have a brief look at the CTRW (and Lévy walk) model with multiple internal states and try to explain the concepts, including the transitions of internal states. For the detailed discussions of these models, please see Chap. 5. Here we only consider a finite number of internal states, and denote the number as N, that is we have N pairs of distributions of waiting time and jump length (or N pairs of velocity distributions and walking time distributions for Lévy walk). Besides, we also need an initial distribution to determine which internal state the particle will choose to stay in for the first step. And a transition matrix with dimension of $N \times N$ is also needed, denoted as M. Actually if we consider the transitions of the internal states with time evolution as a stochastic process, it will be Markovian [Feller (1971)]. Without loss of generality, we denote the elements of matrix M as $m_{i,j}$ (the element of M in i-th row and j-th column) representing the probability of the transition from the i-th internal state to the j-th one. Then the particle will repeat the process from the j-th internal state to the other one according to the transition matrix M. Here we provide a specific example to clarify the process with internal states.

Example 1.1. We consider CTRW model with two internal states. That is we have two pairs of waiting time and jump length distributions, such as $\phi_1(t)$ and $\lambda_1(x)$ representing the first pair of waiting time and jump length distributions respectively, while the second pair denoted as $\phi_2(t)$ and $\lambda_2(x)$. For the first step, these two random variables (waiting time and jump length) obey the corresponding densities determined by a given initial distribution. Specifically, if we choose the initial distribution as $(1/3, 2/3)$ which represents the probability of choosing the first internal state is $1/3$ while choosing the second one is $2/3$ for the first step. Then for a given transition matrix

$$M = \begin{pmatrix} p & 1-p \\ 1-q & q \end{pmatrix},$$

it represents if the particle stays in the first internal state in the previous step, then it will continue staying in it with probability of p and transit to the second internal state with probability $(1-p)$. On the other hand, if the particle stays in the second internal state, then it will go to the first or stay in the second internal state with the probabilities of $(1 - q)$ or q. Let's consider a special case of $p = q = 0$. Then from the transition matrix we can conclude the particle must transit from one internal state to the other one. In practice, depending on the specific problem, we can construct the corresponding transition matrix. Some of the examples of transition matrices have been shown in [Feller (1971)].

For the Fokker-Planck equations, Feynman-Kac equations governing the PDFs of position or functional of CTRW model with multiple internal states, the representation of PDF of Lévy walk with multiple internal states in Fourier-Laplace domain, and more applications of processes with multiple internal states, we will detailedly discuss them in Chap. 5.

In the rest of this book, we will derive the governing equations of the various statistical observations including, position, functional of particles' trajectories, first passage time, fraction of occupation time, etc, for the stochastic models introduced in this chapter.

Chapter 2

Fokker-Planck Equations

Let us now consider the Fokker-Planck equation, which is a partial differential equation that describes the time evolution of the PDF of the positions of particles, and was introduced in Fokker's thesis and independently obtained by Max Planck. It was first applied to study Brownian motion problem, and now has wider applications in the fields of natural sciences, including solid-state physics, anomalous diffusion, chemical physics, quantum theory, and theoretical biology. The aim of this chapter is to introduce fractional derivative and integral and discuss the Fokker-Planck equation.

2.1 Fractional Derivative and Integral

For presenting the governing equations of the statistical observables of the stochastic models introduced in Chap. 1, some non-local operators are needed, for example fractional derivative and integral. In 1695, Leibniz wrote a letter to L'Hospital and discussed whether or not the meaning of derivatives with integer orders could be generalized to derivatives with non-integer orders. L'Hospital was somewhat curious about the problem and asked a simple question in reply: "What if the order will be 1/2?". Leibniz in a re-reply letter dated September 30 of the same year, anticipated: It will lead to a paradox, from which one day useful consequences will be drawn. The date September 30, 1695, is regarded as the exact birthday of the fractional calculus. In the following centuries, the theories of fractional calculus (fractional derivatives and fractional integrals) underwent a significant and even heated development, primarily contributed by pure, not applied, mathematicians. Along this way, it must have entered the minds of more than one enterprising mathematician that differential equations in which the derivatives were of fractional order were conceivable. Only in the last few decades, however, applied scientists and engineers did realize that such fractional differential equations provided a natural framework for the discussion of various kinds of questions, such as viscoelastic systems, electrode-electrolyte polarization, etc. For details, see [Podlubny (1999); Oldham and Spanier (1974); Deng and Zhang (2019)]; and for minor auxiliary examples, see the introductions of [Diethelm *et al.* (2005)].

The anomalous kinetic dynamics described in terms of fractional differential equations is widely studied. In the following, we will introduce some types of fractional derivatives: the Grünwald-Letnikov derivative, the Riemann-Liouville derivative, and fractional substantial derivative. Especially, we compare them with the classical derivative.

2.1.1 *Grünwald-Letnikov Fractional Derivative*

Now we consider the Grünwald-Letnikov derivative, namely the result of operating on a function with fractional operator as the limit of a certain sum. The Grünwald-Letnikov derivative is a basic extension of the derivative in fractional calculus. Let us consider a continuous function $f(t)$. The first-order derivative of the function $f(t)$ is defined by

$$f'(t) = \lim_{h \to 0} \frac{f(t) - f(t-h)}{h}. \tag{2.1}$$

Performing the above operation twice yields the second order derivative

$$
\begin{aligned}
f''(t) &= \frac{d^2 f(t)}{dt^2} \\
&= \lim_{h \to 0} \frac{f'(t) - f'(t-h)}{h} \\
&= \lim_{h \to 0} \frac{1}{h} \left(\frac{f(t) - f(t-h)}{h} - \frac{f(t-h) - f(t-2h)}{h} \right) \\
&= \lim_{h \to 0} \frac{f(t) - 2f(t-h) + f(t-2h)}{h^2}.
\end{aligned}
$$

Iterating this operation yields an expression for the n-th order derivative of a function $f(t)$, namely

$$f^{(n)}(t) = \frac{d^n f(t)}{dt^n} = \lim_{h \to 0} \frac{1}{h^n} \sum_{j=0}^{n} (-1)^j \binom{n}{j} f(t - jh) \tag{2.2}$$

with

$$\binom{n}{j} = \frac{n!}{j!(n-j)!} = \frac{\Gamma(n+1)}{\Gamma(j+1)\Gamma(n-j+1)}.$$

Now Eq. (2.2) can be generalized to non-integer number with the help of the generalized binomial formula

$$_aD_t^\alpha f(t) = \lim_{\substack{h\to 0 \\ nh=t-a}} \frac{1}{h^\alpha} \sum_{j=0}^{n} (-1)^j \binom{\alpha}{j} f(t-jh), \tag{2.3}$$

where

$$\binom{\alpha}{j} = \frac{\Gamma(\alpha+1)}{\Gamma(j+1)\Gamma(\alpha-j+1)}$$

and a is a constant. Similar to Eq. (2.3), the corresponding fractional integral is

$$_aD_t^{-\alpha} f(t) = \lim_{\substack{h\to 0 \\ nh=t-a}} h^\alpha \sum_{j=0}^{n} \frac{\Gamma(\alpha+j)}{j!\Gamma(\alpha)} f(t-jh). \tag{2.4}$$

We can see that Eqs. (2.3) and (2.4) are very appealing in the sense that the equations make less requirements on $f(t)$. Besides, Eq. (2.3) and Eq. (2.4) can recover the integer order derivative and integral, respectively.

2.1.2 *Riemann-Liouville Fractional Derivative*

Note that Grünwald-Letninkov fractional derivative is not convenient to deal with the physical systems since it's defined by series. Equation (2.4) can be rewritten as an integral form. When $\alpha = 1$, there exists

$$_aD_t^{-1} f(t) = \lim_{\substack{h\to 0 \\ nh=t-a}} h \sum_{j=0}^{n} f(t-jh) = \int_0^{t-a} f(t-\tau)d\tau = \int_a^t f(\tau)d\tau,$$

where we use the relation $n = (t-a)/h \to \infty$ as $h \to 0$. By induction, we have

$$_aD_t^{-n} f(t) = \frac{1}{\Gamma(n)} \int_a^t (t-\tau)^{n-1} f(\tau)d\tau. \tag{2.5}$$

Inspired by Eq. (2.5), we can generalize the positive integer order integral into positive real order, and when $t > a$ fractional order integral can be defined as

$$_aD_t^{-\alpha} f(t) = \frac{1}{\Gamma(\alpha)} \int_a^t (t-\tau)^{\alpha-1} f(\tau)d\tau. \tag{2.6}$$

This form of the fractional integral is referred as the Riemann-Liouville fractional integral. It implies that $_aD_t^{-\alpha}$ is a non-local operator, which captures the information of the past.

If $f(t)$ has $(n+1)$-th order derivative, the Grünwald-Letnikov derivative has its equivalent definition

$$_a\mathcal{D}_t^\alpha f(t) = \left(\frac{d}{dt}\right)^{n+1} \frac{1}{\Gamma(n+1-\alpha)} \int_a^t (t-\tau)^{n-\alpha} f(\tau) d\tau \qquad (2.7)$$

with $n < \alpha < n+1$, being called α-th order Riemann-Liouville fractional derivative.

One of the most important properties of the Riemann-Liouville fractional derivative is that for $\alpha > 0$ and $t > a$

$$_a\mathcal{D}_t^\alpha(_a\mathcal{D}_t^{-\alpha} f(t)) = f(t).$$

It implies that the Riemann-Liouville fractional differential operator is a left inverse to the Riemann-Liouville fractional integral operator with the same order α.

In order to have a good understanding of the fractional derivative, we further consider the special case: $f(t) = t^\nu$. The most often used version occurs when $a = 0$. After some simple calculations, Eq. (2.6) gives

$$_0\mathcal{D}_t^{-\alpha} t^\nu = \frac{\Gamma(\nu+1)}{\Gamma(\nu+\alpha+1)} t^{\nu+\alpha}, \nu > -1. \qquad (2.8)$$

Utilizing Eq. (2.7), there exists

$$_0\mathcal{D}_t^\alpha t^\nu = \frac{\Gamma(\nu+1)}{\Gamma(\nu-\alpha+1)} t^{\nu-\alpha}, \nu > -1. \qquad (2.9)$$

As expected, when $\alpha = 1, 2, \ldots, n$, Eq. (2.8) agrees with the classical integral. Equation (2.7) is called semi-derivative when $\alpha = 1/2$; further if $\nu = 0$ in Eq. (2.9), there exists

$$_0\mathcal{D}_t^{1/2} 1 = \frac{1}{\sqrt{\pi t}}.$$

We can see that the Riemann-Liouville fractional derivative of a constant does not vanish. Furthermore, when $f(t) = (t-a)^\nu$, with ν being a real number, the corresponding Riemann-Liouville fractional derivative and integral are presented by

$$_a\mathcal{D}_t^\alpha (t-a)^\nu = \frac{\Gamma(1+\nu)}{\Gamma(1+\nu-\alpha)} (t-a)^{\nu-\alpha} \qquad (2.10)$$

and

$$_a\mathcal{D}_t^{-\alpha} (t-a)^\nu = \frac{\Gamma(1+\nu)}{\Gamma(1+\nu+\alpha)} (t-a)^{\nu+\alpha}, \qquad (2.11)$$

respectively. Note that Eqs. (2.10) and (2.11) are valid for $\nu > -1$, being the same as Eqs. (2.8) and (2.9). Nowadays there are many different forms of fractional derivative or integral operators (see below or above), but the Riemann-Liouville operator is still the most frequently used when solving physical models.

2.1.3 *Fractional Substantial Derivative*

Now we want to introduce the fractional substantial derivative, which appears in the model describing the functional distribution. The functional A is defined as

$$A = \int_0^t U(x(\tau))d\tau, \tag{2.12}$$

where U is a prescribed function and $x(t)$ is the trajectory of the particle. Based on Eq. (2.12), many interesting applications have been dug out. If one is concerned about the time spent by a particle in a given domain, it can be made by taking $U = 1$ if the particle lies in the domain and 0 otherwise [Luchinin and Dolin (2014); Bar-Haim and Klafter (1998); Wu *et al.* (2016)]. The functional is also used to study nuclear magnetic resonance (NMR) [Grebenkov (2007)]; under the influence of inhomogeneous magnetic $U(x(t))$, the total phase accumulated along the trajectory of a nucleus during the time from 0 to t is taken as

$$\vartheta(t) = \gamma \int_0^t U(x(\tau))d\tau,$$

where γ is the nuclear gyromagnetic ratio; and $U(x)$ is respectively specified as x and x^2 to calculate the macroscopic measured signal $E = \langle \exp(i\vartheta(t)) \rangle$; then, NMR indirectly encodes information regarding the trajectory of the particles. If $U = \delta(x - a)$, the functional A denotes the local time at the fixed level a, which is an important quality in probability [Pitman (1999)]. And another instance used in finance is $U(x) = \exp(x)$, which illustrates the price of an Asian stock option in the Black-Scholes framework [Fischer Black (1973)]. Interestingly, these functionals A and their related variants are also used as powerful tools in mathematics and physics, governed by Feynman-Kac equation, which allows us to study the functionals in a quantum mechanical framework [Majumdar (2005); Kac (1949); Perret *et al.* (2015); Baule and Friedrich (2006); Majumdar and Bray (2002); Carmi *et al.* (2010)].

The governing equation of the functional of trajectory of particle is derived in [Carmi *et al.* (2010)] which uses the fractional substantial derivative, defined by

$$D_t^{1-\alpha} f(x,p,t) = \frac{1}{\Gamma(\alpha)} \left[\frac{\partial}{\partial t} + pU(x) \right] \int_0^t \frac{\exp(-(t-\tau)pU(x))}{(t-\tau)^{1-\alpha}} f(x,p,\tau)d\tau \tag{2.13}$$

with $t > 0$ and $0 < \alpha < 1$. As mentioned, here $U(x)$ is a prescribed function in Eq. (2.12). It implies that, due to slowing decay kernel $(t-\tau)^{\alpha-1}$, the evolution of $f(x,p,t)$ is non-Markovian and depends on the entire history. In particular, when $p = 0$, Eq. (2.13) reduces to the Riemann-Liouville fractional derivative operator. With the wide applications of the fractional substantial derivative, it seems urgent to analyze its properties. In Laplace space, Eq. (2.13) can be expressed by a simple form

$$D_t^{1-\alpha}\widehat{f}(x,p,s) = [s + pU(x)]^{1-\alpha}\widehat{f}(x,p,s). \qquad (2.14)$$

Similarly, the corresponding fractional substantial integral is

$$D_t^{-\alpha}f(x,p,t) = \frac{1}{\Gamma(\alpha)}\int_0^t \frac{\exp(-(t-\tau)pU(x))}{(t-\tau)^{1-\alpha}}f(x,p,\tau)d\tau \qquad (2.15)$$

with $0 < \alpha < 1$. See discussions and numerical examples in Refs. [Carmi *et al.* (2010); Chen and Deng (2015); Deng *et al.* (2015); Wu *et al.* (2016); Wang and Deng (2018)].

2.1.4 *Laplace Transform of Fractional Derivative*

Before introducing the formula of the Laplace transform of the fractional derivative, we consider the Laplace transform of n-th integer order derivative of the function $g(t)$. Using integration by parts, there exists

$$\mathcal{L}[g^{(n)}(t)] = s^n\widehat{g}(s) - \sum_{k=0}^{n-1} s^{n-k-1}g^{(k)}(0). \qquad (2.16)$$

Now we proceed with the Laplace transform of Riemann-Liouville fractional integral of order $\alpha > 0$ with $a = 0$. Due to the special structure of the fractional derivative, the convolution property of the Laplace transform will make the problem easy when calculating the Laplace transform. Note that for a function $f(\tau)$ defined on $\tau \geq 0$ the Riemann-Liouville fractional integral reduces to

$$_0D_t^{-\alpha}f(t) = \frac{1}{\Gamma(\alpha)}\int_0^t (t-\tau)^{\alpha-1}f(\tau)d\tau = \frac{1}{\Gamma(\alpha)}t^{\alpha-1} *_t f(t), \qquad (2.17)$$

where $*_t$ represents the convolution of functions w.r.t. t. Using convolution theorem of Laplace transform yields

$$\mathcal{L}[_0D_t^{-\alpha}f(t)] = \mathcal{L}\left[\frac{1}{\Gamma(\alpha)}t^{\alpha-1}\right]\mathcal{L}[f(t)] = s^{-\alpha}\widehat{f}(s). \qquad (2.18)$$

Based on the definition of the Riemann-Liouville fractional derivative Eq. (2.8), the Laplace transform of the Riemann-Liouville fractional derivative of order α is

$$\mathcal{L}[_0\mathcal{D}_t^{\alpha}f(t)] = s^{\alpha}\widehat{f}(s) - \sum_{j=0}^{n-1} s^j (_0\mathcal{D}_t^{\alpha-j-1}f(t))|_{t=0} \qquad (2.19)$$

with $n-1 < \alpha < n$. More details about Eq. (2.19) can be seen in [Oldham and Spanier (1974)]. In the particular case of $0 < \alpha < 1$, the Laplace transform of Riemann-Liouville fractional derivative is

$$\mathcal{L}[_0\mathcal{D}_t^\alpha f(t)] = s^\alpha \widehat{f}(s). \tag{2.20}$$

2.2 Derivation of Fractional Fokker-Planck Equation

In the last decades, anomalous diffusion has been widely studied by different kinds of models. One of the basic models is Fokker-Planck equation, describing the distribution of positions of particles. In the following, we derive the Fokker-Planck equation with a position-dependent external force field $F(x) = -V'(x)$ and a non-local transfer kernel which decays slowly in space and time [Metzler *et al.* (1999b)].

Following [Weiss (1994)], we start from the assumptions that the particles have to leave their current positions during the fixed time interval and they can just go to their nearest neighbors. Then we have the discrete master equation

$$p_j(t + \triangle t) = A_{j-1}p_{j-1}(t) + B_{j+1}p_{j+1}(t), \tag{2.21}$$

where A_{j-1} and B_{j+1} represent the probabilities of moving to right and left, respectively. Note that $A_j + B_j = 1$ for all j and $p_j(t + \triangle t)$ is the probability that the particle is on site j after a jump during $\triangle t$.

Assuming that $\triangle x$ and $\triangle t$ tend to 0 but the ratio $\triangle x^2/\triangle t$ is a finite number. The coefficient in the Fokker-Planck operator is given by

$$\frac{V'(x)}{m\eta_1} = \lim_{\triangle x \to 0, \triangle t \to 0} \frac{\triangle x}{\triangle t}(B(x) - A(x)) \tag{2.22}$$

with m and η_1 being the mass of the diffusion particle and the friction coefficient, respectively. Note that the generalized master equation follows

$$p_j(t + \triangle t) = \sum_{n=1}^{\infty} A_{j,n}p_{j-n}(t) + \sum_{n=1}^{\infty} B_{j,n}p_{j+n}(t),$$

where the terms $A_{j,n}$ and $B_{j,n}$ satisfy the normalized condition $\sum_{n=1}^{\infty}(A_{j,n} + B_{j,n}) = 1$. We further introduce the transfer kernel $\mathcal{K}(x,y)$ depicting the distance between departure site y and site x, which is given by

$$\mathcal{K}(x,y) \equiv w(x-y)\left(A(y)\theta(x-y) + B(y)\theta(y-x)\right), \tag{2.23}$$

where $A(y)$ and $B(y)$ are the probabilities of turning right and left, respectively, and $\theta(x)$ is the Heaviside theta function. Here $w(x)$ is the

PDF of step length. Based on the above equations, the continuous time and continuous space version of the generalized master equation is

$$p(x,t) = \int_{-\infty}^{\infty} \int_{0}^{t} \phi(t-t')\mathcal{K}(x,y)p(y,t')dt'dy + \Phi(t)p_0(x), \qquad (2.24)$$

where $\mathcal{K}(x,y)$ is the memory function and $\phi(t-t')$ describes the distribution of the time intervals between two events. Taking Laplace transform w.r.t. t, Eq. (2.24) gives

$$\widehat{p}(x,s) = \int_{-\infty}^{\infty} \widehat{\phi}(s)\mathcal{K}(x,y)\widehat{p}(y,s)dy + \widehat{\Phi}(s)p_0(x). \qquad (2.25)$$

Now we further perform the Fourier transform w.r.t. x, which gives

$$\widetilde{\widehat{p}}(k,s) = \int_{0}^{\infty} \cos(kx)w(x)dx\widehat{\phi}(s)\widetilde{\widehat{p}}(k,s) + \widehat{\Phi}(s)\widetilde{p}_0(k)$$
$$+ i\int_{0}^{\infty} \sin(kx)w(x)dx\widehat{\phi}(s)\mathcal{F}[(A(x)-B(x))\widehat{p}(x,s)]. \qquad (2.26)$$

Rewriting above equation leads to

$$s\widetilde{\widehat{p}}(k,s) - \widetilde{p}_0(k) = s\widehat{\phi}(s)\int_{0}^{\infty} \cos(kx)w(x)dx\widetilde{\widehat{p}}(k,s) - \widehat{\phi}(s)\widetilde{p}_0(k)$$
$$+ is\widehat{\phi}(s)\int_{0}^{\infty} \sin(kx)w(x)dx\mathcal{F}[(A(x)-B(x))\widehat{p}(x,s)]. \qquad (2.27)$$

With the help of this equation, studying of the distribution of the position $x(t)$ for numerous $w(x)$ and $\phi(t)$ reduces to the inversion of Eq. (2.27).

We now focus on the case of $0 < \alpha < 1$ and $1 < \beta < 2$, i.e., the waiting time distribution $\phi(t) \sim t^{-1-\alpha}$ and the jump length distribution $w(x) \sim 2Ax^{-\beta-1}$ with $x > 0$. According to Eq. (1.32), the cosine transform of $w(x)$ is

$$\int_{0}^{\infty} \cos(kx)w(x)dx \sim 1 - A_\beta|k|^\beta. \qquad (2.28)$$

For the sine transform, we get

$$\int_{0}^{\infty} \sin(kx)w(x)dx \sim A_{\beta,sin}k, \qquad (2.29)$$

with $A_{\beta,sin}$ being a constant determined by the exact form $w(x)$, for example, we can expand $\sin(kx)$ in its Taylor series to get the leading term, i.e., $A_{\beta,sin} = \int_0^\infty xw(x)dx$. Utilizing Eqs. (2.27) and (2.28) gives [Metzler et al. (1999b)]

$$\frac{\partial}{\partial t}p(x,t) = {}_0\mathcal{D}_t^{1-\alpha}\left[\frac{\partial}{\partial x}\frac{A_{\beta,sin}}{b_\alpha}[B(x)-A(x)] + \frac{A_\beta}{b_\alpha}\nabla_x^\beta\right]p(x,t), \qquad (2.30)$$

where $\mathcal{F}[\nabla_x^\beta p(x,t)] = -|k|^\beta \widetilde{p}(k,t)$. Besides, $A_{\beta,sin}(B(x) - A(x))/b_\alpha$ and $A_\beta/(b_\alpha)$, are customarily called the drift and diffusion coefficients, respectively; and there exists

$$- \frac{\partial}{\partial x} \frac{F(x)}{k_B T} = \frac{A_{\beta,sin}}{b_\alpha} (B(x) - A(x)), \tag{2.31}$$

where k_B and T are the Boltzmann constant and temperature, respectively. When $A(x) = B(x) = 1/2$, Eq. (2.30) reduces to

$$\frac{\partial}{\partial t} p(x,t) = \frac{A_\beta}{b_\alpha} \, _0\mathcal{D}_t^{1-\alpha} \nabla^\beta p(x,t). \tag{2.32}$$

If the second moment of jump length is finite, Eq. (2.32) becomes

$$\frac{\partial}{\partial t} p(x,t) = \frac{A_2}{b_\alpha} \, _0\mathcal{D}_t^{1-\alpha} \frac{\partial^2}{\partial x^2} p(x,t). \tag{2.33}$$

2.3 Solution of Fractional Fokker-Planck Equation

Now we introduce a method given by [Barkai (2001)] to find the solution of fractional Fokker-Planck equation. This ingenious method is based on an integral transform mapping a Gaussian type of distribution onto fractional diffusion. This similar method is investigated by Bouchaud [Bouchaud and Georges (1990)], Klafter [Klafter and Zumofen (1994)] in the discussion of the CTRW model; see also the related literature [Saichev and Zaslavsky (1997)].

2.3.1 Integral Form of the Solution for Fokker-Planck Equation

This subsection is to establish the relationship between the solution $p_1(x,t)$ of the ordinary Fokker-Planck equation and the solution $p_\alpha(x,t)$ of the time fractional Fokker-Planck equation, i.e.

$$p_\alpha(x,t) = \int_0^\infty n(\tau,t)p_1(x,\tau)d\tau, \tag{2.34}$$

where $0 < \alpha < 1$ and

$$n(\tau,t) = \frac{d}{d\tau}\left[1 - \ell_\alpha\left(\frac{t}{\tau^{1/\alpha}}\right)\right]$$
$$= \frac{1}{\alpha}\frac{t}{\tau^{1+1/\alpha}}\ell_\alpha\left(\frac{t}{\tau^{1/\alpha}}\right) \tag{2.35}$$

with $\ell_\alpha(\tau)$ being the one sided Lévy stable distribution with index α. In Laplace space, one sided Lévy distribution $\ell_\alpha(\tau)$ is

$$\int_0^\infty \exp(-s\tau)\ell_\alpha(\tau)d\tau = \exp(-s^\alpha).$$

Note that the validity of Eq. (2.34) is under the assumption that the governing equations of $p_\alpha(x,t)$ and $p_1(x,t)$ are both with the free boundary condition and the same initial condition. Besides, there exists similar transform to Eq. (2.34) in high dimensions. Let $p(x,t) \geq 0$ be a normalized density to find a particle at x at time t. Then we have $\int_{-\infty}^\infty p(x,t)dx = 1$. When $\beta > 2$ and $\alpha > 1$, the Gaussian Markovian type of diffusion in an external field $F(x)$ is shown as follows:

$$\frac{\partial p_1(x,t)}{\partial t} = K_{1,2}\mathcal{L}_{NFP}p_1(x,t), \qquad (2.36)$$

where the Fokker-Planck operator

$$\mathcal{L}_{NFP} = -\frac{\partial}{\partial x}\frac{F(x)}{k_BT} + \frac{\partial^2}{\partial x^2},$$

and $K_{1,2}$ and k_B are the diffusion coefficient and Boltzmann constant, respectively. When $\alpha < 1$ and $\beta > 2$, rewriting Eq. (2.30) gives

$$\frac{\partial p_\alpha(x,t)}{\partial t} = K_{\alpha,2}\,_0\mathcal{D}_t^{1-\alpha}\mathcal{L}_{NFP}p_\alpha(x,t), \qquad (2.37)$$

where $K_{\alpha,2}$ is the generalized diffusion coefficient. Note that when $F(x) = 0$, Eq. (2.37) coincides with the Schneider-Wyss fractional equation. In Laplace space, $t \to s$, Eq. (2.37) becomes

$$s\widehat{p}(x,s) - \delta(x - x_0) = K_{\alpha,2}s^{1-\alpha}L_{NFP}\widehat{p}(x,s), \qquad (2.38)$$

where $\delta(x - x_0)$ is the initial condition and we assume free boundary conditions. We would like to recall some properties of Eq. (2.37):

- In the presence of an external time independent binding field, the stationary solution is the Boltzmann distribution.
- Generalized Einstein relation is consistent with linear response theory [Kubo (1966)].
- When $\alpha \to 1$, the standard Smolochwski Fokker-Planck Eq. (2.36) is recovered.

We now turn the discussion to the solution of Eq. (2.37), namely to prove Eq. (2.34). Without loss of generalization, we take $K_{1,2} = K_{\alpha,2} = 1$. Our aim is to find an appropriate solution $n(\tau,t)$.

Performing Laplace transform of Eq. (2.34), and using the normalizations of $p_\alpha(x, t)$ and $p_1(x, t)$ lead to

$$\int_0^\infty \widehat{n}(\tau, s) d\tau = 1/s, \qquad (2.39)$$

which implies that $n(\tau, t)$ is normalized since $\int_0^\infty n(\tau, t) d\tau = 1$. Substituting Eq. (2.34) into Eq. (2.38), there exists

$$s \int_0^\infty \widehat{n}(u, s)\widehat{p}_1(x, s) ds - \delta(x - x_0) = s^{1-\alpha} \int_0^\infty \widehat{n}(u, s)\mathcal{L}_{NFP}\widehat{p}_1(x, s) ds. \qquad (2.40)$$

According to Eq. (2.36), integrating by parts leads to

$$s \int_0^\infty \widehat{n}(u, s)\widehat{p}_1(x, s) ds - \delta(x - x_0) = s^{1-\alpha}[\widehat{n}(\infty, s)p_1(x, \tau = \infty)$$
$$- \widehat{n}(0, s)\widehat{p}_1(x, \tau = 0)] - s^{1-\alpha} \int_0^\infty \frac{\partial}{\partial \tau}\widehat{n}(\tau, s)p_1(x, \tau) d\tau. \qquad (2.41)$$

We can check that $\widehat{n}(s, \infty) = 0$ and $p_1(x, \tau = \infty)$ in Eq. (2.41) is the stationary solution of the standard Fokker-Planck equation. Recall that $p_1(x, \tau = 0) = \delta(x - x_0)$, Eq. (2.41) yields

$$\int_0^\infty \left[s\widehat{n}(\tau, s) + s^{1-\alpha}\frac{\partial}{\partial \tau}\widehat{n}(\tau, s) \right] p_1(x, \tau) d\tau = [1 - s^{1-\alpha}\widehat{n}(0, s)]\delta(x - x_0). \qquad (2.42)$$

Thus we find $\widehat{n}(\tau, s)$ satisfying Eq. (2.42) by solving the following equations:

$$\widehat{n}(0, s) = s^{\alpha-1} \qquad (2.43)$$

and

$$\frac{\partial}{\partial \tau}\widehat{n}(\tau, s) = -s^\alpha \widehat{n}(\tau, s), \qquad (2.44)$$

which implies

$$\widehat{n}(\tau, s) = s^{\alpha-1} \exp(-s^\alpha \tau). \qquad (2.45)$$

Taking inverse Laplace transform yields Eq. (2.35). Finally, we find the solution of Eq. (2.37) by the method of integral transform.

2.3.2 *Solution for Force Free Fractional Diffusion*

Now we show that the solution given by Eq. (2.34) agrees with the known solution of a fractional diffusion equation in m dimensions

$$\frac{\partial p(\mathbf{r}, t)}{\partial t} = {}_0\mathcal{D}_t^{1-\alpha}\nabla^2 p(\mathbf{r}, t).$$

Note that the solution of Eq. (2.36) is Gaussian. From Eq. (2.34), we have

$$p(\mathbf{r}, t) = \int_0^\infty n(\tau, t) \frac{1}{(4\pi\tau)^{m/2}} \exp\left(-\frac{|\mathbf{r}|^2}{4\tau}\right) d\tau,$$

where the initial condition is delta distribution $\delta(x)$. Taking Laplace transform w.r.t. the observation time t leads to

$$\widehat{p}(\mathbf{r}, s) = \frac{s^{\alpha-1}}{(2\pi)^{m/2}} \left(\frac{|\mathbf{r}|}{s^{2/\alpha}}\right)^{1-m/2} K_{m/2-1}(\mathbf{r}s^{2/\alpha}) \qquad (2.46)$$

with $K_{m/2-1}$ being the second kind of Bessel function [Schneider and Wyss (1989)]. The inverse Laplace transform of Eq. (2.46) gives the final solution

$$p(\mathbf{r}, t) = \frac{1}{\alpha\pi^{m/2}|\mathbf{r}|^m} H_{1,2}^{2,0}\left(2^{-2/\alpha}|\mathbf{r}|^{2/\alpha}t^{-1}\Big|_{(m/2,1/\alpha),(1,1/\alpha)}^{(1,1)}\right),$$

where $H_{1,2}^{2,0}$ is the Fox function [Mathai and Saxena (1978)]. According to the series expansion of the Fox function, for small $\xi = |\mathbf{r}|^2/t^\alpha$ with $m = 3$, there exists

$$p(\mathbf{r}, t) = \frac{1}{4\pi t^{3\alpha/2}\xi^{1/2}} \sum_{j=0}^\infty \frac{(-1)^j \xi^{j/2}}{j!\Gamma(1-\alpha(1+j/2))}. \qquad (2.47)$$

It implies that for $m = 3$, $\alpha \neq 1$, and $|\mathbf{r}| \to 0$, the solution $p(\mathbf{r}, t)$ diverges like $p(\mathbf{r}, t) \sim 1/|\mathbf{r}|$.

2.3.3   *Solution for Biased Fractional Wiener Process*

We now discuss a biased one-dimensional time fractional diffusion process with a uniform force $F(x) = F > 0$, the corresponding first moment of which grows more slowly than linearly with the observation time t, namely $\langle x(t)\rangle \sim FK_\alpha t^\alpha/(T\Gamma(1+\alpha))$, with K_α being a generalized diffusion coefficient.

The solution of the classical Fokker-Planck equation Eq. (2.36) is

$$p_1(x, t) = \frac{1}{\sqrt{4\pi t}} \exp\left(-\frac{(x - Ft/T)^2}{4t}\right).$$

In Laplace space, the solution $\widehat{p}(x, s)$ of Eq. (2.37) is

$$\widehat{p}(x, s) = \frac{s^{\alpha-1}T/(FK_\alpha)}{\sqrt{1 + 4s^\alpha T^2/(F^2 K_\alpha)}}$$

$$\times \exp\left(\frac{F(x - \sqrt{1 + 4s^\alpha T^2/(F^2 K_\alpha)}|x|)}{2T}\right). \qquad (2.48)$$

In the long time limit, i.e., $s \to 0$, there exists

$$\widehat{p}(x, s) \sim \begin{cases} \frac{T s^{\alpha-1}}{F K_\alpha} \exp\left(-\frac{T s^\alpha}{F K_\alpha} x\right), & x > 0; \\ \frac{T s^{\alpha-1}}{F K_\alpha} \exp\left(-\frac{F}{T}|x|(1 + \frac{T^2}{F^2 K_\alpha} s^\alpha)\right), & x < 0. \end{cases} \tag{2.49}$$

Due to the effect of the force, in the long time limit, the particles will only move in the one side of the axis. This can be checked as follows: $\int_{-\infty}^{0} \widehat{p}(x, s) dx \sim s^{\alpha-1}$, which means that $\int_{-\infty}^{0} p(x, t) dx \sim 1/t^\alpha$. Taking inverse Laplace transform w.r.t. s, the asymptotic behavior of $p(x, t)$ gives

$$p(x, t) \sim \begin{cases} \frac{t}{\alpha (F/T/K_\alpha)^{1/\alpha} \tau x^{1+1/\alpha}} \ell_\alpha\left(\frac{t}{(F/T/K_\alpha)^{1/\alpha} \tau x^{1/\alpha}}\right), & x > 0; \\ 0, & x < 0. \end{cases} \tag{2.50}$$

It can be checked that the normalized condition of Eq. (2.50) is satisfied. Another interesting result is $\lim_{x \to 0} p(x, t)$. Using the Tauberian theorem, from Eq. (2.49), we can find that $p(0, t) \sim t^{-\alpha}$. When $F = 0$, $p(0, t) \sim t^{-\alpha/2}$. As expected, it can be seen that the particles leave rapidly from the origin position with the help of force.

2.3.4 *Solution Obtained by Separation of Variables*

Here we consider using the separate variables to find the solution of the fractional Fokker-Planck equation with time-independent forces. For simplification, we discuss the one-dimensional fractional Fokker-Planck equation

$$\frac{\partial}{\partial t} p(x, t) = {}_0\mathcal{D}_t^{1-\alpha} L_{FP} p(x, t) \tag{2.51}$$

with

$$L_{FP} = \frac{\partial}{\partial x} \frac{V'(x)}{m \eta_\alpha} + K_\alpha \frac{\partial^2}{\partial x^2}, \tag{2.52}$$

where K_α and η_α represent the generalized diffusion constant and friction, respectively; m denotes the mass of the diffusing particle. Here the dimension of η_α is $\sec^{\alpha-2}$ and an extension of the Fokker-Planck equation to higher dimensional one is made by replacing the spatial derivatives in the one-dimensional Fokker-Planck operator with the corresponding Laplace operators. Note that a generalization of the Einstein relation $K_\alpha = k_B T/(m \eta_\alpha)$ holds for the generalized coefficients K_α and η_α. This relation can be obtained when the stationary state is reached and the stationary solution is given by

$$\frac{V'(x)}{m \eta_\alpha} W_{st} + K_\alpha W'_{st} = 0.$$

We now investigate the solution $p(x,t)$ of Eq. (2.51) with the presence of an arbitrary external force field $F(x)$. In order to obtain a formal solution of Eq. (2.51), we suppose that $p(x,t)$ can be expressed as

$$p(x,t) = \sum_{n=0}^{\infty} h_n(x_0,t_0)p_n(x,t) \tag{2.53}$$

and

$$p_n(x,t) = T_n(t)\psi_n(x), \tag{2.54}$$

where $h_n(x_0,t_0)$ is governed by the initial conditions. Now Eq. (2.51) becomes

$$\psi_n(x)\frac{\partial}{\partial t}T_n(t) =_0\mathcal{D}_t^{1-\alpha}T_n(t)L_{FP}\psi_n(x). \tag{2.55}$$

Rewriting Eq. (2.55) gives the two eigenequations

$$\begin{cases} \frac{dT_n}{dt} = -\lambda_{n,\alpha}\ _0\mathcal{D}_t^{1-\alpha}T_n(t); \\ L_{FP}\psi_n(x) = -\lambda_{n,\alpha}\psi_n(x), \end{cases} \tag{2.56}$$

where $\lambda_{n,\alpha}$ and $\psi_n(x)$ are a group of eigenvalues and eigenfunctions of the operator L_{FP} with $n = 0,1,2,....$ Then from the first equation of Eq. (2.56), there exists

$$T_n(t) = E_\alpha(-\lambda_{n,\alpha}t^\alpha) \equiv \sum_{j=0}^{\infty} \frac{(-\lambda_{n,\alpha}t^\alpha)^j}{\Gamma(1+\alpha j)}. \tag{2.57}$$

For the case of $\alpha = 1$, the corresponding $T_n(t)$ is $\exp(-\lambda_{n,1}t)$. Furthermore, $\lambda_{n,\alpha}$ is related to the standard eigenvalue $\lambda_{n,1}$, obeying $\lambda_{n,\alpha} = (\eta_1/\eta_\alpha)\lambda_{n,1}$. In order to have a deep understanding of solution, we study short and long time behaviors of $T_n(t)$, respectively. When $t \to 0$, Eq. (2.57) gives

$$T_n(t) \sim \exp\left(-\frac{\lambda_{n,\alpha}t^\alpha}{\Gamma(1+\alpha)}\right). \tag{2.58}$$

It shows that $\lim_{t\to 0} T_n(t) = 1$, independent of α. On the other hand, if $t \to \infty$, there exists

$$T_n(t) \sim \frac{1}{\Gamma(1-\alpha)\lambda_{n,\alpha}t^\alpha}. \tag{2.59}$$

We can see that with the increase of the observation time t, the behaviors of $T_n(t)$ change from exponential form to the heavy-tailed power law tending to 0 slowly.

Utilizing Eqs. (2.56) and (2.57), the full solution of Eq. (2.51) is given by the sum over all eigenvalues

$$
\begin{aligned}
p(x,t|x_0,0) = {} & \exp\left(\frac{V(x_0)}{2k_BT} - \frac{V(x)}{2k_BT}\right) \sum_n \left(\exp\left(\frac{V(x)}{2k_BT}\right)\psi_n(x)\right. \\
& \left. \times \exp\left(\frac{V(x_0)}{2k_BT}\right)\psi_n(x_0)E_\alpha(-\lambda_{n,\alpha}t^\alpha)\right).
\end{aligned}
\tag{2.60}
$$

Here x_0 is the initial position of the particle. Arranging the eigenvalues in increasing order, namely, $0 \le \lambda_{0,\alpha} < \lambda_{1,\alpha} < \lambda_{2,\alpha} < ...$, then we can see that the first eigenvalue should be 0 if and only if Eq. (2.51) has a stationary solution.

Chapter 3

Feynman-Kac Equations

In this chapter, we discuss different kinds of Feynman-Kac equations. Generally speaking, Feynman-Kac equations describe the PDFs of the functionals of the particle trajectories and thus have wide applications in different research fields.

3.1 Brownian Functionals

We start with the Brownian functionals which are the functionals of Brownian trajectories. As shown in [Majumdar (2005)], a Brownian functional is defined as

$$A = \int_0^t U[x(\tau)]d\tau, \tag{3.1}$$

where $x(\tau)$ is a Brownian path starting from x_0 at $\tau = 0$ and propagating up to time $\tau = t$ and $U(x)$ is a specified function. Depending on the specific practical problems, we can freely choose the forms of the function $U(x)$. Then we would like to give some interesting examples regarding the crucial applications of Brownian functionals.

Brownian functionals appear in a wide range of problems [Majumdar (2005)], for example, in probability theory, the occupation time spent by a Brownian motion above the origin within a time window of size t [Lévy (1940)],

$$A = \int_0^t \theta[x(\tau)]d\tau. \tag{3.2}$$

In finance, a typical stock price $S(t)$ is sometimes modelled by the exponential of a Brownian motion, $S(t) = e^{-\beta x(t)}$, where β is a constant. An important quantity is the integrated stock price up to some target time t

[Yor (2001)], namely

$$A = \int_0^t e^{-\beta x(\tau)} d\tau. \tag{3.3}$$

Moreover, regarding the stochastic behaviour of daily temperature data, when the deviation from its average is assumed to be a simple Brownian motion $x(t)$ in a harmonic potential (the Ornstein-Uhlenbeck process), the so-called heating degree days are of interest and the functional is defined as

$$A = \int_0^t x(\tau)\theta(x(\tau)) d\tau, \tag{3.4}$$

which measures the integrated excess temperature (here $\theta(x)$ is the Heaviside theta function as denoted in Sec. 2.2) up to time t [Majumdar and Bray (2002)]. The total area (unsigned) under a Brownian motion is defined as

$$A = \int_0^t |x(\tau)| d\tau. \tag{3.5}$$

And this functional is first studied in the context of economics [Cifarelli and Regazzini (1975)], later extensively by probabilists [Shepp (1982)], and also studied by physicists in the context of electron-electron and phase coherence [Altshuler *et al.* (1982)].

3.2 Fractional Feynman-Kac Equations

While the Brownian functionals have a lot of applications, with the widely observed non-Brownian motions the non-Brownian functionals naturally attract the interests of scientists, which is defined as

$$A = \int_0^t U[x(\tau)] d\tau \tag{3.6}$$

with $x(t)$ being the trajectory of the particle and $U(x)$ being some prescribed function. In [Turgeman *et al.* (2009)], the authors present a rather general framework for non-Brownian functionals, and derive the Feynman-Kac equations for an important class of stochastic processes. Specifically, they derive the forward and backward fractional Feynman-Kac equations based on a CTRW model.

In [Turgeman *et al.* (2009)], the authors consider a random walk on a one-dimensional lattice with the lattice spacing a. Particles only jump to the nearest left or right neighbors with equal probabilities. And the waiting time distributions between successive jumps are IID which has the same power law PDF form

$$\phi(\tau) \sim \frac{b_\alpha \tau^{-(1+\alpha)}}{|\Gamma(-\alpha)|}, \qquad (3.7)$$

with b_α being a constant and $0 < \alpha < 1$. Under this condition, the mean waiting time is infinite which leads to anomalous diffusion. The renewal process of the CTRW model comes as follows. First the particle waits on the initial point x_0 for a random time τ drawn from the distribution $\phi(\tau)$ and then jumps to the position $x_0 + a$ with probability $1/2$, or alternatively, to the position $x_0 - a$ with the same probability $1/2$. After that, the process is renewed and repeated.

3.2.1 *Forward Fractional Feynman-Kac Equation*

For the non-Brownian functional, to start with, let $G(x, A, t)$ be the joint PDF of finding the particle at position x with the functional value A at time t. Assume the particle performed its last jump at time $t - \tau$. According to the CTRW model, the particle is at position x with the functional value A at time t if it arrived at position x with the functional value $A - \tau U(x)$ at time $t - \tau$ and its waiting time exceeds the length τ. Let $Q_n(x, A, t)dxdA$ be the probability of the particle arriving exactly in $(x, x + dx)$ with the functional value $(A, A + dA)$ after n jumps at time t. Denote

$$\Phi(\tau) = 1 - \int_0^\tau \phi(\tau')\, d\tau' \qquad (3.8)$$

as the survival probability, namely, the probability for not moving in the time interval $(t - \tau, t)$. Consequently, one can get the relation

$$G(x, A, t) = \int_0^t \Phi(\tau) \sum_{n=0}^\infty Q_n[x, A - \tau U(x), t - \tau]d\tau. \qquad (3.9)$$

If the particle reaches exactly the point x after $n + 1$ steps, it should reach one of the nearest neighbor points $x - a$ or $x + a$ after n steps, with equal probabilities. Similarly, in order to have the functional value A after $n + 1$ steps, the particle is supposed to have the functional value $A - \tau U(x - a)$, or $A - \tau U(x + a)$, respectively, after n steps. Here τ is the random waiting time between the n steps and $n + 1$ steps, which is distributed as $\phi(\tau)$. Naturally, this derivation leads to the recursive relation

$$Q_{n+1}(x, A, t) = \frac{1}{2} \int_0^t \phi(\tau)Q_n[x - a, A - \tau U(x - a), t - \tau]d\tau$$
$$\qquad (3.10)$$
$$+ \frac{1}{2} \int_0^t \phi(\tau)Q_n[x + a, A - \tau U(x + a), t - \tau]d\tau.$$

Suppose the functionals only have positive support. Naturally one can conduct the Laplace transform $A \to p$, and particularly,

$$\int_0^\infty Q_n[x, A - U(x)\tau, t - \tau]e^{-pA}dA = e^{-pU(x)\tau}Q_n(x, p, t - \tau), \quad (3.11)$$

using the property of Laplace transform. Thus, conducting Laplace transform $A \to p$ of both sides of Eq. (3.9) gives

$$G(x, p, t) = \int_0^t \Phi(\tau) \sum_{n=0}^\infty e^{-pU(x)\tau}Q_n(x, p, t - \tau)d\tau. \quad (3.12)$$

Next step is to introduce the Laplace transform w.r.t. time, $t \to s$, of both sides of Eq. (3.12). Using the convolution theorem of Laplace transform, one obtains

$$\hat{G}(x, p, s) = \sum_{n=0}^\infty \frac{1 - \hat{\phi}[s + pU(x)]}{s + pU(x)}\hat{Q}_n(x, p, s), \quad (3.13)$$

where

$$\hat{\phi}(s) = \int_0^\infty \phi(t)\exp(-st)dt \quad (3.14)$$

is the Laplace transform of the waiting time distribution. One can further perform Fourier transform $x \to k$ on Eq. (3.13), which yields

$$\tilde{\hat{G}}(k, p, s) = \sum_{n=0}^\infty \frac{1 - \hat{\phi}\left[s + pU\left(-i\frac{\partial}{\partial k}\right)\right]}{s + pU\left(-i\frac{\partial}{\partial k}\right)}\tilde{\hat{Q}}_n(k, p, s). \quad (3.15)$$

Similarly, performing the Laplace transforms $A \to p$ as well as $t \to s$ and Fourier transform $x \to k$ sequentially on the both sides of Eq. (3.10) yields

$$\tilde{\hat{Q}}_{n+1}(k, p, s) = \cos(ka)\hat{\phi}\left[s + pU\left(-i\frac{\partial}{\partial k}\right)\right]\tilde{\hat{Q}}_n(k, p, s). \quad (3.16)$$

Note that the initial condition for $n = 0$ is given by

$$\tilde{\hat{Q}}_0(k, p, s) = 1, \quad (3.17)$$

since

$$Q_0(x, A, t) = \delta(x)\delta(A)\delta(t), \quad (3.18)$$

namely, the particle start from $x = 0$ at $t = 0$ with $A = 0$. Combining Eq. (3.15) and Eq. (3.16), it can be obtained that

$$\tilde{\hat{G}}(k, p, s) = \frac{1 - \hat{\phi}[s + pU(-i\frac{\partial}{\partial k})]}{s + pU(-i\frac{\partial}{\partial k})} \cdot \frac{1}{1 - \cos(ka)\hat{\phi}[s + pU(-i\frac{\partial}{\partial k})]}. \quad (3.19)$$

When $p = 0$, according to the definition of the Laplace transform,

$$\tilde{\hat{Q}}(k, p = 0, s) = \int_0^\infty \tilde{Q}(k, A, s)dA = \tilde{Q}(k, s). \qquad (3.20)$$

Moreover, substituting $p = 0$ into Eq. (3.19), one can easily recover the well-known Montroll-Weiss equation [Metzler and Klafter (2000); Montroll and Weiss (1965)]

$$\tilde{Q}(k, s) = \frac{1 - \hat{\phi}(s)}{s} \frac{1}{1 - \cos(ka)\hat{\phi}(s)}. \qquad (3.21)$$

For the assumed power law waiting time distribution, one can find its Laplace transform in the small s expansion

$$\hat{\phi}(s) = 1 - b_\alpha s^\alpha + \cdots . \qquad (3.22)$$

And for simplicity, denote the anomalous diffusion coefficient as

$$K_\alpha = \frac{a^2}{2b_\alpha}. \qquad (3.23)$$

Next step is to substitute Eq. (3.22) into Eq. (3.19) and thus in the long time and large scale limits, one finds

$$\tilde{\hat{G}}(k, p, s) \sim \left[s + pU\left(-i\frac{\partial}{\partial k}\right) \right]^{\alpha-1} \frac{1}{K_\alpha k^2 + \left[s + pU\left(-i\frac{\partial}{\partial k}\right) \right]^\alpha}. \qquad (3.24)$$

Rearrange the above expression and invert it back to the time-space domain. We get the forward fractional Feynman-Kac equation

$$\frac{\partial G(x, p, t)}{\partial t} = K_\alpha \frac{\partial^2}{\partial x^2} D_t^{1-\alpha} G(x, p, t) - pU(x)G(x, p, t), \qquad (3.25)$$

where the fractional substantial derivative operator $D_t^{1-\alpha}$ is defined by Eq. (2.13) on page 29.

3.2.2 *Backward Fractional Feynman-Kac Equation*

In practice, the backward Feynman-Kac equation turns out to be very useful. Let $G_{x_0}(A, t)$ be the PDF of the functional A, when the process starts at the initial position x_0. According to the CTRW model, the particle jumps to either $x_0 + a$ or $x_0 - a$ with equal probabilities, when the very first renewal happens at time τ satisfying $\tau < t$. Alternatively, the particle does not move at all during the whole measurement time $(0, t)$. Translating this observation to an equation, there exists [Turgeman *et al.* (2009)]

$$G_{x_0}(A, t) = \frac{1}{2} \int_0^t \phi(\tau) G_{x_0+a}[A - \tau U(x_0), t - \tau] d\tau$$
$$+ \frac{1}{2} \int_0^t \phi(\tau) G_{x_0-a}[A - \tau U(x_0), t - \tau] d\tau \qquad (3.26)$$
$$+ \Phi(t)\delta[A - U(x_0)t].$$

Then using the Laplace-Fourier transform technique similar to that used in the derivation of the forward Feynman-Kac equation, one finds, in the continuum limit, the backward fractional Feynman-Kac equation

$$\frac{\partial}{\partial t} G_{x_0}(p, t) = K_\alpha D_t^{1-\alpha} \frac{\partial^2}{\partial x_0^2} G_{x_0}(p, t) - pU(x_0) G_{x_0}(p, t). \qquad (3.27)$$

3.2.3   *Distribution of Occupation Times*

The occupation time in half space is defined as

$$t^+ = \int_0^t \theta[x(\tau)] d\tau, \qquad (3.28)$$

and its distribution was first calculated by Lamperti [Lamperti (1958)]. In [Turgeman *et al.* (2009)], the authors also consider this functional as an application of the derived backward Feynman-Kac equation. Substitute $\theta(x_0)$ of Eq. (3.28) into Eq. (3.27), and solve the backward Feynman-Kac equation separately for $x_0 > 0$ and $x_0 < 0$ with the continuities of the solution and its first derivative regarding x_0 (in the Laplace domain). Denote the fraction of time spent in half space as

$$p^+ = \frac{t^+}{t}. \qquad (3.29)$$

For initial position $x_0 = 0$, it is found that, after inversion of the solution to the time domain, the PDF of p^+ is

$$f(p^+) = \frac{\sin \frac{\pi \alpha}{2}}{\pi} \times \frac{(p^+)^{\alpha/2-1} (1 - p^+)^{\alpha/2-1}}{(1 - p^+)^\alpha + (p^+)^\alpha + 2(1 - p^+)^{\alpha/2} (p^+)^{\alpha/2} \cos \frac{\pi \alpha}{2}}. \qquad (3.30)$$

The above Lamperti PDF has two peaks at $p^+ = 1$ and $p^+ = 0$. Its minimum is achieved when $p^+ = 1/2$. The expectation is

$$\langle p^+ \rangle = \frac{1}{2}. \qquad (3.31)$$

In the limit $\alpha \to 0$, two delta functions on $p^+ = 1$ and $p^+ = 0$ are obtained, which indicates that the particle is localized in either $x > 0$ or $x < 0$ for the whole observation time. For $\alpha \to 1$, the well-known arcsine law of Lévy is recovered [Majumdar (2005)].

In [Carmi *et al.* (2010); Carmi and Barkai (2011)] the authors continue to investigate the different scenarios of the fractional Feynman-Kac equations and their multiple applications, including but not limited to the first passage time, the maximal displacement, weak ergodicity breaking, and the hitting probability.

3.3 Tempered Fractional Feynman-Kac Equations

Tempered anomalous diffusion describes the very slow transition from anomalous to normal diffusion, and it has many applications in physical, biological, and chemical processes [Cartea and del Castillo-Negrete (2007); Bruno *et al.* (2004); Baeumer and Meerschaert (2010); Meerschaert *et al.* (2014); Stanislavsky *et al.* (2008)]. The tempered anomalous diffusion can be described by the CTRW model with truncated power law waiting time and/or jump length distribution(s). In [Wu *et al.* (2016)], the authors take the tempered power law function as the waiting time distribution in the CTRW model and derive the forward and backward Feynman-Kac equations governing the distribution of the functionals of the tempered anomalous diffusion. The derivations include several cases: random walk on lattice, random walk on lattice with forces, and random walk with (tempered) power law jump distribution. Besides the derivations of the tempered fractional Feynman-Kac equations, several concrete examples of the functionals of the tempered anomalous diffusion are analytically and explicitly analyzed.

3.3.1 *Model and Tempered Dynamics*

The CTRW model with tempered power law waiting time distribution is used as the underlying process leading to tempered anomalous diffusion. First, one can consider the CTRW on an infinite one-dimensional lattice with the unique spacing distance a. And a particle is only allowed to jump to its nearest neighbors. When the external potential is placed on the lattice, one can assume that the probabilities of jumping left $L(x)$ and right $R(x)$ depend on the external force $F(x)$ at the position x. If $F(x)=0$, then

$$R(x) = L(x) = \frac{1}{2}. \tag{3.32}$$

The waiting times between two successive jump events are IID random variables with exponentially truncated stable distribution (ETSD) $\psi(t, \lambda)$, and are independent of the external force. It characterizes the slow transition from the anomalous to normal diffusion, controlled by the parameter λ. This ETSD is useful for rigorous analysis of diffusion behavior because it is an infinitely divisible distribution, and thus its distribution or characteristic function can be explicitly derived. The Laplace transform for $\phi(t, \lambda)$ is given by [Feller (1971)]

$$e^{\hat{\eta}(s,\lambda)} = \int_0^{+\infty} \phi(t, \lambda)e^{-st}dt, \tag{3.33}$$

where

$$\hat{\eta}(s, \lambda) = -B_\alpha(\lambda + s)^\alpha + B_\alpha\lambda^\alpha. \tag{3.34}$$

Hence, the Laplace transform (for small s and λ) of ETSD $\phi(t, \lambda)$ results in

$$\hat{\phi}(s, \lambda) \simeq 1 - B_\alpha(\lambda + s)^\alpha + B_\alpha\lambda^\alpha. \tag{3.35}$$

The process starts at $x = x_0$, and the particle waits at x_0 for a random time t drawn from the distribution $\phi(t, \lambda)$. After that, it jumps to either $x_0 + a$ (with probability $R(x_0)$) or $x_0 - a$ (with probability $L(x_0)$). Then, the process is renewed in the same way.

3.3.2 *Tempered Fractional Feynman-Kac Equations of Random Walk on a One-Dimensional Lattice*

Consider the CTRW on a lattice with

$$R(x) = L(x) = \frac{1}{2}. \tag{3.36}$$

Let $G(x, A, t)$ be the joint PDF of finding the particle at position x and time t with the functional value A. Here the functional A is defined in Eq. (3.6). And $\tilde{\hat{G}}(k, p, s)$ is the Fourier transform $x \to k$, and Laplace transforms $t \to s$, $A \to p$ of $G(x, A, t)$. In this subsection, based on the CTRW model describing the tempered anomalous diffusion, the forward and backward tempered fractional Feynman-Kac equations are derived. We start with the derivation of the forward tempered fractional Feynman-Kac equation for the random walk on a lattice. Substituting the asymptotic form of $\hat{\phi}(s, \lambda)$ into Eq. (3.19) and using

$$\cos(ka) \simeq 1 - \frac{a^2k^2}{2}, \tag{3.37}$$

for the long wavelength $k \to 0$ corresponding to large x (or the continuous limit $a \to 0$) as is well known, then we have

$$\tilde{\hat{G}}(k, p, s) \simeq \frac{b_\alpha[\lambda + s + pU(-i\frac{\partial}{\partial k})]^\alpha - b_\alpha \lambda^\alpha}{s + pU(-i\frac{\partial}{\partial k})}$$

$$\times \frac{1}{\frac{a^2 k^2}{2} + b_\alpha[\lambda + s + pU(-i\frac{\partial}{\partial k})]^\alpha - b_\alpha \lambda^\alpha}. \quad (3.38)$$

After some rearrangements, we invert it to the space-time domain $k \to x$ and $s \to t$ and the tempered fractional Feynman-Kac equation is obtained

$$\frac{\partial}{\partial t} G(x, p, t) = \left[\lambda^\alpha D_t^{1-\alpha,\lambda} - \lambda\right] \left[G(x, p, t) - e^{-pU(x)t} \delta(x)\right]$$

$$- pU(x)G(x, p, t) + K_\alpha \frac{\partial^2}{\partial x^2} D_t^{1-\alpha,\lambda} G(x, p, t), \quad (3.39)$$

with the initial condition

$$G(x, A, t = 0) = \delta(x)\delta(A), \quad (3.40)$$

or

$$G(x, p, t = 0) = \delta(x), \quad (3.41)$$

where $\delta(\cdot)$ is the Dirac delta function. And the constant

$$K_\alpha = \frac{a^2}{2b_\alpha}, \quad (3.42)$$

with units m^2/sec^α, is finite for $a \to 0$ and $b_\alpha \to 0$. This is a generalized Einstein relation for tempered motion. In Laplace space, we have

$$D_t^{1-\alpha,\lambda} \to [\lambda + s + pU(x)]^{1-\alpha}, \quad (3.43)$$

and in t space, the tempered fractional substantial derivative is

$$D_t^{1-\alpha,\lambda} G(x, p, t) =$$

$$\frac{1}{\Gamma(\alpha)} \left[\lambda + pU(x) + \frac{\partial}{\partial t}\right] \int_0^t \frac{e^{-(t-\tau)\cdot(\lambda + pU(x))}}{(t - \tau)^{1-\alpha}} G(x, p, \tau) d\tau. \quad (3.44)$$

Thus, due to the long waiting times, the evolution of $G(x, p, t)$ is non-Markovian and depends on the entire history. Next is to list several special cases of the forward tempered fractional Feynman-Kac equation. First of all, when λ is finite and $\alpha = 1$, the corresponding equation is

$$\frac{\partial}{\partial t} G(x, p, t) = K_1 \frac{\partial^2}{\partial x^2} G(x, p, t) - pU(x)G(x, p, t), \quad (3.45)$$

which is simply the classical Feynman-Kac equation [Kac (1949)]. Thus, exponential truncation has no effect on normal diffusion. As is well known the Feynman-Kac equation is the imaginary time Schrödinger equation, where $U(x)$ serves as the potential field. When $\lambda = 0$, Eq. (3.39) reduces to the imaginary time fractional Schrödinger equation [Turgeman *et al.* (2009)], namely, the fractional Feynman-Kac equation

$$\frac{\partial}{\partial t}G(x,p,t) = K_\alpha \frac{\partial^2}{\partial x^2} D_t^{1-\alpha} G(x,p,t) - pU(x)G(x,p,t). \qquad (3.46)$$

Then we continue with the derivation of the backward tempered Feynman-Kac equation for the random walk on a lattice. In some cases we may be just interested in the distribution of A, so integrating $G(x,A,t)$ over all x is necessary. Therefore, it would be convenient to obtain an equation for $G_{x_0}(A,t)$, which is the PDF of the functional A at time t for a process starting at x_0. According to the CTRW model, the particle starts at $x = x_0$; after its first jump at time τ, it is at either $x_0 + a$ or $x_0 - a$. Alternatively, the particle doesn't move at all during the measurement time $(0, t)$. Translating this process to an equation with the tempered power law waiting time distribution, there exists

$$\begin{aligned}
G_{x_0}(A,t) = \ &\frac{1}{2}\int_0^t \phi(\tau,\lambda)G_{x_0+a}[A-\tau U(x_0), t-\tau]d\tau \\
&+ \frac{1}{2}\int_0^t \phi(\tau,\lambda)G_{x_0-a}[A-\tau U(x_0), t-\tau]d\tau \\
&+ \Phi(t,\lambda)\delta[A-U(x_0)t],
\end{aligned} \qquad (3.47)$$

where $\tau U(x_0)$ is the contribution to A from the pausing time on x_0 in the time interval $(0,\tau)$; and the probability that particle remains motionless on its initial location is

$$\Phi(t,\lambda) = 1 - \int_0^t \phi(\tau,\lambda)d\tau. \qquad (3.48)$$

The Laplace transform of $\Phi(t,\lambda)$ follows from the form for the Laplace transform of an integral and reads

$$\hat{\Phi}(s,\lambda) = \frac{1 - \hat{\phi}(s,\lambda)}{s}. \qquad (3.49)$$

Here $\hat{\phi}(s,\lambda)$ is also given by Eq. (3.35). Taking Laplace transforms $A \to p$ and $t \to s$, as well as Fourier transform $x_0 \to k$, we have

$$\begin{aligned}
\tilde{\hat{G}}_k(p,s) = \ &\left\{1 - b_\alpha\left[\lambda + pU\left(-i\frac{\partial}{\partial k}\right) + s\right]^\alpha + b_\alpha \lambda^\alpha\right\}\cos(ka)\tilde{\hat{G}}_k(p,s) \\
&+ \frac{b_\alpha[\lambda + pU\left(-i\frac{\partial}{\partial k}\right) + s]^\alpha - b_\alpha\lambda^\alpha}{pU\left(-i\frac{\partial}{\partial k}\right) + s}\delta(k).
\end{aligned}$$

$$(3.50)$$

Rearranging the expressions and taking approximation $k \to 0$, $\cos(ka) \simeq 1 - \frac{a^2 k^2}{2}$ in the last equation we find

$$\left[\lambda + pU \left(-i \frac{\partial}{\partial k} \right) + s \right]^{1-\alpha} \frac{a^2 k^2}{2b_\alpha} \tilde{G}_k(p, s) - \delta(k)$$

$$+ \left\{ \left[\lambda + pU \left(-i \frac{\partial}{\partial k} \right) + s \right] - \lambda^\alpha \left[\lambda + pU \left(-i \frac{\partial}{\partial k} \right) + s \right]^{1-\alpha} \right\} \tilde{G}_k(p, s)$$

$$= \frac{\lambda - \lambda^\alpha [\lambda + pU(-i \frac{\partial}{\partial k}) + s]^{1-\alpha}}{pU(-i \frac{\partial}{\partial k}) + s} \delta(k).$$

$$(3.51)$$

Inverting to the space-time domain $s \to t$ and $k \to x_0$ similar to that used in the derivation of the forward equation, in the continuum limit, we get the backward tempered fractional Feynman-Kac equation

$$\frac{\partial}{\partial t} G_{x_0}(p, t) = \left[\lambda^\alpha D_t^{1-\alpha, \lambda} - \lambda \right] \left[G_{x_0}(p, t) - e^{-pU(x_0)t} \right] - pU(x_0) G_{x_0}(p, t)$$

$$+ K_\alpha D_t^{1-\alpha, \lambda} \frac{\partial^2}{\partial x_0^2} G_{x_0}(p, t).$$

$$(3.52)$$

The initial condition is

$$G_{x_0}(A, t = 0) = \delta(A), \qquad (3.53)$$

or

$$G_{x_0}(p, t = 0) = 1. \qquad (3.54)$$

The symbol $D_t^{1-\alpha, \lambda}$ is the tempered fractional substantial derivative, in the Laplace domain, defined as

$$D_t^{1-\alpha, \lambda} \to [\lambda + s + pU(x_0)]^{1-\alpha}. \qquad (3.55)$$

Notice that here, this operator appears to the left of the Laplacian $\frac{\partial^2}{\partial x_0^2}$ in Eq. (3.52), in contrast to the forward equation (3.39). When $\lambda = 0$, Eq. (3.52) turns to the backward fractional Feynman-Kac equation [Turgeman et al. (2009)]

$$\frac{\partial}{\partial t} G_{x_0}(p, t) = K_\alpha D_t^{1-\alpha} \frac{\partial^2}{\partial x_0^2} G_{x_0}(p, t) - pU(x_0) G_{x_0}(p, t). \qquad (3.56)$$

3.3.3   Tempered Fractional Feynman-Kac Equations of Random Walk with Forces

In this subsection we consider the CTRW on lattice but with forces, which means the probabilities of jumping left $L(x)$ and right $R(x)$ are no longer equal. Assuming the system is coupled to a heat bath at temperature T and stays in detailed balance, there exists

$$L(x) \exp\left[-\frac{V(x)}{k_B T}\right] = R(x-a) \exp\left[-\frac{V(x-a)}{k_B T}\right]. \tag{3.57}$$

For small a, expanding $R(x)$, $L(x)$, and the exponential function leads to

$$R(x) \simeq \frac{1}{2}\left[1 + \frac{aF(x)}{2k_B T}\right] \tag{3.58}$$

and

$$L(x) \simeq \frac{1}{2}\left[1 - \frac{aF(x)}{2k_B T}\right], \tag{3.59}$$

where

$$F(x) = -V'(x). \tag{3.60}$$

At first, we investigate the derivation of the forward tempered fractional Feynman-Kac equation with forces. For the long wavelength $k \to 0$, using

$$\sin(ka) \simeq ka \tag{3.61}$$

and following Eq. (18) in [Carmi and Barkai (2011)] lead to

$$\tilde{\hat{G}}(k,p,s) = \frac{1 - \hat{\phi}[s + pU(-i\frac{\partial}{\partial k}), \lambda]}{s + pU(-i\frac{\partial}{\partial k})}$$

$$\times \frac{1}{1 - \left[1 - \frac{a^2 k^2}{2} + i(ka)\frac{aF(-i\frac{\partial}{\partial K})}{2k_B T}\right]\hat{\phi}[s + pU(-i\frac{\partial}{\partial k}), \lambda]}. \tag{3.62}$$

Substituting $\hat{\phi}(s, \lambda)$ into Eq. (3.62) and rearranging the equation, we obtain

$$\frac{a^2}{2b_\alpha}\left\{k^2 - ik\frac{F\left(-i\frac{\partial}{\partial k}\right)}{k_B T}\right\}\left[\lambda + s + pU\left(-i\frac{\partial}{\partial k}\right)\right]^{1-\alpha}\tilde{\hat{G}}(k,p,s)$$

$$+ \left[\lambda + s + pU\left(-i\frac{\partial}{\partial k}\right)\right]\tilde{\hat{G}}(k,p,s) - 1$$

$$- \lambda^\alpha\left[\lambda + s + pU\left(-i\frac{\partial}{\partial k}\right)\right]^{1-\alpha}\tilde{\hat{G}}(k,p,s)$$

$$= \frac{\lambda - \lambda^\alpha[\lambda + s + pU(-i\frac{\partial}{\partial k})]^{1-\alpha}}{s + pU(-i\frac{\partial}{\partial k})}. \tag{3.63}$$

Inverting $k \to x$, $s \to t$, then

$$\frac{\partial}{\partial t}G(x,p,t) = \left[\lambda^{\alpha}D_t^{1-\alpha,\lambda} - \lambda\right]\left[G(x,p,t) - e^{-pU(x)t}\delta(x)\right] - pU(x)G(x,p,t)$$
$$+ K_{\alpha}\left[\frac{\partial^2}{\partial x^2} - \frac{\partial}{\partial x}\frac{F(x)}{k_B T}\right]D_t^{1-\alpha,\lambda}G(x,p,t).$$

(3.64)

If $F(x) = 0$, then Eq. (3.64) turns to Eq. (3.39). When $\lambda = 0$, Eq. (3.64) changes to

$$\frac{\partial}{\partial t}G(x,p,t) = K_{\alpha}\left[\frac{\partial^2}{\partial x^2} - \frac{\partial}{\partial x}\frac{F(x)}{k_B T}\right]D_t^{1-\alpha}G(x,p,t) - pU(x)G(x,p,t),$$

(3.65)

which is exactly the same as Eq. (22) given in [Carmi and Barkai (2011)]. Next step is to derive the backward tempered fractional Feynman-Kac equation with forces. As mentioned in the previous subsection, if we are just interested in the distribution of the functional A, the backward Feynman-Kac equation should be more useful and convenient. For $\tilde{\hat{G}}_k(p,s)$, the following equation holds [Carmi and Barkai (2011)]

$$\tilde{\hat{G}}_k(p,s) = \hat{\phi}\left[pU\left(-i\frac{\partial}{\partial k}\right) + s, \lambda\right] \cdot \left[\cos(ka) - \frac{aF(-i\frac{\partial}{\partial k})}{2k_B T}i\sin(ka)\right]\tilde{\hat{G}}_k(p,s)$$
$$+ \hat{\Phi}\left[pU\left(-i\frac{\partial}{\partial k}\right) + s, \lambda\right]\delta(k).$$

(3.66)

Substituting the specific forms of $\hat{\Phi}(s)$ and $\hat{\phi}(s,\lambda)$ into Eq. (3.66), and after some rearrangements, we have

$$\frac{a^2}{2b_{\alpha}}\left[\lambda + pU\left(-i\frac{\partial}{\partial k}\right) + s\right]^{1-\alpha}\left[k^2 + \frac{F(-i\frac{\partial}{\partial k})}{k_B T}(ik)\right]\tilde{\hat{G}}_k(p,s)$$
$$+ \left[\lambda + pU\left(-i\frac{\partial}{\partial k}\right) + s\right]\tilde{\hat{G}}_k(p,s) - \delta(k)$$
$$- \lambda^{\alpha}\left[\lambda + pU\left(-i\frac{\partial}{\partial k}\right) + s\right]^{1-\alpha}\tilde{\hat{G}}_k(p,s)$$
$$= \frac{\lambda - \lambda^{\alpha}[\lambda + pU(-i\frac{\partial}{\partial k}) + s]^{1-\alpha}}{pU(-i\frac{\partial}{\partial k}) + s}\delta(k).$$

(3.67)

Taking inversions of the above equation, $k \to x_0$ and $s \to t$, we get

$$\frac{\partial}{\partial t}G_{x_0}(p,t) = \left[\lambda^{\alpha}D_t^{1-\alpha,\lambda} - \lambda\right]\left[G_{x_0}(p,t) - e^{-pU(x_0)t}\right] - pU(x_0)G_{x_0}(p,t)$$
$$+ K_{\alpha}D_t^{1-\alpha,\lambda}\left[\frac{\partial^2}{\partial x_0^2} + \frac{F(x_0)}{k_B T}\frac{\partial}{\partial x_0}\right]G_{x_0}(p,t).$$

(3.68)

If $F(x) = 0$, then Eq. (3.68) is exactly the same as Eq. (3.52). For $\lambda = 0$, Eq. (3.68) reduces to

$$\frac{\partial}{\partial t} G_{x_0}(p, t) = K_\alpha D_t^{1-\alpha} \left[\frac{\partial^2}{\partial x_0^2} + \frac{F(x_0)}{k_B T} \frac{\partial}{\partial x_0} \right] G_{x_0}(p, t) - pU(x_0) G_{x_0}(p, t).$$

(3.69)

3.3.4   *Distribution of Occupation Time in Half Space*

In probability theory, an important object of interest is the occupation time, i.e., the time spent by a stochastic motion above the origin within a time window of size t. Thus define the occupation time in the half space $x > 0$ as

$$t^+ = A = \int_0^t \theta[x(\tau)] d\tau,$$

(3.70)

i.e., $U(x) = \theta(x) = 1$ for $x \geq 0$ and is zero otherwise. For example, for Brownian motion the PDF of t^+ is the famous arcsine distribution. In order to find the PDF of the occupation time, here we consider the backward fractional Feynman-Kac equation Eq. (3.52) with regular jump length distribution in Laplace s space. When $x_0 < 0$, we have

$$- K_\alpha (\lambda + s)^{1-\alpha} \frac{\partial^2}{\partial x_0^2} \hat{G}_{x_0}(p, s) + (\lambda + s) \hat{G}_{x_0}(p, s) - 1$$

$$= \lambda^\alpha (\lambda + s)^{1-\alpha} \hat{G}_{x_0}(p, s) + \frac{\lambda - \lambda^\alpha (\lambda + s)^{1-\alpha}}{s}.$$

(3.71)

And for $x_0 > 0$, correspondingly,

$$- K_\alpha (\lambda + s + p)^{1-\alpha} \frac{\partial^2}{\partial x_0^2} \hat{G}_{x_0}(p, s) + (\lambda + s + p) \hat{G}_{x_0}(p, s) - 1$$

$$= \lambda^\alpha (\lambda + s + p)^{1-\alpha} \hat{G}_{x_0}(p, s) + \frac{\lambda - \lambda^\alpha (\lambda + s + p)^{1-\alpha}}{s + p}.$$

(3.72)

Rewriting the above equations leads to

$$\hat{G}_{x_0}(p, s) = \begin{cases} K_\alpha \frac{1}{(\lambda+s)^\alpha - \lambda^\alpha} \frac{\partial^2}{\partial x_0^2} \hat{G}_{x_0}(p, s) + \frac{1}{s}, & x_0 < 0; \\ \\ K_\alpha \frac{1}{(\lambda+s+p)^\alpha - \lambda^\alpha} \frac{\partial^2}{\partial x_0^2} \hat{G}_{x_0}(p, s) + \frac{1}{s+p}, & x_0 > 0. \end{cases}$$

(3.73)

Solving the equations in each half space individually and requiring that $\hat{G}_{x_0}(p, s)$ is finite for $|x_0| \to \infty$ lead to

$$\hat{G}_{x_0}(p, s) = \begin{cases} C_0 \exp\left(x_0 \sqrt{\frac{(\lambda+s)^\alpha - \lambda^\alpha}{K_\alpha}} \right) + \frac{1}{s}, & x_0 < 0; \\ \\ C_1 \exp\left(-x_0 \sqrt{\frac{(\lambda+s)^\alpha - \lambda^\alpha}{K_\alpha}} \right) + \frac{1}{s+p}, & x_0 > 0. \end{cases}$$

(3.74)

The particle can never arrive at $x > 0$ for $x_0 \to -\infty$; thus $G_{x_0}(T^+, t) = \delta(t^+)$ and $\hat{G}_{x_0}(p, s) = \frac{1}{s}$, in conformity to Eq. (3.74). Likewise, for $x_0 \to +\infty$, the particle is never at $x < 0$ and thus $G_{x_0}(t^+, t) = \delta(T^+ - t)$ and $\hat{G}_{x_0}(p, s) = \frac{1}{s+p}$, as expected in Eq. (3.74). Then demanding that $\hat{G}_{x_0}(p, s)$ and its first derivative are continuous at $x_0 = 0$, yields a pair of equations about C_0 and C_1, i.e.,

$$\begin{cases} C_0 + \frac{1}{s} = C_1 + \frac{1}{s+p}; \\ \\ C_0 \sqrt{(\lambda + s)^\alpha - \lambda^\alpha} = -C_1 \sqrt{(\lambda + s + p)^\alpha - \lambda^\alpha}. \end{cases} \tag{3.75}$$

By solving these equations, we get

$$\begin{cases} C_0 = -\dfrac{p\sqrt{(\lambda+s+p)^\alpha - \lambda^\alpha}}{s(s+p)(\sqrt{(\lambda+s+p)^\alpha - \lambda^\alpha} + \sqrt{(\lambda+s)^\alpha - \lambda^\alpha})}; \\ \\ C_1 = \dfrac{p\sqrt{(\lambda+s)^\alpha - \lambda^\alpha}}{s(s+p)(\sqrt{(\lambda+s+p)^\alpha - \lambda^\alpha} + \sqrt{(\lambda+s)^\alpha - \lambda^\alpha})}. \end{cases} \tag{3.76}$$

Assume that the particle starts at $x_0 = 0$. Substituting $x_0 = 0$ in Eq. (3.74), then

$$\hat{G}_0(p, s) = C_0 + \frac{1}{s} = C_1 + \frac{1}{s+p}, \tag{3.77}$$

namely,

$$\hat{G}_0(p, s) = \frac{s\sqrt{(\lambda + s + p)^\alpha - \lambda^\alpha} + (s + p)\sqrt{(\lambda + s)^\alpha - \lambda^\alpha}}{s(s + p)(\sqrt{(\lambda + s + p)^\alpha - \lambda^\alpha} + \sqrt{(\lambda + s)^\alpha - \lambda^\alpha})}, \tag{3.78}$$

which describes the PDF of t^+ and is valid for all times. Specially, if $\alpha = 1$, then

$$\hat{G}_0(p, s) = s^{-1/2}(s + p)^{-1/2}. \tag{3.79}$$

This can be inverted to give the equilibrium PDF of the occupation fraction t^+/t,

$$f_{t^+/t}(x) = \frac{1}{\pi\sqrt{x(1 - x)}},$$

which is the arcsine law [Majumdar and Comtet (2002)].

3.3.5 *Distribution of First Passage Time*

As is well known, the first passage time can be the time t_f which a particle starting at $x_0 = -b$ ($b > 0$) takes to hit $x = 0$ for the first time [Redner (2001)] and is widely applied in physics and other disciplines. A relationship between the distribution of first passage time and the occupation time functional is [Kac (1951)]

$$P\{t_f > t\} = P\left\{\max_{0 \leq \tau \leq t} x(\tau) < b\right\} = \lim_{p \to \infty} G_{x_0}(p, t), \qquad (3.80)$$

where $G_{x_0}(p, t)$ describes the Laplace transform of the PDF of functional t^+. For $x_0 = -b$ and $p \to \infty$, according to Eqs. (3.74) and (3.76), we have

$$\lim_{p \to \infty} \hat{G}_{-b}(p, s) = \frac{1}{s} - \frac{1}{s} \exp\left(-b\sqrt{\frac{(\lambda + s)^\alpha - \lambda^\alpha}{K_\alpha}}\right). \qquad (3.81)$$

In accordance with the definition of the first passage time, its PDF satisfies

$$f(t) = \frac{\partial}{\partial t}(1 - P\{t_f > t\}) = -\frac{\partial}{\partial t} \lim_{p \to \infty} G_{-b}(p, t). \qquad (3.82)$$

Hence, in the Laplace space, we have

$$\hat{f}(s) = -s \lim_{p \to \infty} \hat{G}_{-b}(p, s) + 1$$

$$= \exp\left(-b\sqrt{\frac{(\lambda + s)^\alpha - \lambda^\alpha}{K_\alpha}}\right). \qquad (3.83)$$

Expanding Eq. (3.83) in small s,

$$\hat{f}(s) \simeq 1 - b\sqrt{\frac{\alpha \lambda^{\alpha-1} s}{K_\alpha}}. \qquad (3.84)$$

Taking inverse Laplace transform for long time, $s \to t$, we have

$$f(t) \simeq \frac{b}{|\Gamma(-\frac{1}{2})|}\sqrt{\frac{\alpha \lambda^{\alpha-1}}{K_\alpha}} t^{-\frac{3}{2}}, \qquad (3.85)$$

for all $\alpha \in (0, 1)$, which coincides with the famous $t^{-\frac{3}{2}}$ decay law [Redner (2001)] of a one-dimensional random walk and decreases with the increasing of λ. Hence,

$$P\{t_f > t\} = \int_t^\infty f(t_f) dt_f \simeq \frac{b}{\sqrt{\pi}}\sqrt{\frac{\alpha \lambda^{\alpha-1}}{K_\alpha}} t^{-1/2}. \qquad (3.86)$$

However, if $s \to \infty$, corresponding to small t, from Eq. (3.83), we have

$$\hat{f}(s) \simeq \exp\left(-\frac{b}{\sqrt{K_\alpha}} s^{\frac{\alpha}{2}}\right). \tag{3.87}$$

In t space, the above equation tends to be the one sided Lévy laws $l_{\alpha/2}(t)$. Hence $f(t)$ decays very fast to zero when $t \to 0$ and behaves as $t^{-1-\alpha/2}$ for short but not too short time. When $\lambda = 0$, then waiting times are power law distributed. Equation (3.83) becomes

$$\hat{f}(s) = \exp\left(-\frac{b}{\sqrt{K_\alpha}} s^{\frac{\alpha}{2}}\right). \tag{3.88}$$

In t space, Eq. (3.88) is the one sided Lévy laws $l_{\alpha/2}(t)$. And then $f(t)$ decays very fast to zero when $t \to 0$. For $t \to \infty$, $f(t)$ behaves as $t^{-(1+\alpha/2)}$, which is in agreement with the results given in [Barkai (2001)], indicating that $\langle t_f \rangle$ is infinite for all $\alpha \in (0,1)$.

3.3.6 *Distribution of Maximal Displacement*

Now we develop another application of Eq. (3.74). Let

$$x_m \equiv \max_{0 \le \tau \le t} x(\tau). \tag{3.89}$$

Then

$$P\{x_m < b\} = \lim_{p \to \infty} G_{x_0}(p, t). \tag{3.90}$$

From the last subsection we have, for $x_0 = -b$ (Eq. (3.81)),

$$P\{x_m < b\} = \frac{1}{s} - \frac{1}{s}\exp\left(-b\sqrt{\frac{(\lambda+s)^\alpha - \lambda^\alpha}{K_\alpha}}\right). \tag{3.91}$$

Then the PDF of x_m is

$$\hat{p}(x_m, s) = \frac{1}{s}\sqrt{\frac{(\lambda+s)^\alpha - \lambda^\alpha}{K_\alpha}} \exp\left(-x_m\sqrt{\frac{(\lambda+s)^\alpha - \lambda^\alpha}{K_\alpha}}\right). \tag{3.92}$$

When $\lambda = 0$, the above equation becomes

$$\hat{p}(x_m, s) = \frac{1}{s}\sqrt{\frac{s^\alpha}{K_\alpha}} \exp\left(-x_m\sqrt{\frac{s^\alpha}{K_\alpha}}\right). \tag{3.93}$$

Inverting $s \to t$, $x_m > 0$, we have

$$p(x_m, t) = \sqrt{\frac{8}{\alpha^2 K_\alpha}} \frac{t}{\left(x_m\sqrt{\frac{2}{K_\alpha}}\right)^{1+\frac{2}{\alpha}}} l_{\frac{\alpha}{2}}\left[\frac{t}{\left(x_m\sqrt{\frac{2}{K_\alpha}}\right)^{\frac{2}{\alpha}}}\right]. \tag{3.94}$$

The PDF is in agreement with the result of [Schehr and Le Doussal (2010)], derived via a renormalization group method.

3.3.7 Fluctuations of Occupation Fraction

In this subsection, we focus on the occupation fraction

$$\varepsilon \equiv \frac{t^+}{t}. \tag{3.95}$$

As shown in the previous subsection, it is hard to analytically invert Eq. (3.78), which governs the distribution $\hat{G}_0(p, s)$ of the occupation time in the half space. But it can be effective to calculate its moments, according to

$$\langle (t^+)^n \rangle_s = (-1)^n \frac{\partial^n}{\partial p^n} \hat{G}_0(p, s)|_{p=0}. \tag{3.96}$$

The first moment is calculated as

$$\langle t^+ \rangle_s = -\frac{\partial}{\partial p} \hat{G}_0(p, s)|_{p=0} = \frac{1}{2s^2}. \tag{3.97}$$

Performing the inverse Laplace transform, we have

$$\langle t^+ \rangle = \frac{t}{2}, \tag{3.98}$$

or

$$\langle \varepsilon \rangle_t = \frac{1}{2}, \tag{3.99}$$

which coincides with the results of $\lambda = 0$. That is to say, exponential tempering has no influence on the first moment of the occupation time, as expected from symmetry.

For the second moment, we have

$$\langle (t^+)^2 \rangle_s = \frac{\partial^2}{\partial p^2} \hat{G}_0(p, s)|_{p=0} = \frac{1}{s^3} - \frac{\alpha(s+\lambda)^{\alpha-1}}{4s^2[(s+\lambda)^\alpha - \lambda^\alpha]}. \tag{3.100}$$

Inverting both sides of Eq. (3.100) to the time domain leads to

$$\langle (t^+)^2 \rangle \simeq \frac{t^2}{2} - \frac{\alpha}{4} t * e^{-\lambda t} E_{\alpha,1}[\lambda^\alpha t^\alpha], \tag{3.101}$$

where the symbol '$*$' describes the convolution operator

$$f(t) * g(t) = \int_0^t f(t - \tau) g(\tau) d\tau, \tag{3.102}$$

and we used the Laplace transform relation

$$\int_0^\infty e^{-st} E_{\alpha,1}(at^\alpha) dt = \frac{s^{\alpha-1}}{s^\alpha - a}. \tag{3.103}$$

Here $E_{\alpha,1}(z)$ is the Mittag-Leffler function, defined as

$$E_{\alpha,1}(z) = \sum_{n=0}^{\infty} \frac{z^n}{\Gamma(1+\alpha n)}. \tag{3.104}$$

When $s \to \infty$, λ can be ignored in Eq. (3.100). Then we obtain

$$\langle (t^+)^2 \rangle_s \simeq \frac{4-\alpha}{4s^3}. \tag{3.105}$$

Hence, when $t \to 0$, the inverse Laplace transform of above equation gives that

$$\langle (t^+)^2 \rangle \simeq \frac{4-\alpha}{8} t^2. \tag{3.106}$$

In fact, this result can also be derived with a different method. When $t \to 0$, we have

$$E_{\alpha,1}(\lambda^\alpha t^\alpha) \simeq 1 + \frac{\lambda^\alpha t^\alpha}{\Gamma(1+\alpha)}. \tag{3.107}$$

Consequently, we can calculate that

$$
\begin{aligned}
& t * e^{-\lambda t} E_{\alpha,1}(\lambda^\alpha t^\alpha) \\
& \simeq \int_0^t (t-\tau) e^{-\lambda \tau} \left[1 + \frac{\lambda^\alpha \tau^\alpha}{\Gamma(1+\alpha)} \right] d\tau \\
& = \frac{e^{-\lambda t} + \lambda t - 1}{\lambda^2} + \frac{e^{-\lambda t} t^{\alpha+1} \lambda^{\alpha-1}}{\Gamma(\alpha+1)} \\
& \quad + \frac{t\lambda^\alpha - (\alpha+1)\lambda^{\alpha-1}}{\Gamma(\alpha+1)} \int_0^t e^{-\lambda \tau} \tau^\alpha d\tau.
\end{aligned} \tag{3.108}
$$

As $t \to 0$, the second and the third terms are zeros. Meanwhile, we use the approximation

$$e^{-\lambda t} \simeq 1 - \lambda t + \frac{\lambda^2 t^2}{2}. \tag{3.109}$$

Therefore, we have

$$t * e^{-\lambda t} E_{\alpha,1}(\lambda^\alpha t^\alpha) \simeq \frac{t^2}{2}. \tag{3.110}$$

Substituting it into Eq. (3.101), we obtain the same result.

Finally, we obtain the fluctuations of the occupation fraction

$$\langle (\Delta\varepsilon)^2 \rangle_t = \langle \varepsilon^2 \rangle_t - \langle \varepsilon \rangle_t^2 \simeq \frac{1-\alpha/2}{4}. \tag{3.111}$$

This is the expected result as [Carmi and Barkai (2011)], since for short time λ has no effect on the process and the PDF Eq. (3.78) is Lamperti's law with index $\alpha/2$. As $s \to 0$, expanding Eq. (3.100) in small s leads to

$$\langle (t^+)^2 \rangle_s \simeq \frac{1}{s^3} - \frac{1}{4s^3} = \frac{3}{4s^3}. \tag{3.112}$$

Taking inverse Laplace transform, we find that when $t \to \infty$,

$$\langle (t^+)^2 \rangle \simeq \frac{3}{8} t^2.$$

Then, when $t \to \infty$,

$$\langle (\triangle \varepsilon)^2 \rangle_t \simeq \frac{1}{8}. \tag{3.113}$$

3.4   Feynman-Kac Equations Revisited: Langevin Picture

The CTRW models and Langevin equations are popular microscopic models to mathematically model stochastic dynamics in the natural world. Under the framework of CTRW, the Feynman-Kac equations governing the PDFs of the functionals have been derived, including those of the paths of anomalous diffusion [Turgeman *et al.* (2009); Wu *et al.* (2016); Carmi *et al.* (2010)] and reaction diffusion processes [Hou and Deng (2018)].

As is well known, Langevin equations are more convenient when we are concerned with the effect of an external field and/or stochastic noises generated from a fluctuating environment. In [Wang *et al.* (2018b)], the authors derive the corresponding forward Feynman-Kac equations governing the PDFs of the functionals of paths of the Langevin system with both space- and time-dependent force fields and arbitrary multiplicative noise, and the backward version is proposed for a system with arbitrary additive noise or multiplicative Gaussian white noise together with a force field.

3.4.1   *Forward Feynman-Kac Equation*

Consider the dynamical system with a fluctuating environment described by the overdamped Langevin equation:

$$\dot{x}(t) = f(x(t), t) + g(x(t), t)\xi(t), \tag{3.114}$$

where $x(t)$ is the particle coordinate, $f(x, t)$ is the force field, $\xi(t)$ is the Lévy noise resulting from a fluctuating environment, and $g(x, t)$ is the multiplicative noise term. The Lévy noise $\xi(t)$ is used, which is the formal time derivative of its corresponding Lévy process $\eta(t)$. That is to say, the increment of $\eta(t)$ could be defined as the time integral of $\xi(t)$,

$$\delta\eta(t) = \eta(t + \tau) - \eta(t) = \int_t^{t+\tau} \xi(t')dt'. \tag{3.115}$$

Similarly, the increment of $x(t)$ is defined as

$$\delta x(t) = x(t + \tau) - x(t). \tag{3.116}$$

For the particle trajectory $x(t)$ undergoing the Langevin system Eq. (3.114) during a time interval τ $(\tau \to 0)$, its increment satisfies

$$\delta x(t) = f(x(t), t)\tau + g(x(t), t)\delta\eta(t), \tag{3.117}$$

in the Itô interpretation [Itô (1950); Risken (1989)]. Because of the stationary increment property of the Lévy process, we know that $\delta\eta(t)$ has the same distribution as $\eta(\tau)$ with characteristic function denoted by [Applebaum (2009)]:

$$\langle e^{-ik\eta(\tau)} \rangle = e^{\tau\phi_0(k)}, \tag{3.118}$$

where the Lévy exponent $\phi_0(k)$ characterizes the jump structure of the Lévy noise $\xi(t)$. For a specific Lévy noise, it has the specific form that

$$\phi_0(k) = -k^2 \tag{3.119}$$

for Gaussian white noise and

$$\phi_0(k) = -|k|^\beta \tag{3.120}$$

for non-Gaussian β-stable Lévy noise.

In order to obtain the joint PDF of position x and functional A at time t, $G(x, A, t)$, we define its Fourier transforms $x \to k$, $A \to p$ as

$$\tilde{G}(k, p, t) = \int_{-\infty}^{\infty} \int_{-\infty}^{\infty} e^{-ikx-ipA}G(x, A, t)dxdA,$$

and we write it in the usual way,

$$\tilde{G}(k, p, t) = \langle e^{-ikx(t)}e^{-ipA(t)} \rangle. \tag{3.121}$$

Being similar to the increment $\delta x(t)$ in Eq. (3.117), one has the increment

$$\delta A(t) = A(t + \tau) - A(t) = U(x(t))\tau, \tag{3.122}$$

during the time interval τ $(\tau \to 0)$. Then we consider the increment of $G(x, A, t)$ in Fourier space,

$$\delta\tilde{G}(k, p, t) := \tilde{G}(k, p, t + \tau) - \tilde{G}(k, p, t), \tag{3.123}$$

which can be written as

$$\delta\tilde{G}(k,p,t) = \langle e^{-ikx(t+\tau)-ipA(t+\tau)} \rangle - \langle e^{-ikx(t)-ipA(t)} \rangle. \tag{3.124}$$

Substituting the increments $\delta x(t)$ as well as $\delta A(t)$ into Eq. (3.124) and taking $\tau \to 0$, we obtain

$$
\begin{aligned}
\delta\tilde{G}(k,p,t) = {} & \langle e^{-ikx(t)-ipA(t)} \big(e^{-ikg(x(t),t)\delta\eta(t)} - 1 \big) \rangle \\
& - ik\tau \langle e^{-ikx(t)-ipA(t)} f(x(t),t) \rangle \\
& - ip\tau \langle e^{-ikx(t)-ipA(t)} U(x(t)) \rangle.
\end{aligned}
\tag{3.125}
$$

Note that the first term on the right hand side of Eq. (3.125) denotes the joint PDF $G(x,A,t)$ and the PDF of the noise increment $\delta\eta(t)$, and $\delta\eta(t)$ is independent of particle trajectory $x(t)$. The characteristic function of the noise increment $\delta\eta(t)$ in Eq. (3.118) gives that

$$\lim_{\tau\to0} \frac{1}{\tau} \langle (e^{-ikg(x(t),t)\delta\eta(t)} - 1) \rangle = \phi_0(kg(x(t),t)). \tag{3.126}$$

The second and third terms in Eq. (3.125) are exactly the Fourier transforms of a compound function regrading $G(x,A,t)$, namely,

$$ik\langle e^{-ikx(t)-ipA(t)} f(x(t),t) \rangle = \mathcal{F}_x\mathcal{F}_A\left\{ \frac{\partial}{\partial x} f(x,t)G(x,A,t) \right\}, \tag{3.127}$$

and

$$ip\langle e^{-ikx(t)-ipA(t)} U(x(t)) \rangle = ip\mathcal{F}_x\mathcal{F}_A\left\{ U(x)G(x,A,t) \right\}. \tag{3.128}$$

Based on Eqs. (3.126), (3.127) and (3.128), dividing Eq. (3.125) by τ and taking the limit $\tau \to 0$, we obtain the forward Feynman-Kac equation in Fourier space:

$$
\begin{aligned}
\frac{\partial\tilde{G}(k,p,t)}{\partial t} = {} & \mathcal{F}_x\{\phi_0(kg(x,t))G(x,p,t)\} \\
& - \mathcal{F}_x\left\{ \frac{\partial}{\partial x} f(x,t)G(x,p,t) + ipU(x)G(x,p,t) \right\}.
\end{aligned}
\tag{3.129}
$$

Once the form of $\phi_0(kg(x,t))$ is given for a specific noise, the corresponding forward Feynman-Kac equation in x space is obtained.

If the deterministic time variable in Langevin equation Eq. (3.114) is replaced by a positive non-decreasing one-dimensional Lévy process, called subordinator [Applebaum (2009)], then the subordinated stochastic process could be described by the following coupled Langevin equation

$$
\begin{aligned}
\dot{x}(s) &= f(x(s),T(s)) + g(x(s),T(s))\xi(s), \\
\dot{T}(s) &= \zeta(s).
\end{aligned}
\tag{3.130}
$$

Here we adopt the fully skewed α-stable Lévy noise $\zeta(s)$ with $0 < \alpha < 1$, which is independent of the arbitrary Lévy noise $\xi(s)$. Then the combined process is defined as

$$y(t) = x(S(t)) \qquad (3.131)$$

with the inverse α-stable subordinator $S(t)$, which is the first passage time of the α-stable subordinator $\{T(s), s \geq 0\}$ and defined [Piryatinska *et al.* (2005); Magdziarz *et al.* (2006)] as $S(t) = \inf_{s>0}\{s : T(s) > t\}$. Note that the time-dependent force f and multiplicative noise term g should depend on the physical time $T(s)$, rather than the operation time s, due to the physical interpretation [Magdziarz *et al.* (2008); Heinsalu *et al.* (2007)]. Denote the corresponding functional of process $y(t)$ as

$$W(t) = \int_0^t U(y(t'))dt'. \qquad (3.132)$$

Then the forward Feynman-Kac equation of the joint PDF $G(y, W, t)$ in Fourier space $(y \to k, W \to p)$ is

$$\frac{\partial \tilde{G}(k,p,t)}{\partial t} = \mathcal{F}_y\{\phi_0(kg(y,t))D_t^{1-\alpha}G(y,p,t)\}$$
$$- \mathcal{F}_y\left\{ \frac{\partial}{\partial y}f(y,t)D_t^{1-\alpha}G(y,p,t) + ipU(y)G(y,p,t) \right\}, \qquad (3.133)$$

which recovers Eq. (3.129) when $\alpha = 1$.

The next step is to consider some particular cases of the derived equations above. To start with, let $p = 0$ in Eq. (3.129). In this case, $G(x, p = 0, t) = \int_0^\infty G(x, A, t)dA$ reduces to $G(x, t)$, the marginal PDF of finding the particle at position x at time t. Correspondingly, the forward Feynman-Kac equation Eq. (3.129) reduces to the generalized Fokker-Planck equation [Denisov *et al.* (2009)], where three kinds of noises (Gaussian white noise, Poisson white noise and Lévy stable noise) are considered for the specific forms of this equation. If the noise $\xi(t)$ is the Gaussian white noise in Eq. (3.133), for arbitrary $f(x, t)$ and $g(x, t)$, we get the forward Feynman-Kac equation:

$$\frac{\partial G(y,p,t)}{\partial t} = \left[-\frac{\partial}{\partial y}f(y,t) + \frac{\partial^2}{\partial y^2}g^2(y,t) \right]$$
$$\times D_t^{1-\alpha}G(y,p,t) - ipU(y)G(y,p,t). \qquad (3.134)$$

This equation is consistent with the forward Feynman-Kac equation with inverse α-stable subordinator proposed in [Cairoli and Baule (2017)] by Langevin-type approach. Especially when $g(x, t) \equiv 1$, it recovers the equation in [Carmi and Barkai (2011)] by CTRW models. If the noise $\xi(t)$ is the non-Gaussian β-stable noise in Eq. (3.133), for arbitrary $f(x,t)$ and $g(x,t)$, the forward Feynman-Kac equation becomes

$$\frac{\partial G(y,p,t)}{\partial t} = \left[-\frac{\partial}{\partial y} f(y,t) + \nabla_y^\beta |g(y,t)|^\beta \right]$$
$$\times D_t^{1-\alpha} G(y,p,t) - ipU(y)G(y,p,t),$$

(3.135)

where ∇_y^β is the Riesz space fractional derivative operator with Fourier symbol $-|k|^\beta$ [Wu *et al.* (2016); Carmi *et al.* (2010)]; and in y space,

$$\nabla_y^\beta h(y) = -\frac{-\infty D_y^\beta h(y) + {}_y D_\infty^\beta h(y)}{2\cos(\beta\pi/2)},$$

where for $n-1 < \beta < n$,

$$-\infty D_y^\beta h(y) = \frac{1}{\Gamma(n-\beta)} \frac{d^n}{dy^n} \int_{-\infty}^y \frac{h(y')}{(y-y')^{\beta+1-n}} dy',$$

$${}_y D_\infty^\beta h(y) = \frac{(-1)^n}{\Gamma(n-\beta)} \frac{d^n}{dy^n} \int_y^\infty \frac{h(y')}{(y'-y)^{\beta+1-n}} dy'.$$

This equation extends Eq. (3.134) to Lévy stable noise, denoting the heavy-tailed jump length in CTRW models.

Note that if the functional A is positive at any time t, the Fourier transform $A \to p$ will be replaced by the Laplace transform

$$G(x,p,t) = \int_0^\infty e^{-pA} G(x,A,t) dA.$$

(3.136)

Under this condition, the corresponding forward Feynman-Kac equation will be obtained by replacing ip with p in Eq. (3.133).

3.4.2 *Backward Feynman-Kac Equation*

As presented above, the forward Feynman-Kac equation Eq. (3.135), now we focus on deriving the backward Feynman-Kac equation governing $G_{x_0}(A,t)$—the PDF of functional A at time t, given that the process has started at x_0. In the following derivation, we consider this stochastic process

$$\dot{x}(t) = f(x(t)) + g(x(t))\xi(t),$$

(3.137)

where $\xi(t)$ is also a Lévy noise. Note that here the functions f and g do not explicitly depend on the time variable t. If not, the time-dependent force field (or the multiplicative term) induces a different displacement for a particle located at the same position but different time. In this case, it is difficult to let the functional A only depend on the initial position x_0 without using the information of the whole path $x(t)$.

Different from the increment δA considered in the forward Feynman-Kac equation, here we should build the relation between A and x_0 as, during the time interval τ ($\tau \to 0$),

$$A(t+\tau)|_{x_0} = \int_0^\tau U(x(t'))dt' + \int_\tau^{t+\tau} U(x(t'))dt'$$
$$= U(x_0)\tau + A(t)|_{x(\tau)}, \qquad (3.138)$$

where $A(t+\tau)|_{x_0}$ denotes the functional A at time $t+\tau$ with the initial position x_0. Let $t=0$ in Eq. (3.117) and $x(\tau)$ can be expressed as

$$x(\tau) = x_0 + f(x_0)\tau + g(x_0)\eta(\tau). \qquad (3.139)$$

Expressing $G_{x_0}(A,t)$ in the Fourier space as

$$G_{x_0}(p,t) = \langle e^{-ipA(t)|_{x_0}} \rangle,$$

we could get the form of $G_{x_0}(p,t+\tau)$ from Eq. (3.138) as

$$G_{x_0}(p,t+\tau) = \langle \langle e^{-ipA(t)|_{x(\tau)}} \rangle \rangle e^{-ipU(x_0)\tau}, \qquad (3.140)$$

where the internal angular brackets denote the average of $A(t)|_{x(\tau)}$ and the external ones are the average of $\eta(\tau)$.

Then the increment $\delta G_{x_0}(p,t)$ can be expressed as

$$\delta G_{x_0}(p,t) := G_{x_0}(p,t+\tau) - G_{x_0}(p,t)$$
$$= \langle \langle e^{-ipA(t)|_{x(\tau)}} \rangle \rangle e^{-ipU(x_0)\tau} - \langle e^{-ipA(t)|_{x_0}} \rangle.$$

Taking $\tau \to 0$, and omitting the higher order term of τ, we get

$$\delta G_{x_0}(p,t) = \langle \langle e^{-ipA(t)|_{x(\tau)}} \rangle \rangle - \langle e^{-ipA(t)|_{x_0}} \rangle$$
$$- ipU(x_0)\tau \langle e^{-ipA(t)|_{x_0}} \rangle, \qquad (3.141)$$

where the last term on the right hand side equals to $-ipU(x_0)\tau G_{x_0}(p,t)$. Next, we will deal with the first two terms on the right hand side of Eq. (3.141) carefully by keeping the terms containing $\mathcal{O}(\tau)$, denoting the same order of τ, but removing the terms of $o(\tau)$, signifying the higher order. Taking Fourier transform $x_0 \to k_0$ in Eq. (3.141), then $\langle e^{-ipA(t)|_{x_0}} \rangle$ becomes $\tilde{G}_{k_0}(p,t)$. But for $\langle \langle e^{-ipA(t)|_{x(\tau)}} \rangle \rangle$, it is not easy to get the form in Fourier space. Hence, we assume that the noise in this system is additive and we take $g(x) \equiv 1$ for simplicity. For the nonconstant $g(x)$ case, one can refer to [Wang *et al.* (2018b)] for details. Denote $T_\eta = \langle e^{-ipA(t)|_{x(\tau)}} \rangle$. Since $g(x) \equiv 1$, Eq. (3.139) becomes

$$x(\tau) = x_0 + f(x_0)\tau + \eta(\tau), \qquad (3.142)$$

where $f(x_0)$ depends on the initial position x_0. Therefore, $x(\tau)$ is not a simple shift of x_0 and we write the Fourier transform ($x_0 \to k_0$) of $\langle T_\eta \rangle$ as

$$\mathcal{F}_{x_0}\{\langle T_\eta \rangle\} = \left\langle \int_{-\infty}^{\infty} e^{-ik_0 x(\tau)} T_\eta e^{ik_0(f(x_0)\tau + \eta(\tau))} dx_0 \right\rangle.$$

Then we turn dx_0 into $dx(\tau)$ and get

$$\mathcal{F}_{x_0}\{\langle T_\eta \rangle\} = \left\langle \int_{-\infty}^{\infty} e^{-ik_0 x(\tau)} T_\eta e^{ik_0(f(x_0)\tau + \eta(\tau))} dx(\tau) \right\rangle$$
$$- \left\langle \int_{-\infty}^{\infty} e^{-ik_0 x(\tau)} T_\eta e^{ik_0(f(x_0)\tau + \eta(\tau))} \frac{df(x_0)}{dx_0} \tau dx_0 \right\rangle. \tag{3.143}$$

Since all x_0 and $f(x_0)$ are multiplied by τ in Eq. (3.143), replacing all x_0 by $x(\tau)$ in Eq. (3.143) yields higher-order terms of τ, which can be omitted. Recall

$$e^{ik_0 f(x_0)\tau} \simeq 1 + ik_0 f(x_0)\tau, \tag{3.144}$$

and then the first term on the right hand side of Eq. (3.143) reduces to

$$\left\langle \int_{-\infty}^{\infty} e^{-ik_0 x(\tau)} T_\eta e^{ik_0 \eta(\tau)} dx(\tau) \right\rangle + ik_0 \tau \left\langle \int_{-\infty}^{\infty} e^{-ik_0 x(\tau)} T_\eta f(x(\tau)) dx(\tau) \right\rangle,$$

where the second term equals to

$$\tau \mathcal{F}_{x_0}\left\{ \frac{\partial}{\partial x_0} f(x_0) G_{x_0}(p,t) \right\}. \tag{3.145}$$

The second term on the right hand side of Eq. (3.143) gives

$$-\tau \left\langle \int_{-\infty}^{\infty} e^{-ik_0 x(\tau)} T_\eta \frac{df(x(\tau))}{dx(\tau)} dx(\tau) \right\rangle = -\tau \mathcal{F}_{x_0}\left\{ \frac{df(x_0)}{dx_0} G_{x_0}(p,t) \right\}.$$

Therefore, the Fourier transform of $\langle\langle e^{-ipA(t)|_{x(\tau)}} \rangle\rangle - \langle e^{-ipA(t)|_{x_0}} \rangle$ in Eq. (3.141), with $x(\tau)$ replaced by y, reduces to

$$\left\langle \int_{-\infty}^{\infty} e^{-ik_0 y} T_\eta (e^{ik_0 \eta(\tau)} - 1) dy \right\rangle + \tau \mathcal{F}\left\{ f(x_0) \frac{\partial G_{x_0}(p,t)}{\partial x_0} \right\},$$

i.e.,

$$\tau \phi_0(-k_0) \tilde{G}_{k_0}(p,t) + \tau \mathcal{F}_{x_0}\left\{ f(x_0) \frac{\partial G_{x_0}(p,t)}{\partial x_0} \right\}.$$

Dividing Eq. (3.141) by τ and taking the limit $\tau \to 0$, we obtain the backward Feynman-Kac equation in Fourier space:

$$\frac{\partial \tilde{G}_{k_0}(p,t)}{\partial t} = \phi_0(-k_0) \tilde{G}_{k_0}(p,t)$$
$$+ \mathcal{F}_{x_0}\left\{ f(x_0) \frac{\partial G_{x_0}(p,t)}{\partial x_0} - ipU(x_0) G_{x_0}(p,t) \right\}. \tag{3.146}$$

If the noise $\xi(t)$ is Gaussian white noise, then $\phi_0(-k_0) = -k_0^2$ and we get the backward Feynman-Kac equation:

$$\frac{\partial G_{x_0}(p,t)}{\partial t} = \frac{\partial^2}{\partial x_0^2} G_{x_0}(p,t) + f(x_0)\frac{\partial}{\partial x_0} G_{x_0}(p,t) - ipU(x_0)G_{x_0}(p,t),$$

$$(3.147)$$

which is the same as the backward Feynman-Kac equation with $\alpha = 1$ proposed in [Carmi and Barkai (2011)] in the CTRW framework. Here, α is the exponent characterizing the waiting time PDF in CTRW models or the subordinator PDF in the Langevin system.

If the noise $\xi(t)$ is non-Gaussian β-stable noise, i.e., $\phi_0(-k_0) = -|k_0|^\beta$, then the backward Feynman-Kac equation becomes

$$\frac{\partial G_{x_0}(p,t)}{\partial t} = \nabla_{x_0}^\beta G_{x_0}(p,t) + f(x_0)\frac{\partial}{\partial x_0} G_{x_0}(p,t) - ipU(x_0)G_{x_0}(p,t). \quad (3.148)$$

This is an extension for the backward Feynman-Kac equation derived in the CTRW framework [Carmi *et al.* (2010)], where the jump length obeys heavy-tailed distribution but without a force field $f(x)$.

When $g(x)$ is not a constant, we assume $\xi(t)$ to be Gaussian white noise and derive the backward Feynman-Kac equation

$$\frac{\partial G_{x_0}(p,t)}{\partial t} = g^2(x_0)\frac{\partial^2}{\partial x_0^2} G_{x_0}(p,t) + f(x_0)\frac{\partial}{\partial x_0} G_{x_0}(p,t) - ipU(x_0)G_{x_0}(p,t),$$

$$(3.149)$$

which goes back to Eq. (3.147) when $g(x_0) \equiv 1$.

3.4.3 *Distribution of Occupation Time in Positive Half Space*

We first discuss the occupation time in $x > 0$ for a particle moving freely but with a multiplicative Gaussian white noise in a box $[-L, L]$ and $L > 0$. Particularly, we take $U(x_0)$ in Eq. (3.149) to be $\theta(x_0)$, and then get occupation time of a particle in the positive half space as

$$T^+(t) = \int_0^t \theta[x(t')]dt'. \quad (3.150)$$

In this case, $T^+(t)$ is always positive. Thus, we replace the Fourier transform by Laplace transform in Eq. (3.149) and remove i in it. To find the distribution of $T^+(t)$, we take the Laplace transform of the backward Feynman-Kac equation Eq. (3.149) ($t \to s$):

$$s\hat{G}_{x_0}(p,s) - 1 = g^2(x_0)\frac{\partial^2}{\partial x_0^2}\hat{G}_{x_0}(p,s) + f(x_0)\frac{\partial}{\partial x_0}\hat{G}_{x_0}(p,s) - pU(x_0)\hat{G}_{x_0}(p,s).$$

$$(3.151)$$

To consider the effect of multiplicative noise, we specify

$$f(x_0) = 0 \qquad (3.152)$$

and

$$g(x_0) = aL - x_0 \qquad (3.153)$$

with $a > 1$ to keep $g(x_0)$ positive. The constant aL in $g(x_0)$ measures the intensity of the additive component of the random force. Hence, Eq. (3.151) becomes

$$(aL - x_0)^2 \frac{\partial^2 \hat{G}_{x_0}(p, s)}{\partial x_0^2} - (s + pU(x_0))\hat{G}_{x_0}(p, s) = -1.$$

With a variable substitution $y = aL - x_0 > 0$, the celebrated Euler equation is obtained:

$$y^2 \frac{\partial^2 \hat{\bar{G}}_y(p, s)}{\partial y^2} - (s + p\bar{U}(y))\hat{\bar{G}}_y(p, s) = -1,$$

which can be solved by another variable substitution $y = e^t$. Finally, we get the solutions of Eq. (3.151) in two half-spaces, respectively,

$$\hat{G}_{x_0}(p, s)$$
$$= \begin{cases} C_1(aL - x_0)^{\lambda_1} + C_2(aL - x_0)^{\lambda_2} + \frac{1}{s+p} & x_0 > 0 \\ C_3(aL - x_0)^{\lambda_3} + C_4(aL - x_0)^{\lambda_4} + \frac{1}{s} & x_0 < 0, \end{cases} \qquad (3.154)$$

where

$$\lambda_{1,2} = \frac{1 \mp \sqrt{1 + 4(s + p)}}{2}, \quad \lambda_{3,4} = \frac{1 \mp \sqrt{1 + 4s}}{2}. \qquad (3.155)$$

Assume the reflecting boundary condition to Eq. (3.154), i.e.,

$$\left. \frac{\partial \hat{G}_{x_0}(p, s)}{\partial x_0} \right|_{x_0 = \pm L} = 0. \qquad (3.156)$$

The two conditions Eq. (3.156) together with two other conditions ($\hat{G}_{x_0}(p, s)$ and its derivative are continuous at $x_0 = 0$) can determine the four coefficients C_{1-4} in Eq. (3.154). Then we get the final solution $\hat{G}_{x_0}(p, s)$ at $x_0 = 0$:

$$\hat{G}_0(p, s) = \frac{p}{s(p + s)} \cdot \frac{F_1 F_2}{F_3 F_4 - F_1 F_2} + \frac{1}{s}, \qquad (3.157)$$

where

$$F_1 = a^{\lambda_4} - \frac{\lambda_4}{\lambda_3}(a+1)^{\lambda_4 - \lambda_3} a^{\lambda_3},$$

$$F_2 = \lambda_2 [a^{\lambda_2} - (a-1)^{\lambda_2 - \lambda_1} a^{\lambda_1}],$$

$$F_3 = \lambda_4 [a^{\lambda_4} - (a+1)^{\lambda_4 - \lambda_3} a^{\lambda_3}],$$

$$F_4 = a^{\lambda_2} - \frac{\lambda_2}{\lambda_1}(a-1)^{\lambda_2 - \lambda_1} a^{\lambda_1}.$$

Equation (3.157) is the PDF of T^+ in Laplace space, but it cannot be inverted easily. Nevertheless, the first moment of the occupation time $T^+(t)$ can be computed by taking the inverse Laplace transform [Klafter and Sokolov (2011)] of

$$\langle T^+(s) \rangle = - \left. \frac{\partial \hat{G}_0(p, s)}{\partial p} \right|_{p=0}.$$

By this formula, from Eq. (3.157), one can get

$$\langle T^+(s) \rangle = -\frac{1}{s^2} \cdot \left. \frac{F_1 F_2}{F_3 F_4 - F_1 F_2} \right|_{p=0}. \tag{3.158}$$

For long time, i.e., $s \ll 1$, $(\lambda_1 = \lambda_3 \sim -s, \lambda_2 = \lambda_4 \sim 1)$,

$$\langle T^+(t) \rangle \simeq \frac{a+1}{2a} t. \tag{3.159}$$

For short time, i.e., $s \gg 1$, $(\lambda_1 = \lambda_3 \sim -\sqrt{s}, \lambda_2 = \lambda_4 \sim \sqrt{s})$,

$$\langle T^+(t) \rangle \simeq \frac{1}{2} t. \tag{3.160}$$

It can be seen that for both long time and short time, $\langle T^+(t) \rangle$ scales asymptotically as t. Therefore, it is natural to consider the PDF of the occupation fraction

$$T_f \equiv \frac{T^+}{t}. \tag{3.161}$$

For long time, i.e., $s \ll 1$, together with $p \ll 1$ due to the scale of $T^+(t)$, we have $\lambda_1 \sim -(s+p), \lambda_2 \sim 1, \lambda_3 \sim -s, \lambda_4 \sim 1$ from Eq. (3.155) and $F_1 \sim (a+1)/s$, $F_2 \sim 1$, $F_3 \sim -1$, $F_4 \sim (a-1)/(s+p)$, which gives the asymptotic expression of Eq. (3.157):

$$\hat{G}_0(p, s) \simeq \frac{2a}{2as + (a+1)p}.$$

By inverting the scaling form of a double Laplace transform in [Godrèche and Luck (2001)], after some calculations, using the nascent delta function:

$$\lim_{\epsilon \to 0} \frac{\epsilon}{\pi(x^2 + \epsilon^2)} = \delta(x),$$

we obtain the PDF of T_f:

$$G(T_f) \simeq \frac{r}{T_f} \cdot \delta(T_f - r) \overset{d}{=} \delta(T_f - r), \qquad (3.162)$$

where $r = \frac{a+1}{2a}$ and '$\overset{d}{=}$' denotes identical distribution. Note that the PDF of T_f in Eq. (3.162) is normalized. Especially, T_f reduces to a deterministic event for large t, occurring at r with probability 1. But the value r depends on a. When a is sufficiently large, this value will approach $1/2$. For short time, i.e., $s \gg 1$, we have $\lambda_1 \sim -\sqrt{s+p}$, $\lambda_2 \sim \sqrt{s+p}$, $\lambda_3 \sim -\sqrt{s}$, $\lambda_4 \sim \sqrt{s}$ from Eq. (3.155) and $F_1 \sim (a+1)^{2\sqrt{s}} a^{-\sqrt{s}}$, $F_2 \sim \sqrt{s+p} a^{\sqrt{s+p}}$, $F_3 \sim -\sqrt{s}(A+1)^{2\sqrt{s}} a^{-\sqrt{s}}$, $F_4 \sim a^{\sqrt{s+p}}$, which result in the asymptotic expression of Eq. (3.157):

$$\hat{G}_0(p, s) \simeq -\frac{p}{s(p+s)} \cdot \frac{\sqrt{s+p}}{\sqrt{s} + \sqrt{s+p}} + \frac{1}{s}.$$

Then we obtain the PDF of T_f:

$$G(T_f) \simeq \frac{1}{\pi} \cdot \frac{1}{\sqrt{x}\sqrt{1-x}}, \qquad (3.163)$$

which is consistent with the classical Brownian functional [Majumdar (2005)]. This result is as expected since for short time the particle does not interact with the boundaries and behaves like a free particle. Furthermore, if the time t is sufficiently small, such that $x \ll aL$, then the multiplicative noise term approximates an additive noise term aL, so the PDFs of occupation fractions T_f in cases $g(x) = aL \pm x$ both become the Lamperti PDF [Wang *et al.* (2018b)].

3.4.4 *Distribution of First Passage Time*

We further investigate the first passage time t_f based on the results of the occupation time in half box. Assume a particle moves freely in the box $[-L, L]$. By definition, t_f denotes the time it takes a particle starting at $x_0 = -bL$, $0 < b < 1$ to reach $x = 0$ for the first time [Redner (2001)]. The distribution of t_f can be obtained from the occupation time distribution by using an identity due to Kac [Kac (1951)]. Taking $x_0 = -bL$ in Eq. (3.154) in the previous subsection, we get

$$\hat{G}_{-bL}(p, s) = \frac{p}{s(p+s)} \cdot \frac{F_{1b}F_2}{F_3F_4 - F_1F_2} + \frac{1}{s}, \qquad (3.164)$$

where $F_1, \cdots, F_4$ are the same as the ones in Eq. (3.157) and

$$F_{1b} = (a+b)^{\lambda_4} - \frac{\lambda_4}{\lambda_3}(a+1)^{\lambda_4-\lambda_3}(a+b)^{\lambda_3}.$$

When $p \to \infty$, we consider the long-time behaviour (i.e., $s \to 0$) and have $\lambda_1 \sim -\sqrt{p}, \lambda_2 \sim \sqrt{p}, \lambda_3 \sim -s, \lambda_4 \sim 1$. Substituting λ_{1-4} into Eq. (3.164) yields

$$\lim_{p \to \infty} \hat{G}_{-bL}(p, s) \simeq \ln\left(1 + \frac{b}{A}\right) - \frac{b}{1 + A} =: C_{Ab},$$

which is a constant only depending on a and b.

Since the PDF of the first passage time satisfies

$$f(t) = \frac{\partial}{\partial t}[1 - P\{t_f > t\}], \tag{3.165}$$

and we have the PDF of t_f in Laplace s space

$$\hat{f}(s) \simeq 1 - C_{Ab}s \simeq e^{-C_{Ab}s},$$

and thus in the time domain,

$$f(t_f) \simeq \delta(t_f - C_{Ab}).$$

This means that the first passage time is a deterministic event, occurring at C_{Ab} with probability 1. Furthermore, for $0 < b < 1 < A$, C_{Ab} is monotonously increasing of b but decreasing of A, being the same as physical intuition.

3.4.5 *Area under Random Walk Curve*

Now we turn to one application of the derived Feynman-Kac equation. Assume $U(x) = x$ and we have the functional

$$A_x = \int_0^t x(t')dt', \tag{3.166}$$

which denotes the total area under the curve of the trajectory $x(t)$ [Friedrich *et al.* (2006); Grebenkov (2007)]. This functional A_x is also related to the phase accumulated by spins in an NMR experiment [Grebenkov (2007)]. Since the analytical solutions of $G_{x_0}(p, t)$ in Eq. (3.148) cannot be easily obtained due to the Riesz space fractional derivative operator ∇_x^β, we resort to the forward Feynman-Kac equation Eq. (3.135) by integrating the solution $G(x, p, t)$ over x with initial position x_0 to get the marginal PDF of $G_{x_0}(p, t)$.

In the case of a harmonic potential, where $V(x, t) = bx^2/2$ $(b > 0)$, $(f(x, t) = -\partial V(x, t)/\partial x = -bx)$ and $g(x, t) \equiv 1$, $U(x) = x$, $\alpha = 1$, the forward Feynman-Kac equation Eq. (3.129) takes the form

$$\frac{\partial \tilde{G}(k, p, t)}{\partial t} + (bk - p)\frac{\partial}{\partial k}\tilde{G}(k, p, t) = \phi_0(k)\tilde{G}(k, p, t).$$

Its general solution is given as follows [Polyanin *et al.* (2002)]:

$$\tilde{G}(k,p,t) = \exp\left[\int_0^k \frac{\phi_0(z)}{bz-p}dz + c_1\right] \times \Psi\left[\frac{1}{b}\ln|bk-p| - t + c_2\right], \quad (3.167)$$

where c_1, c_2 are constants and $\Psi(x)$ is an arbitrary function. Using the initial condition $\tilde{G}(k,p,0) = 1$ (the particle starts at $x_0 = 0$), we get

$$\Psi\left[\frac{1}{b}\ln|bk-p| + c_2\right] = \exp\left[-\int_0^k \frac{\phi_0(z)}{bz-p}dz - c_1\right]. \quad (3.168)$$

Then replacing k by $l(k) := \frac{bk-p}{be^{bt}} + \frac{p}{b}$ in Eq. (3.168) yields

$$\Psi\left[\frac{1}{b}\ln|bk-p| - t + c_2\right] = \exp\left[-\int_0^{l(k)} \frac{\phi_0(z)}{bz-p}dz - c_1\right].$$

Substituting this result into Eq. (3.167), we obtain

$$\tilde{G}(k,p,t) = \exp\left[\int_{l(k)}^k \frac{\phi_0(z)}{bz-p}dz\right].$$

Let $k = 0$ and we get the PDF of functional A_x in Fourier space ($A_x \to p$):

$$G(p,t) := \tilde{G}(k,p,t)|_{k=0} = \exp\left[\int_{\frac{p}{b}(1-e^{-bt})}^0 \frac{\phi_0(z)}{bz-p}dz\right]. \quad (3.169)$$

Now we discuss the specific dynamical behaviour of functional A_x with Lévy β-stable noise

$$\phi_0(k) = -|k|^\beta. \quad (3.170)$$

With a variable substitution $z = \frac{p}{b}(1-e^{-bt})y$, Eq. (3.169) can be represented in the form

$$\ln G(p,t) = -C_b(t)\left(\frac{1-e^{-bt}}{b}\right)^{\beta+1}|p|^\beta, \quad (3.171)$$

where $C_b(t)$ is independent of p [Gradshteyn and Ryzhik (1980)]:

$$C_b(t) = \int_0^1 \frac{y^\beta}{1-(1-e^{-bt})y}dy$$
$$= B(\beta+1,1) \cdot {}_2F_1(1,\beta+1;\beta+2;1-e^{-bt}),$$

where ${}_2F_1(\cdot)$ is a hypergeometric function. According to Eq. (3.171), the functional A_x also obeys the Lévy β-stable distribution. Then we further consider the coefficient in front of $|p|^\beta$ in Eq. (3.171). For long time $t \to \infty$, we find that

$$C_b(t) = \int_0^1 \frac{y^\beta}{1-(1-e^{-bt})y}dy \simeq bt,$$

since this integral scales as bt in both two extreme cases ($\beta = 0$ and $\beta = 2$). Substituting it into Eq. (3.171) yields

$$G(p,t) \simeq \exp(-b^{-\beta}t|p|^{\beta}) \qquad \text{as} \quad t \to \infty. \tag{3.172}$$

For short time $t \to 0$, $_2F_1(1, \beta + 1; \beta + 2; 1 - e^{-bt}) \sim 1$, and thus

$$G(p,t) \simeq \exp\left(-\frac{t^{\beta+1}}{\beta+1}|p|^{\beta}\right) \qquad \text{as} \quad t \to 0. \tag{3.173}$$

For the special case $\beta = 2$, i.e., Gaussian white noise, by the formula

$$\langle A_x^2 \rangle = \frac{\partial^2}{\partial p^2}G(p,t)\bigg|_{p=0} \,,$$

we get

$$\langle A_x^2 \rangle \simeq 2b^{-2}t, \qquad \text{as} \quad t \to \infty, \tag{3.174}$$

and

$$\langle A_x^2 \rangle \simeq \frac{2}{3}t^3, \qquad \text{as} \quad t \to 0, \tag{3.175}$$

which are verified by numerical simulations [Wang *et al.* (2018b)]. The functional A_x exhibits a crossover between different scaling regimes (from t^3 to t). When the particle begins its movement from the origin, i.e., $x \ll 1$, the effect of force ($f = -bx$) can be omitted. As time goes on, this effect is getting bigger, and eventually it produces the multi-scale phenomenon. On the contrary, for the case without the force field f, i.e., $b = 0$, it is equivalent to $b \to 0$ for any t from Eq. (3.171). Then only the single-scale phenomenon $\langle A_x^2 \rangle \simeq \frac{2}{3}t^3$ can be observed, which is consistent with [Carmi *et al.* (2010)] by taking $\alpha = 1$ there. As for the general case $0 < \beta < 2$, the MSD of A_x diverges [Metzler and Klafter (2000)]: $\langle A_x^2 \rangle \to \infty$. The fractional moments can be written as

$$\langle |A_x|^{\delta} \rangle \propto \tilde{t}^{\delta/\beta}, \tag{3.176}$$

where $0 < \delta < \beta < 2$. From Eq. (3.172) and Eq. (3.173), one can get that in Eq. (3.176) $\tilde{t}$ should be $t^{\beta+1}$ for the short time and t for the long time, respectively. So we rescale the fractional moments and get the pseudo second moment $[A_x^2] \propto \tilde{t}^{2/\beta}$. An alternative method is to consider the $(A_x - t)$ scaling relations, or to measure the width of the PDF $G(A_x, t)$ rather than its variance [Metzler and Klafter (2000)]. More precisely, consider the particle in an imaginary growing box [Jespersen *et al.* (1999)] and define

$$\langle A_x^2 \rangle_L := \int_{L_1 t^{1/\beta}}^{L_2 t^{1/\beta}} A_x^2 G(A_x, t)dA_x \simeq \tilde{t}^{2/\beta},$$

where L_1 and L_2 are chosen to adapt the scaling regimes in Eq. (3.172) and Eq. (3.173), i.e., for long time $-L_1 = L_2 = \sqrt{2b^{-\beta}}$ while for short time $-L_1 = L_2 = \sqrt{2/(1+\beta)}$.

Chapter 4

Aging Fokker-Planck and Feynman-Kac Equations

One of the most omnipresent phenomena in nature is aging, being clearly a process of time t. The models naturally coming into our mind are the stochastic processes, the leading examples of which are the renewal processes. As the generalization of Poisson process [Papoulis (1984)], we mainly focus on the renewal processes with IID holding times between any adjacent two renewals, and the distribution of holding times are power law with divergent first moment. Unlike Poisson process, in this case, the power law renewal process is no longer Markovian, which leads to nonstationarity, exhibiting aging behaviors [Barkai and Cheng (2003)]. An important result drawn from aging systems with power law waiting times is the fact that in a growing fraction of trajectories no jump occurs within the observation window from the aging time t_a to $t_a + t$, implying that a large number of individual particles are split into mobile and immobile ones [Metzler et $al.$ (2014); Schulz et $al.$ (2014, 2013)]. Furthermore, aging also affects other behaviors of anomalous diffusion, such as, the second moment of fluctuation of occupation fraction, the first passage time [Metzler et $al.$ (2014)], moments of the number of renewal events [Deng et $al.$ (2016)], and so on. In this chapter, our aim is to discuss the corresponding aging Fokker-Planck and Feynman-Kac equations.

4.1 Aging CTRW

For the power law renewal process, if there is a jump in each renewal, we get the compound power law renewal process; the size of the jump is generally an IID random variable with a specified probability distribution. This compound renewal process can effectively characterize anomalous diffusion [Krüsemann et $al.$ (2016)], whose MSD is a nonlinear function of time t, i.e., $\langle x^2(t) \rangle \simeq K_\alpha t^\alpha$, $\alpha \neq 1$. For the real physical process, the observation time should not be exactly the starting time of the process. Monthus and Bouchaud [Monthus and Bouchaud (1996)] introduce a CTRW framework, which can be used to study aging behaviors. It is called generalized CTRW or aging continuous time random walk (ACTRW) [Barkai and Cheng (2003)], where the observation time does not start at time $t = 0$ but at some later instant time $t = t_a > 0$ and t_a denotes the aging time; see Fig. 4.1.

First, we briefly outline the main ingredients in the ACTRW. The ACTRW modifies the statistic of the time interval for the first jump, namely, the waiting time PDF of the first jump, which is denoted by $\omega(t_a, t)$. It describes a CTRW process having the aging time interval $(0, t_a)$, while t_a corresponds to the initial observation time. Aging means that the number of renewals in the time interval $(t_a, t_a + t)$ depends on the aging time t_a, even when the former is long. Thus ACTRW and CTRW generally exhibit different behaviors.

More concretely, both ACTRW and CTRW are the following renewal process: a walker is trapped at the origin for time $\tau_1 = t_1$, then makes a jump and the displacement is x_1; the walker is further trapped at x_1 for time $\tau_2 = t_2 - t_1$, and then jumps to a new position and its step length is x_2; this process is then renewed. So they are characterized by a set of waiting times $\{\tau_1, \tau_2, \ldots, \tau_n, \ldots\}$ and displacements $\{x_1, x_2, \ldots, x_n, \ldots\}$, respectively. The difference between them is that for CTRW, all τ_i are IID random variables with a common PDF $\phi(\tau)$, while for ACTRW, the distribution of τ_1 is different from all the other τ_i, i.e., τ_1 has PDF $\omega(t_a, \tau_1)$ and all the other τ_i are IID with a common PDF $\phi(\tau)$; for the ACTRW process, the observation time of the random walk starts from the time t_a, therefore $\omega(t_a, \tau_1)$ may depend on the aging time of the process t_a.

When $t_a = 0$, we have $\omega(t_a, \tau_1) = \phi(\tau_1)$, which agrees with the well known Montroll-Weiss nonequilibrium process. In order to investigate ACTRW, we should first discuss the aging renewal process [Barkai and Cheng (2003)]. In what follows, we suppose that $p_{N_a}(t_a, t)$ is the probability of the renewal process $N_a(t_a, t)$, where $N_a(t_a, t) = N(t_a + t) - N(t_a)$ and $N(t)$ denotes the number of renewals taking place by time t, i.e., $N_a(t_a, t)$ is the number of renewals in time interval $(t_a, t_a + t)$ for a process starting at time zero. Our aim is to discuss the properties of the renewal process in the time interval $(t_a, t_a + t)$.

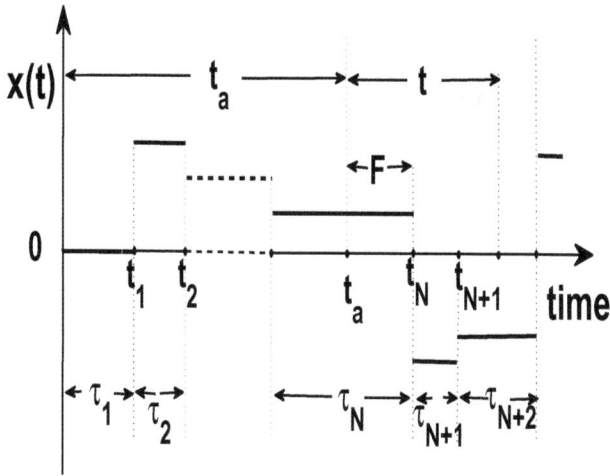

Fig. 4.1 Comparison between CTRW and ACTRW model. Here the particle starts to move from $x = 0$ at time $t = 0$. F represents the first waiting time, namely $t_N - t_a$. Though the process starts at $t = 0$, we begin to observe the process at time $t = t_a$, and after time t, we finish observing the process.

4.2 Aging Renewal Theory

We discuss the aging effects of the renewal processes with exponentially tempered power law waiting time PDF

$$\varphi(t) = \ell_\alpha(t)e^{\lambda^\alpha - \lambda t} \sim \frac{1}{-\Gamma(-\alpha)}t^{-(1+\alpha)}e^{-\lambda t}, \qquad (4.1)$$

where $0 < \alpha < 1$, and $\lambda > 0$ generally is a small parameter. As previously mentioned, $\ell_\alpha(t)$ is the one sided Lévy distribution. The semi-heavy tails and none scale-free waiting time properties of $\varphi(t)$ play a particularly prominent role in diffusion phenomena.

From Eq. (4.1), it can be noted that if $1 \ll t \ll 1/\lambda$, $\varphi(t) \sim t^{-(1+\alpha)}$, while if $t \gg 1/\lambda$, $\varphi(t) \sim \exp(-\lambda t)$. For the random variables generated by Eq. (4.1), Fig. 4.2 shows that the maximum and range of fluctuations vary dramatically with the change of λ. The introduced tempering forces the renewal process to converge from non-Gaussian to Gaussian. But the convergence is very slow, requiring a long time to find the trend. So, with

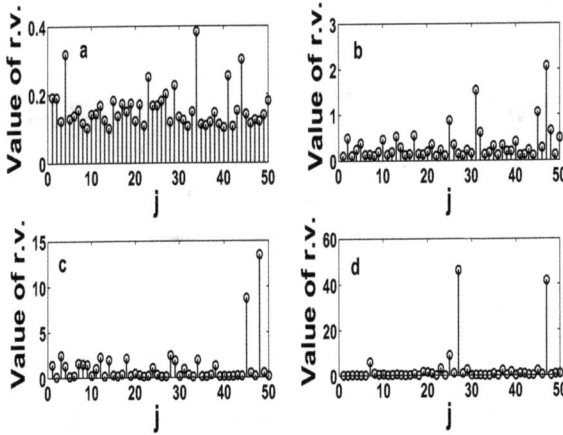

Fig. 4.2 Random variables (r.v.) generated by Eq. (4.1) with $\alpha = 0.5$. And the parameter λ is chosen, respectively, as (a) $\lambda = 10$, (b) $\lambda = 10^{-1}$, (c) $\lambda = 10^{-3}$, (d) $\lambda = 10^{-5}$; and $j = 1, 2, \ldots, 50$ correspond to the first, second, $\ldots$, 50-th variable, respectively. With the decrease of λ, the 'large' jumps are found.

the time passed by, both of the non-Gaussian and Gaussian processes can be described.

According to the renewal theory developed by Godrèche and Luck [Godrèche and Luck (2001)],

$$\widehat{\omega}(u,s) = \frac{1}{1 - \widehat{\phi}(u)} \frac{\widehat{\phi}(u) - \widehat{\phi}(s)}{s - u}, \qquad (4.2)$$

where $\widehat{\omega}(u,s)$ is the double Laplace transforms of the PDF of the first waiting time $\omega(t_a, t)$, and $\widehat{\phi}(u)$ (or $\widehat{\phi}(s)$) is the Laplace transform of $\phi(t)$. This section focuses on taking $\phi(t)$ as (tempered) power law, and its Laplace transform $(t \to s)$ has the asymptotic form

$$\widehat{\phi}(s) = e^{-(s+\lambda)^\alpha + \lambda^\alpha} \sim 1 + \lambda^\alpha - (s+\lambda)^\alpha, \qquad 0 < \alpha < 1. \qquad (4.3)$$

Now we consider to derive $p_{N_a}(t_a, t)$, being the probability of making N_a steps in the interval $(t_a, t_a + t)$. Especially, for $N_a = 0$, there exists

$$\Phi(t_a, t) = 1 - \int_0^t \omega(t_a, \tau) d\tau. \qquad (4.4)$$

Performing Laplace transform w.r.t. t, $t \to s$, yields

$$\widehat{\Phi}(t_a, s) = \frac{1 - \widehat{\omega}(t_a, s)}{s}. \qquad (4.5)$$

Let us now consider the generalized case of $N_a \geq 1$, given as

$$p_{N_a}(t_a, t) = \int_0^t \int_0^\tau w(t_a, y)Q_{N_a-1}(\tau - y)dy\Phi(t - \tau)d\tau. \qquad (4.6)$$

Note that Eq. (4.6) is similar to the probability of taking N_a steps up to time t for CTRW. Due to the special role of the first step, $Q_N(t)$ is replaced by a convolution of $w(t_a, t)$ and $Q_{N_a-1}(t)$. The double Laplace transforms of the PDF of $N_a(t_a, t)$ read, $t_a \to u$, $t \to s$,

$$\widehat{p}_{N_a}(u, s) = \begin{cases} \dfrac{1-s\widehat{w}(u,s)}{su}, & N_a = 0; \\[3mm] \widehat{w}(u, s)\widehat{\phi}^{N_a-1}(s)\dfrac{1-\widehat{\phi}(s)}{s}, & N_a \geq 1. \end{cases} \qquad (4.7)$$

For the particular case $\alpha = 1$, from Eq. (4.3), there exists $\widehat{w}(u, s) \sim 1/u$, and from Eq. (4.7) we can get $p_{N_a}(t_a, t) \sim \delta(N_a - t)$; for this case $p_{N_a}(t_a, t)$ is independent of t_a. Since $\widehat{w}(u, s)$ plays a key role in our discussion, we now derive the analytical formula of $w(t_a, t)$,

$$\begin{aligned} \widehat{w}(u, t) &= \frac{1}{1 - \widehat{\phi}(u)} \exp(ut) \int_t^\infty \exp(-uy)\phi(y)dy \\ &\sim \frac{(u + \lambda)^\alpha}{\lambda^\alpha - (u + \lambda)^\alpha} \frac{\exp(ut)}{\Gamma(-\alpha)}\Gamma(-\alpha, (u + \lambda)t), \end{aligned} \qquad (4.8)$$

where $\Gamma(\alpha, x) = \int_x^\infty \exp(-t)t^{\alpha-1}dt$ is incomplete Gamma function. Using the Laplace transform of incomplete Gamma function [Erdélyi *et al.* (1954)], Eq. (4.8) yields

$$w(t_a, t) = \frac{\exp(-\lambda t)}{-\Gamma(-\alpha)}g(t_a) *_t (\exp(-\lambda t_a) \cdot (t_a + t)^{-\alpha-1}), \qquad (4.9)$$

where '$*_t$' is the convolution operator w.r.t. t, and $g(t_a) = t_a^{\alpha-1}\exp(-\lambda t_a)$ $E_{\alpha,\alpha}(\lambda^\alpha t_a^\alpha)$; Figs. 4.3 and 4.4 confirm the correctness of Eq. (4.9) by calculating the probability of the moving parts of the particles $\int_0^t w(t_a, \tau)d\tau$, and show the trend of $\int_0^t w(t_a, \tau)d\tau$ with time t for different λ and t_a. From the second line of Eq. (4.8), if $u \ll \lambda$, i.e., $t_a \gg 1/\lambda$, Eq. (4.8) can be given by

$$w(t_a, t) \sim \frac{\lambda \exp(-\lambda t)\sin(\pi\alpha)}{\alpha\pi t^\alpha} \int_0^{t_a} \exp(-\lambda\tau)\frac{1}{\tau + t}\tau^\alpha d\tau. \qquad (4.10)$$

If $u \gg \lambda$, i.e., $t_a \ll 1/\lambda$, then there exists $\frac{(u+\lambda)^\alpha}{(u+\lambda)^\alpha - \lambda^\alpha} \to 1$. Eq. (4.8) can be further simplified as

$$w(t_a, t) \sim \frac{\sin(\pi\alpha)\exp(-\lambda(t + t_a))}{\pi}\frac{1}{t + t_a}\left(\frac{t_a}{t}\right)^\alpha; \qquad (4.11)$$

under the further assumption $t \ll 1/\lambda$, i.e., $\lambda t \ll 1$, there exists

$$\omega(t_a, t) \sim \frac{\sin(\pi\alpha)}{\pi} \frac{1}{t + t_a} \left(\frac{t_a}{t}\right)^{\alpha}, \tag{4.12}$$

being the same as the one for the power law waiting time, i.e., $\lambda = 0$. When λ is sufficiently large, Eq. (4.10) plays a dominant role. In the following,

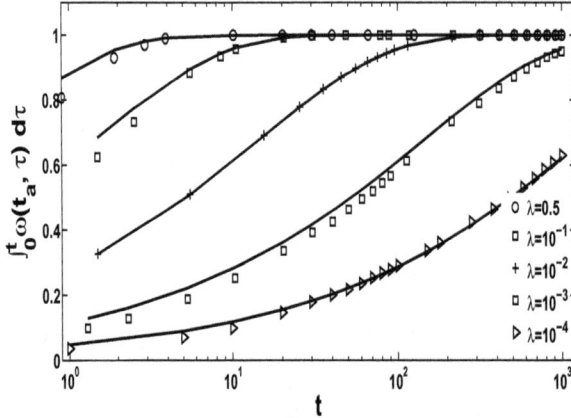

Fig. 4.3 Probability of particles making jumps during the time interval $(t_a, t_a + t)$, i.e., $\int_0^t \omega(t_a, \tau)d\tau$, for $t_a \gg 1/\lambda$. The parameters are taken as $\alpha = 0.6$, $t_a = 10^4$, and $t = 1000$; and the symbolled lines are obtained by averaging 5000 trajectories with different λ. The solid lines from down to up corresponding to the increasing λ are the theoretical results obtained from Eq. (4.9). When λ is large, the probability goes to one quickly.

we analyze the asymptotic form of $\omega(t_a, t)$ with $\alpha \in (0, 1)$. From Eqs. (4.2) and (4.3), there exists

$$\widehat{\omega}(u, s) \sim \frac{(u + \lambda)^{\alpha} - (s + \lambda)^{\alpha}}{(s - u)[\lambda^{\alpha} - (u + \lambda)^{\alpha}]}. \tag{4.13}$$

In the following we consider another important quantity, the survival probability $\Phi(t_a, t)$ [Krüsemann *et al.* (2014)], which gives the probability of making no jump during the interval $(t_a, t_a + t)$. In Laplace space, we have

$$\widehat{\Phi}(u, s) \sim \frac{1}{us} - \frac{(u + \lambda)^{\alpha} - (s + \lambda)^{\alpha}}{s(s - u)[\lambda^{\alpha} - (u + \lambda)^{\alpha}]}. \tag{4.14}$$

It is instructive to consider two different cases. If $u \ll s$, i.e., $t_a \gg t$, there exists

$$\widehat{\Phi}(u, s) \sim \frac{1}{us} - \frac{(s + \lambda)^{\alpha} - \lambda^{\alpha}}{s^2[(u + \lambda)^{\alpha} - \lambda^{\alpha}]}. \tag{4.15}$$

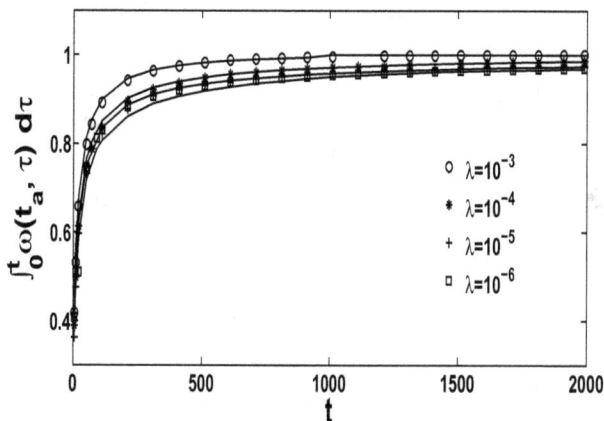

Fig. 4.4 Probability of particles making jumps during the time interval $(t_a, t_a + t)$, i.e., $\int_0^t \omega(t_a, \tau)d\tau$. The parameters are taken as $\alpha = 0.6$, $t_a = 100$; and the symboled lines are obtained by averaging 5000 trajectories with different λ. The solid lines from down to up corresponding to the increasing λ are the theoretical results of Eq. (4.9). When λ is sufficiently small, the distribution is almost the same as pure power law for short times.

For $\lambda \ll s$, i.e., $t_0 \ll t \ll 1/\lambda$, then $(s + \lambda)^\alpha \sim s^\alpha (1 + \lambda/s)^\alpha \sim s^\alpha$. Performing double inverse Laplace transforms of the above equations results in

$$\Phi(t_a, t) \sim 1 - t_a^{\alpha-1} \exp(-\lambda t_a) E_{\alpha,\alpha}(\lambda^\alpha t_a^\alpha)\left(-\lambda^\alpha t + \frac{t^{1-\alpha}}{\Gamma(2-\alpha)}\right). \quad (4.16)$$

For $t_a \ll 1/\lambda$, Eq. (4.16) can be simplified as

$$\Phi(t_a, t) \sim 1 - t_a^{\alpha-1}\left(-\lambda^\alpha t + \frac{t^{1-\alpha}}{\Gamma(2-\alpha)}\right). \quad (4.17)$$

It can be noted that $t^{1-\alpha}/\Gamma(2-\alpha)$ is larger than $\lambda^\alpha t$ in the parenthesis of Eq. (4.17), since $\lambda^\alpha t = (\lambda t)^\alpha t^{1-\alpha} \ll t^{1-\alpha}$. For $t_a \gg 1/\lambda$, from Eq. (4.16), we obtain

$$\Phi(t_a, t) \sim 1 - \frac{t^{1-\alpha}}{\langle \tau \rangle \Gamma(2-\alpha)}, \quad (4.18)$$

being confirmed by Fig. 4.5, i.e., the lines tend to be close for big t_a.

For $t \ll t_a \ll 1/\lambda$, we have $\Phi(t_a, t) \sim 1$, i.e., for small λ, $\phi(t) \sim t^{-1-\alpha}$, the waiting time is generally long. It implies that for a small observation time t, we cannot find movement of the particles. Equation (4.14) can also be rewritten as

$$\hat{\Phi}(u, s) \sim \frac{1}{us} + \frac{1}{s(s-u)} + \frac{(s+\lambda)^\alpha - \lambda^\alpha}{s(u-s)[(u+\lambda)^\alpha - \lambda^\alpha]}. \quad (4.19)$$

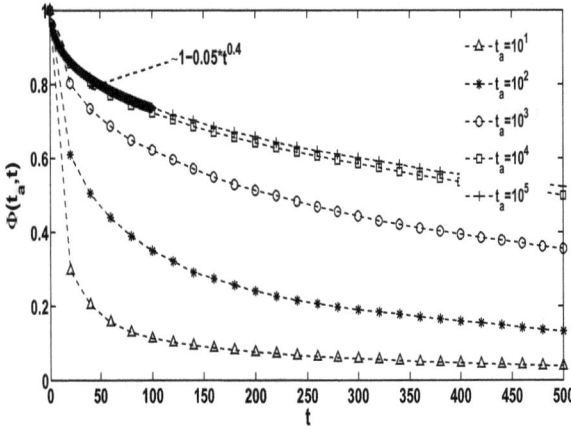

Fig. 4.5 Time evolution of $\Phi(t_a, t)$ with different t_a. The parameters are taken as $\alpha = 0.6$, $\lambda = 10^{-4}$. The lines are obtained by averaging 5000 trajectories. The dashed line, $1 - 0.05 \times t^{0.4}$, is the fitting result for small t and big t_a, which agrees with Eq. (4.18).

For the case $s \ll u$, i.e., $t_a \ll t$, Eq. (4.19) yields

$$\widehat{\Phi}(u, s) \sim \frac{(s + \lambda)^\alpha - \lambda^\alpha}{us[(u + \lambda)^\alpha - \lambda^\alpha]}; \tag{4.20}$$

under the further assumption $s \gg \lambda$, we have

$$\Phi(t_a, t) \sim 1 *_{t_a} g(t_a) \frac{t^{-\alpha}}{\Gamma(1 - \alpha)}. \tag{4.21}$$

When $t_a \ll t \ll 1/\lambda$,

$$\Phi(t_a, t) \sim \frac{\sin(\pi\alpha)}{\pi\alpha} \left(\frac{t}{t_a}\right)^{-\alpha}, \tag{4.22}$$

being the same result for the pure power law case ($\lambda = 0$) and confirmed by Fig. 4.6.

When the aging time is sufficiently long compared to the observation time and λ is small, the probability of making no jump during the time interval $(t_a, t_a + t)$ approaches to one, i.e., the system is completely trapped. On the contrary, if t_a is short, while the observation time is long enough, then the particles are unacted on the aging time. So at least one jump will be made, namely, the possibility of making no jump is zero. Indeed, from Eq. (4.22), it can be easily obtained that $\Phi(t_a, t) \sim 0$ when $t/t_a \to \infty$. Note that $\Phi(t_a, t) \sim 0$ can also be directly obtained from Eq. (4.14) under

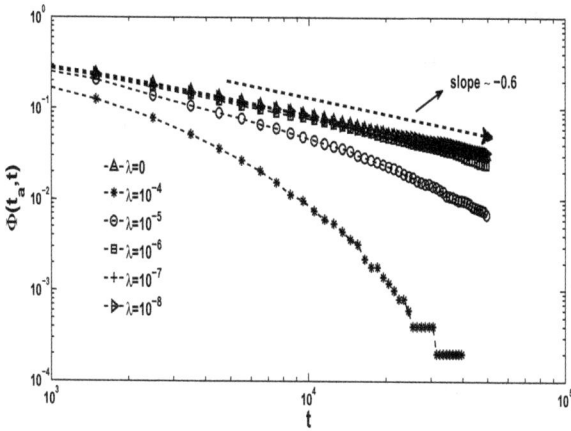

Fig. 4.6   Time evolution of $\Phi(t_a, t)$ with different λ. The parameters are taken as $\alpha = 0.6$, and $t_a = 500$. The lines are obtained by averaging 5000 trajectories. The dashed line with arrow is the indicator of slope -0.6, confirming Eq. (4.22).

the assumption $t \gg t_a$. From Eq. (4.7), we can write the double Laplace transforms of the PDF of $N_a(t_a, t)$ as

$$\widehat{p}_{N_a}(u, s) = \frac{\delta(N_a)}{s}\left[\frac{1}{u} - \widehat{\omega}(u, s)\right] + \widehat{\omega}(u, s)\widehat{\phi}^{N_a - 1}(s)\frac{1 - \widehat{\phi}(s)}{s}. \qquad (4.23)$$

Inserting Eq. (4.3) into the above equation yields

$$\widehat{p}_{N_a}(u, s) \sim \widehat{\omega}(u, s)\exp(-N_a[(s + \lambda)^\alpha - \lambda^\alpha])\frac{(s + \lambda)^\alpha - \lambda^\alpha}{s}$$
$$+ \frac{\delta(N_a)}{s}\left[\frac{1}{u} - \widehat{\omega}(u, s)\right]. \qquad (4.24)$$

From now on, we start to calculate the q-th moment of $N_a(t_a, t)$, which reads

$$\langle(\widehat{N}_a(u, s))^q\rangle \sim \int_0^\infty (N_a(t_a, t))^q p_{N_a}(t_a, t) dN_a$$
$$= \frac{\Gamma(q + 1)\widehat{\omega}(u, s)}{s[(s + \lambda)^\alpha - \lambda^\alpha]^q}. \qquad (4.25)$$

For the cases that $s \gg u$ and $\lambda \ll s$ or $s \ll u$, there exist

$$\langle(\widehat{N}_a(u, s))^q\rangle \sim \begin{cases} \frac{\Gamma(q+1)}{((u+\lambda)^\alpha - \lambda^\alpha)s^{2+\alpha(q-1)}}, & \text{for } u \ll s, \ \lambda \ll s; \\ \\ \frac{\Gamma(q+1)}{(\alpha\lambda^{\alpha-1})^q}\frac{1}{us^{1+q}}, & \text{for } s \ll u. \end{cases} \qquad (4.26)$$

By taking double inverse Laplace transforms, we have

$$\langle(N_a(t_a,t))^q\rangle \sim \begin{cases} \frac{\Gamma(q+1)}{\Gamma(2+\alpha q-\alpha)}g(t_a)t^{\alpha q-\alpha+1}, \text{ for } t \ll t_a, \ \lambda t \ll 1; \\ \left(\frac{t}{\langle\tau\rangle}\right)^q, \qquad\qquad \text{ for } t_a \ll t. \end{cases} \qquad (4.27)$$

It implies that in the long time scale $(t \gg t_a)$ $\langle N_a(t_a,t)\rangle$ scales as t, and the behavior of $\langle(N_a(t_a,t))^q\rangle \sim t^q$, like a Poisson type of renewal process.

Taking $q = 1$ in Eq. (4.25) leads to

$$\langle\widehat{N}_a(u,s)\rangle = \frac{(u+\lambda)^\alpha - (s+\lambda)^\alpha}{s(s-u)[\lambda^\alpha - (u+\lambda)^\alpha][(s+\lambda)^\alpha - \lambda^\alpha]}, \qquad (4.28)$$

which can be rewritten as

$$\langle\widehat{N}_a(u,s)\rangle = \frac{1}{s(u-s)}\left(\frac{1}{(s+\lambda)^\alpha - \lambda^\alpha} - \frac{1}{(u+\lambda)^\alpha - \lambda^\alpha}\right). \qquad (4.29)$$

We will confirm that if both t_a and t are in large scales, $\langle N_a(t_a,t)\rangle \sim t/\langle\tau\rangle$, which is an important result for normal diffusion. For small s and u, using the Taylor expansions $(s+\lambda)^\alpha \sim \lambda^\alpha+\alpha\lambda^{\alpha-1}s$ and $(u+\lambda)^\alpha \sim \lambda^\alpha+\alpha\lambda^{\alpha-1}u$, from Eq. (4.29), we have

$$\langle\widehat{N}_a(u,s)\rangle \sim \frac{1}{\alpha\lambda^{\alpha-1}s^2u}. \qquad (4.30)$$

Performing double inverse Laplace transforms of the above equation yields

$$\langle N_a(t_a,t)\rangle \sim \frac{t}{\alpha\lambda^{\alpha-1}} = \frac{t}{\langle\tau\rangle}, \qquad (4.31)$$

where $\langle\tau\rangle = \alpha\lambda^{\alpha-1}$, being the first moment of tempered power law waiting time.

For the weak aging system, $t \gg t_a$, i.e., $s \ll u$, performing the double inverse Laplace transforms of both sides of Eq. (4.29) yields

$$\langle N_a(t_a,t)\rangle \sim (t^{\alpha-1}e^{-\lambda t}E_{\alpha,\alpha}(\lambda^\alpha t^\alpha)) *_t 1 - (t_a^{\alpha-1}e^{-\lambda t_a}E_{\alpha,\alpha}(\lambda^\alpha t_a^\alpha)) *_{t_a} 1$$
$$\sim (t^{\alpha-1}e^{-\lambda t}E_{\alpha,\alpha}(\lambda^\alpha t^\alpha)) *_t 1.$$
$$(4.32)$$

Equation (4.32) is confirmed by simulating trajectories of the particle; see Fig. 4.7.

For the special case $\lambda = 0$, it can be noted that $\langle N_a(t_a,t)\rangle \sim t^\alpha$. For $t \gg 1/\lambda$, using the asymptotic expansion of Mittag-Leffler function and inserting the corresponding asymptotic behavior into Eq. (4.32), we again obtain $\langle N_a(t_a,t)\rangle \sim \frac{t}{\langle\tau\rangle}$. In the long time scale, the process converges to the Gaussian process, and then the first moment of the number of renewal events grows linearly with the observation time t. For $t \ll 1/\lambda$, from Eq. (4.32) we have

$$\langle N_a(t_a, t) \rangle \sim \frac{1}{\Gamma(1+\alpha)} t^\alpha, \qquad (4.33)$$

which is the same as the pure power law case. It can be seen that when $t \gg t_a$, the first moment of N_a is not relevant to the aging time t_a. From Eqs. (4.31) and (4.33), we can see that λ plays an important role in our discussion as expected.

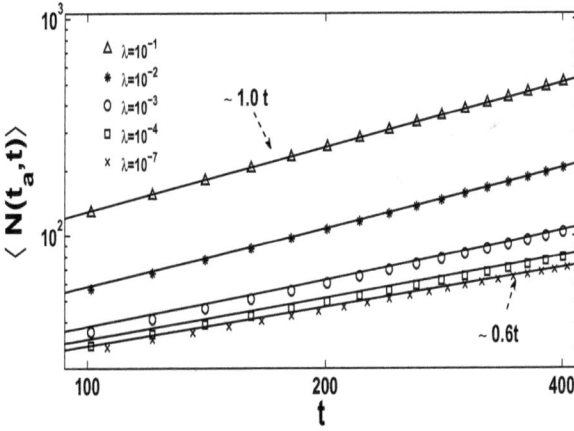

Fig. 4.7 Time evolution of the ensemble average of the renewal times $N_a(t_a, t)$ with the waiting time PDF shown in Eq. (4.1) for weak aging. The parameters are taken as $\alpha = 0.6$ and $t_a = 3$. The solid lines are the analytical results shown in Eq. (4.32) and the symbols are obtained by averaging 5000 trajectories.

While for the strong aging system, $t_a \gg t$, i.e., $u \ll s$, there exists

$$\langle \widehat{N}_a(u, s) \rangle \sim \frac{1}{s^2[(u+\lambda)^\alpha - \lambda^\alpha]}, \qquad (4.34)$$

which yields

$$\langle N_a(t_a, t) \rangle \sim t t_a^{\alpha-1} e^{-\lambda t_a} E_{\alpha,\alpha}(\lambda^\alpha t_a^\alpha), \qquad (4.35)$$

being confirmed by simulating the particle trajectories; see Fig. 4.8. Following the methods used above, for $t_a \gg 1/\lambda$, the term $t_a^{\alpha-1} E_{\alpha,\alpha}(\lambda^\alpha t_a^\alpha) e^{-\lambda t_a}$ tends to $\lambda^{1-\alpha}/\alpha$. Then we have

$$\langle N_a(t_a, t) \rangle \sim \frac{t}{\langle \tau \rangle}, \qquad (4.36)$$

which is verified by numerical simulations, see Fig. 4.9. For $t_a \ll 1/\lambda$, there exists

$$\langle N_a(t_a, t) \rangle \sim \frac{1}{\Gamma(\alpha)} t t_a^{\alpha-1}. \tag{4.37}$$

The above results for the first moment of $N_a(t_a, t)$ can be summarized as:

- if t_a or t is greater than $1/\lambda$, then $\langle N_a(t_a, t) \rangle \sim t/\langle \tau \rangle$;
- for $t \gg t_a$ and $t \ll 1/\lambda$, i.e., $t^{-\alpha-1} \exp(-\lambda t) \sim t^{-\alpha-1}$, $\langle N_a(t_a, t) \rangle \sim t^\alpha$;
- for $t \ll t_a$ and $t_a \ll 1/\lambda$, $\langle N_a(t_a, t) \rangle$ behaves as $t t_a^{\alpha-1}$.

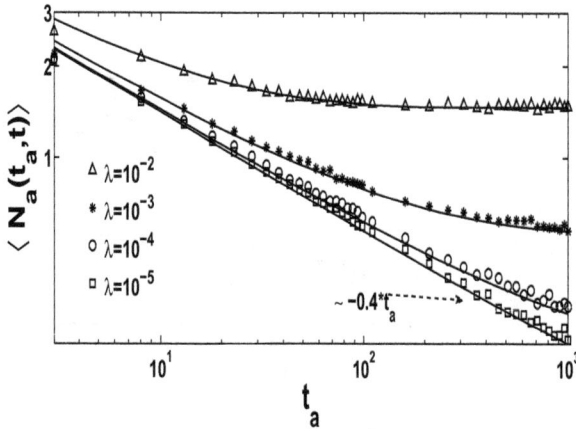

Fig. 4.8 $\langle N_a(t_a, t) \rangle$ versus t_a for different λ. The number of particles is 5000, $t = 5$, $\alpha = 0.6$, and $\lambda = 10^{-2}$, 10^{-3}, 10^{-4}, and 10^{-5}. The solid lines are the analytical results of Eq. (4.35) by choosing different λ, while the symbols represent the corresponding simulation results.

4.3 ACTRW with Tempered Power Law Waiting Time

We now turn to discuss the ACTRW and the MSD for the cases $t_a \ll t$ and $t \ll t_a$. Besides, the numerical simulations confirm the analytical expressions of the MSD and the propagators, respectively.

4.3.1 MSD

After understanding the statistics of the number of renewals, we go further to discuss the ACTRW with the (tempered) power law waiting time

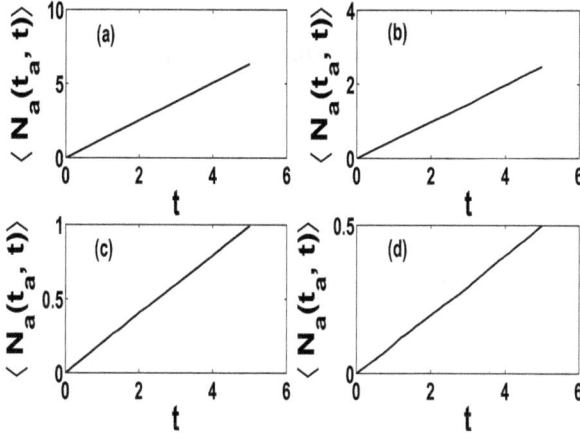

Fig. 4.9 Time evolution of the ensemble average of the renewals $N_a(t_a, t)$ for strong aging ($t_a \gg t$). The parameters $\alpha = 0.5$, $t_a = 10^4$, and $\lambda = 10^{-1}$ for (a); $\lambda = 10^{-2}$ for (b); $\lambda = 10^{-5}$ for (c); $\lambda = 10^{-7}$ for (d). The symbols are obtained by averaging 10^4 trajectories. It can be seen that $\langle N_a(t_a, t) \rangle$ grows linearly with time t for fixed t_a for all values of λ, which confirms the analytical result in Eq. (4.35).

distribution. Here we focus on the symmetric random walk in one spacial dimension, i.e., the distribution of jump lengths $w(x) = w(-x)$; and $M_2 = \int_{-\infty}^{+\infty} x^2 w(x) dx$ is finite. For such a random walk, we denote $p(x, t_a, t)$ as the PDF of particles' position in the decoupled tempered ACTRW with aging time t_a. Then

$$p(x, t_a, t) = \sum_{N_a=0}^{\infty} p_{N_a}(t_a, t) \chi_{N_a}(x), \tag{4.38}$$

where again $p_{N_a}(t_a, t)$ means the probability of jumping N_a steps in the time interval $(t_a, t_a + t)$, and $\chi_{N_a}(x)$ the probability of jumping to the position x after N_a steps. Since the step length is IID with a common PDF $w(x)$, in the Fourier-Laplace domain,

$$\widetilde{\overline{p}}(k, u, s) = \sum_{N_a=0}^{\infty} \widehat{p}_{N_a}(u, s) \widetilde{w}^{N_a}(k), \tag{4.39}$$

where $\widetilde{w}^{N_a}(k)$ means the N_a powers of $\widetilde{w}(k)$; and $\widetilde{w}(k)$ is the Fourier transform of $w(x)$. Inserting Eq. (4.7) into Eq. (4.39) leads to

$$\widetilde{\overline{p}}(k, u, s) = \frac{1 - u\widehat{\omega}(u, s)}{su} + \frac{\widehat{\omega}(u, s)(1 - \widehat{\phi}(s))}{s} \frac{\widetilde{w}(k)}{1 - \widehat{\phi}(s)\widetilde{w}(k)}. \tag{4.40}$$

It can be noticed that the first term of the right hand side of Eq. (4.40) corresponds to the probability of making no step during the interval $(t_a, t_a + t)$; in fact, its double inverse Laplace transforms is $1 - \int_0^t w(t_a, \tau)d\tau$, being exactly the survival probability. Differentiating Eq. (4.40) twice w.r.t. k and setting $k = 0$, we obtain the MSD, i.e.,

$$\langle \widehat{r}^2(u, s) \rangle = \frac{\widehat{\omega}(u, s)M_2}{s[1 - \widehat{\phi}(s)]}. \tag{4.41}$$

For the MSD, we present the results of the weak aging and strong aging systems, respectively, i.e.,

$$\langle \widehat{r}^2(u, s) \rangle \sim \begin{cases} \frac{M_2}{su[(s+\lambda)^\alpha - \lambda^\alpha]}, & \text{for } s \ll u; \\ \\ \frac{M_2}{s^2[(u+\lambda)^\alpha - \lambda^\alpha]}, & \text{for } s \gg u. \end{cases} \tag{4.42}$$

Performing double inverse Laplace transforms of $\langle \widehat{r}^2(u, s) \rangle$ yields

$$\langle r^2(t_a, t) \rangle \sim \begin{cases} M_2 g(t) *_t 1, & \text{for } t \gg t_a; \\ \\ M_2 t g(t_a), & \text{for } t_a \gg t, \end{cases} \tag{4.43}$$

where $g(z) = z^{\alpha-1} \exp(-\lambda z) E_{\alpha,\alpha}(\lambda^\alpha z^\alpha)$; for the simulations, see Figs. 4.10 and 4.11. From Fig. 4.11, we can see the large fluctuations of $\langle r^2(t_a, t) \rangle$ even if the number of trajectories is 10,000. This is because most of particles are trapped in the initial position for $t_a \gg t$ and $t \ll 1/\lambda$, which is consistent with Eq. (4.16). This is related to population splitting [Cherstvy and Metzler (2013)]. Figure 4.12 further confirms that when $t \gg t_a$, $\langle r^2(t_a, t) \rangle$ does not depend on t_a.

It can be noted that when $t \gg t_a$, the MSD has no aging effect; while $t \ll t_a$ ($\lambda t_a \ll 1$), the MSD is deeply affected by the aging time t_a. The surprising result is that $\langle r^2(t_a, t) \rangle \sim \langle N_a(t_a, t) \rangle$, when the second order moment of the jump length is finite. Note that the same things happen for the pure power law waiting time distribution.

4.3.2 *Propagator Function $p(x, t_a, t)$*

We further discuss the propagator function $p(x, t_a, t)$ of the tempered AC-TRW. Omitting the motionless part of Eq. (4.40), taking $w(x)$ as Gaussian, using the formula of Fourier transform $\mathcal{F}[\exp(-a|x|)] = \frac{-2a}{k^2+a^2}$ with $a > 0$, and performing inverse Fourier transform w.r.t. k, there exists

$$\widehat{p}(x, u, s) \sim \frac{\widehat{\omega}(u, s)}{2s} F_1(s, x) \tag{4.44}$$

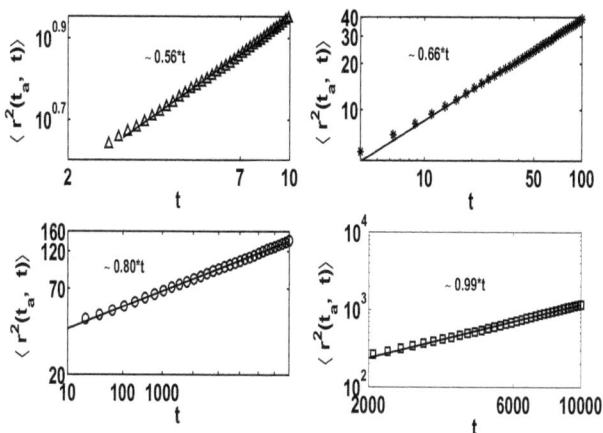

Fig. 4.10 The relation between the $\langle r^2(t_a, t)\rangle$ and the observation time t for $t_a \ll t$ got from the numerical simulations (symbols) and the theoretical results shown in Eq. (4.43) (solid line). The parameters are $\alpha = 0.6$, $\lambda = 10^{-3}$, $t_a = 1$, and the number of trajectories is 5000. With the increase of λ, the behavior of $\langle r^2(t_a, t)\rangle$ indicates the process changes from subdiffusion to normal diffusion.

with

$$F_1(s, x) = \sqrt{\frac{(s+\lambda)^\alpha - \lambda^\alpha}{0.5(1 + \lambda^\alpha - (s+\lambda)^\alpha)}} \exp\left(-|x|\sqrt{\frac{(s+\lambda)^\alpha - \lambda^\alpha}{0.5(1 + \lambda^\alpha - (s+\lambda)^\alpha)}}\right).$$

For $s \ll u$, Eq. (4.44) can be rewritten as

$$\widehat{p}(x, t_a, s) \sim \left(\frac{\lambda^\alpha - (s+\lambda)^\alpha}{2s}[g(t_a) *_{t_a} 1] + 1\right) F_1(s, x). \tag{4.45}$$

From Fig. 4.13, it can be noted that for small λ ($\lambda = 10^{-3}$ or $\lambda = 10^{-4}$) the propagator functions display the characteristics of α-stable distribution; while for large λ, $p(x, t_a, t)$ shows the classical Gaussian behavior.

For $u \gg s$, Eq. (4.40) yields

$$\widehat{p}(x, t_a, s) \sim \frac{(s+\lambda)^\alpha - \lambda^\alpha}{2s^2} g(t_a) F_1(s, x). \tag{4.46}$$

Contrary to Fig. 4.13, Fig. 4.14 displays the behaviors of the α-stable distribution for all kinds of λ.

From the numerical results and the theoretical ones, we can see that the 'α-stable distribution' characteristics can be found for small λ. Both of the distributions for $\lambda = 10^{-3}$ and $\lambda = 10^{-4}$ have the sharp peaks and the tails decay slowly. While for large λ, the top of the distribution for Eq. (4.45)

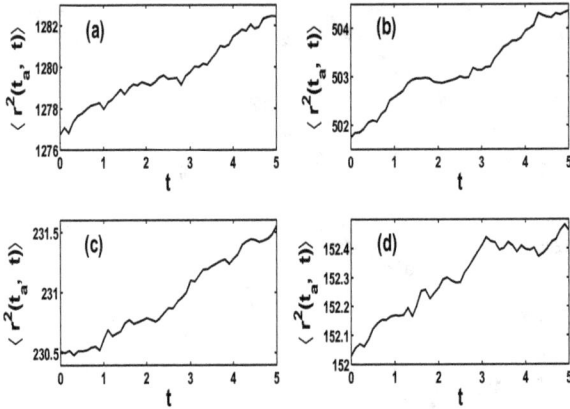

Fig. 4.11 The relation between $\langle r^2(t_a, t) \rangle$ and the observation time t for $t_a \gg t$. It can be noted that $\langle r^2(t_a, t) \rangle$ increases linearly with t for fixed t_a and the fluctuations are large because of population splitting. The parameters $t_a = 10000$, $t = 10$, $\lambda = 10^{-1}$ for (a), $\lambda = 10^{-2}$ for (b), $\lambda = 10^{-5}$ for (c), $\lambda = 10^{-7}$ for (d), $\alpha = 0.6$ and the number of the trajectories is $10,000$.

is smooth, being different from the case of small λ. Therefore, depending on the choice of λ, one can control the behaviors of the propagator.

4.4   Strong Relation between Fluctuation and Response

Now we discuss the aging from a new point of view. Based on the CTRW model, consider such a process: the particles begin to move at time $t = 0$ and undergo unbiased diffusion in the time interval $(0, t_a)$. We suppose an external field is switched on the system starting from t_a. Because of the effect of force, the particles may have small disturbances on the unbiased case, i.e., the responses to the external field. The responses are measured by the ensemble average of the positions. If the averaged response of the particles depends on t_a, the process exhibits aging. Generally speaking, giving some disturbances to a system, some characteristics (parameters of thermodynamics) of the system will change, being called response [Bertin and Bouchaud (2003)]. Under the small disturbance of external field, if the change of the parameter of thermodynamics is proportional to the force of the external field, then it is called linear response. It is important to use drift diffusion to consider aging. Using the method given in [Froemberg

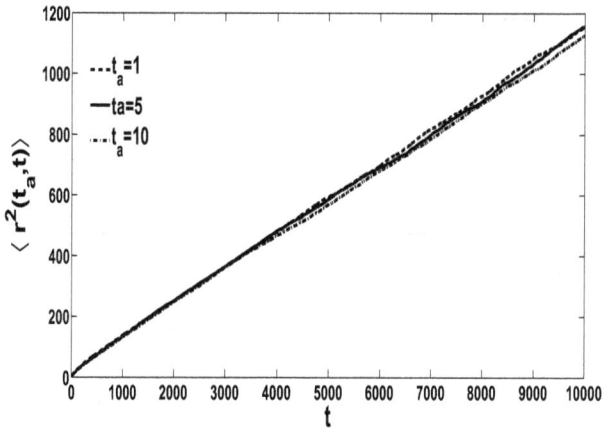

Fig. 4.12   The relation between $\langle r^2(t_a, t) \rangle$ and the observation time t for various t_a with $t \gg t_a$.

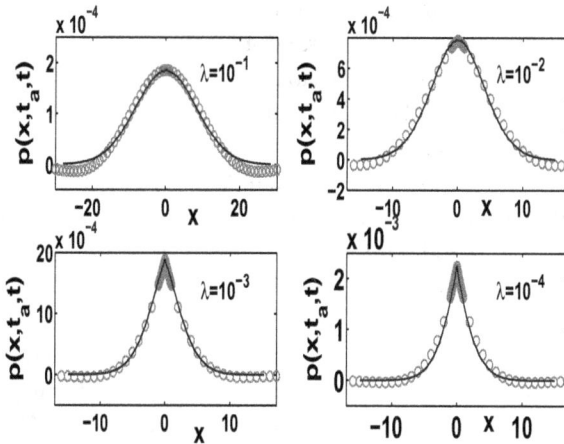

Fig. 4.13   Propagator functions with $t_a = 3$, $\alpha = 0.6$, and $t = 500$ for different λ. The symbols are obtained by calculating Eq. (4.45), and the solid lines are got from 10^4 trajectories of the particles.

and Barkai (2013); Barkai (2007); Allegrini *et al.* (2005); Shemer and Barkai (2009)], we discuss the tempered aging Einstein relation.

Next, we introduce some related variables. Let us consider a simple example of random walk on one-dimensional lattice; the length of the lattice

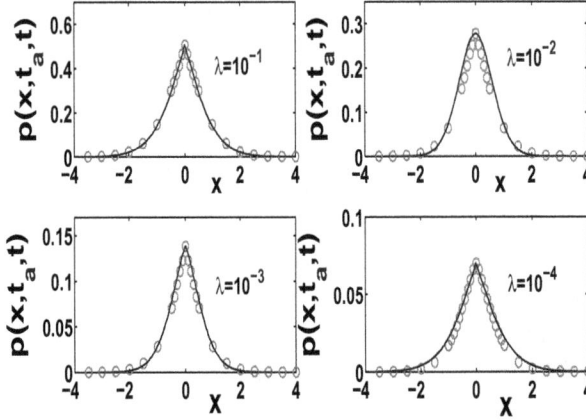

Fig. 4.14 Strong aging case (contrary to Fig. 4.13) with $t_a = 500$ and $t = 3$. The other parameters are the same as Fig. 4.13.

is a, and the particles can only move to their neighboring sites. Waiting times between different steps of the random walk are considered to be independent and have the same distribution $\phi(t)$. We use h to denote the small bias under the influence of the external field; for simplicity, the induced bias is supposed to be independent of the position. Jumps to the right (left) are performed with the probability $1/2 + h/2$ $(1/2 - h/2)$. The particles move unbiasedly till t_a; after that, the impact of the external field occurs, which lasts for time t_b. The total time is $t = t_a + t_b$; $(0, t_a)$ is called aging interval with $h = 0$, i.e., the particles have the same probability to move left or right; and $(t_a, t_a + t_b)$ is called response interval with $0 < h < 1$. Let $x = x_a + x_b$, where $x_a = \sum_{i=1}^{N_a} x_i^a$ is the displacement performed in the aging time interval and $x_b = \sum_{i=1}^{N_b} x_i^b$ is the displacement performed in the response time interval, x_i^a, x_i^b are the step lengths, and N_a, N_b are the number of events happened in the two time intervals, respectively.

We consider the correlation function $\langle (x_a)^2 x_b \rangle$ which shows the impact between $(x_a)^2$ in the aging interval and x_b in the response interval. And define a parameter F_R to show the relation between fluctuation and response [Barkai (2007)],

$$F_R = \frac{\langle (x_a)^2 x_b \rangle}{\langle (x_a)^2 \rangle \langle x_b \rangle} - 1. \tag{4.47}$$

If $F_R = 0$, it shows that x_a^2 and x_b are independent with each other. Using the relation $\langle x_a^2 \rangle = a^2 \langle N_a \rangle$ and $\langle x_b \rangle = ha \langle N_b \rangle$, then F_R can be shown in another way,

$$F_R = \frac{\langle N_a N_b \rangle}{\langle N_a \rangle \langle N_b \rangle} - 1. \tag{4.48}$$

We further introduce $X_{t_a,t_b}(N_a, N_b)$, the probability of occurring N_a events in the aging interval and N_b events in the response interval. Following the result given in [Godrèche and Luck (2001)],

$$X_{t_a,t_b}(N_a, N_b) = \langle I(t_{N_a} < t_a < t_{N_a+1}) I(t_{N_a+N_b} < t_a + t_b < t_{N_a+N_b+1}) \rangle, \tag{4.49}$$

where $I(t_{N_a} < t_a < t_{N_a+1}) = 1$ if the event inside the parenthesis occurs, and 0 otherwise. Using double Laplace transforms, if $N_b = 0$,

$$\widehat{X}_{s,u}(N_a, N_b) = \frac{\widehat{\phi}^{N_a}(s)}{u} \left[\frac{1 - \widehat{\phi}(s)}{s} - \frac{\widehat{\phi}(u) - \widehat{\phi}(s)}{s - u} \right]; \tag{4.50}$$

and if $N_b \geq 1$,

$$\widehat{X}_{s,u}(N_a, N_b) = \frac{\widehat{\phi}^{N_a}(s) \widehat{\phi}^{N_b-1}(u)}{u(s - u)} [1 - \widehat{\phi}(u)][\widehat{\phi}(u) - \widehat{\phi}(s)].$$

Summing N_a, N_b from 0 to ∞ leads to

$$\langle \widehat{N}_a \widehat{N}_b \rangle_{s,u} = \frac{[\widehat{\phi}(u) - \widehat{\phi}(s)]\widehat{\phi}(s)}{u(s - u)(1 - \widehat{\phi}(s))^2(1 - \widehat{\phi}(u))}.$$

Assume that t_a and t_b are sufficient large. Then the above equation can be written as

$$\langle \widehat{N}_a \widehat{N}_b \rangle_{s,u} \sim \frac{1}{u(s - u)(1 - \widehat{\phi}(s))(1 - \widehat{\phi}(u))} - \frac{1}{u(s - u)(1 - \widehat{\phi}(s))^2}. \tag{4.51}$$

We are interested in two different limiting behaviors of Eq. (4.51), namely $t_b \ll t_a$ and $t_a \ll t_b$. For the former case, it corresponds to $s \ll u$. From Eq. (4.51),

$$\langle \widehat{N}_a \widehat{N}_b \rangle_{s,u} \sim -\frac{1}{u^2(1 - \widehat{\phi}(s))(1 - \widehat{\phi}(u))} + \frac{1}{u^2(1 - \widehat{\phi}(s))^2}. \tag{4.52}$$

Taking the double inverse Laplace transform, we have

$$\langle N_a N_b \rangle_{t_a,t_b} \sim t_b g(t_a) *_{t_a} g(t_a) - [t_b *_{t_b} g(t_b)]g(t_a) \tag{4.53}$$

with $g(z) = z^{\alpha-1} E_{\alpha,\alpha}(\lambda^\alpha z^\alpha) \exp(-\lambda z)$ and '$*$' being the Laplace convolution operator defined above. Similarly, if $t_a \ll t_b$, we have

$$\langle N_a N_b \rangle_{t_a,t_b} \sim [1 *_{t_b} g(t_b)][1 *_{t_a} g(t_a)] - [1 *_{t_a} g(t_a) *_{t_a} g(t_a)]. \tag{4.54}$$

Utilizing the relation

$$\langle \widehat{N}_a \rangle_{s,u} = \frac{\widehat{\phi}(s)}{us(1 - \widehat{\phi}(s))} \sim \frac{1}{us(1 - \widehat{\phi}(s))},$$

then we have

$$\langle N_a \rangle_{t_a, t_b} \sim 1 *_{t_a} g(t_a). \tag{4.55}$$

Note that $\langle \widehat{N}_b \rangle_{s,u}$ is the same as Eq. (4.28). With the help of Eqs. (4.53), (4.54), and (4.140), the limiting behaviors of Eq. (4.48) are obtained.

In the following, we further consider the Einstein relation [Froemberg and Barkai (2013); Shemer and Barkai (2009)] for the tempered aging process. Denoting $\langle x(t_a, t_b) \rangle_F$ as the first order moment of the displacement under the influence of a force F, from Eq. (4.35) and $\langle x(t_a, t_b) \rangle_F = ha \langle N_b \rangle$, we get that for $t_a \gg t_b$,

$$\langle x(t_a, t_b) \rangle_F \sim ha \cdot t_b t_a^{\alpha-1} E_{\alpha,\alpha}(\lambda^\alpha t_a^\alpha) \exp(-\lambda t_a). \tag{4.56}$$

Denoting $\langle r^2(t_a, t_b) \rangle_0$ as the MSD of the random walk without external force, from Eq. (4.43) we obtain that for $t_a \gg t_b$,

$$\langle r^2(t_a, t_b) \rangle_0 \sim M_2 t_b t_a^{\alpha-1} E_{\alpha,\alpha}(\lambda^\alpha t_a^\alpha) \exp(-\lambda t_a) \tag{4.57}$$

with $M_2 = 1/2a^2 + 1/2(-a)^2 = a^2$. Under the assumption $h = aF/(2K_b T) \ll 1$, we obtain the following relation being the same as power law case [Barkai and Cheng (2003)],

$$\langle x(t_a, t_b) \rangle_F \sim \frac{F}{2K_b T} \langle r^2(t_a, t_b) \rangle_0. \tag{4.58}$$

4.5 Fokker-Planck Equations for Tempered ACTRW

We now derive the Fokker-Planck equation of the tempered ACTRW, which can be used to solve the tempered aging diffusion problems with different types of boundary and initial conditions. Omitting the motionless part of Eq. (4.40) and taking $w(x)$ as Gaussian distribution (i.e., $\widetilde{w}(k) \sim 1 - \frac{1}{2}k^2$), we have

$$\widetilde{p}(k, u, s) = \frac{\widehat{\omega}(u, s)}{s} \frac{(s + \lambda)^\alpha - \lambda^\alpha}{(s + \lambda)^\alpha - \lambda^\alpha + \frac{1}{2}k^2(1 + \lambda^\alpha - (s + \lambda)^\alpha)}. \tag{4.59}$$

For small s and k, Eq. (4.59) can be rewritten as

$$\left((s + \lambda)^\alpha - \lambda^\alpha + \frac{1}{2}(1 + \lambda^\alpha)k^2 - \frac{1}{2}(s + \lambda)^\alpha k^2 \right) \widetilde{p}(k, u, s)$$

$$= \frac{\widehat{\omega}(u, s)}{s} \left((s + \lambda)^\alpha - \lambda^\alpha \right). \tag{4.60}$$

Performing inverse Fourier transform of Eq. (4.60), and noticing $\mathcal{F}^{-1}[k^2 \widetilde{y}(k)] = -\frac{\partial^2}{\partial x^2} y(x)$, there exists

$$
((s+\lambda)^\alpha - \lambda^\alpha)\widehat{p}(x,u,s) - \frac{1}{2}(1+\lambda^\alpha)\frac{\partial^2}{\partial x^2}\widehat{p}(x,u,s)
$$

$$
+ \frac{1}{2}(s+\lambda)^\alpha \frac{\partial^2}{\partial x^2}\widehat{p}(x,u,s) \qquad (4.61)
$$

$$
= ((s+\lambda)^\alpha - \lambda^\alpha)\frac{\widehat{\omega}(u,s)}{s}\delta(x).
$$

Next we further perform the double inverse Laplace transforms of Eq. (4.61) and obtain the corresponding equation. First we introduce the tempered fractional derivative. For $0 < q < 1$, taking the Laplace transform of the Riemann-Liouville fractional derivative results in

$$
\mathcal{L}[_0\mathcal{D}_t^q y(t)] = s^q \mathcal{L}[y(t)] = s^q \widehat{y}(s). \qquad (4.62)
$$

From Eq. (4.62), there exists

$$
\mathcal{L}[_0\mathcal{D}_t^q(\exp(\lambda t)f(t))] = s^q \widehat{f}(s-\lambda). \qquad (4.63)
$$

Then, we have

$$
\mathcal{L}[\exp(-\lambda t)_0\mathcal{D}_t^q(\exp(\lambda t)f(t))] = (s+\lambda)^q \widehat{f}(s), \qquad (4.64)
$$

which is the Laplace transform of the tempered fractional derivative [Meerschaert and Sikorskii (2012)], defined as

$$
\partial_t^{q,\lambda} f(t) = \exp(-\lambda t)_0\mathcal{D}_t^q(\exp(\lambda t)f(t)).
$$

Using the above equations, the inverse Laplace transform of Eq. (4.61) w.r.t. s reads

$$
(\partial_t^{\alpha,\lambda} - \lambda^\alpha)\widehat{p}(x,u,t) = \frac{1}{2}(1+\lambda^\alpha)\frac{\partial^2}{\partial x^2}\widehat{p}(x,u,t) - \frac{1}{2}\partial_t^{\alpha,\lambda}\frac{\partial^2}{\partial x^2}\widehat{p}(x,u,t)
$$

$$
+ \left(\partial_t^{\alpha,\lambda}\widehat{\omega}(u,t) *_t 1 - \lambda^\alpha \widehat{\omega}(u,t) *_t 1\right)\delta(x);
$$

$$
(4.65)
$$

further performing the inverse Laplace transform of Eq. (4.65) w.r.t. u results in

$$
(\partial_t^{\alpha,\lambda} - \lambda^\alpha)p(x,t_a,t) = \frac{1}{2}(1+\lambda^\alpha)\frac{\partial^2}{\partial x^2}p(x,t_a,t) - \frac{1}{2}\partial_t^{\alpha,\lambda}\frac{\partial^2}{\partial x^2}p(x,t_a,t)
$$

$$
+ \left(\partial_t^{\alpha,\lambda}\omega(t_a,t) *_t 1 - \lambda^\alpha \omega(t_a,t) *_t 1\right)\delta(x).
$$

$$
(4.66)
$$

Equation (4.66) is the Fokker-Planck equation of the Green function $p(x, t_a, t)$ in the case that the waiting time distribution is the tempered power law illustrated in Eq. (4.1).

For the non-tempered case, namely, $\lambda = 0$ and $\phi(t) \sim t^{-1-\alpha}$, taking $\lambda = 0$ in Eq. (4.59) results in

$$\widetilde{\widehat{p}}(k, u, s) \sim \widehat{\omega}(u, s) \frac{s^{\alpha-1}}{s^{\alpha} + \frac{k^2}{2}(1 - s^{\alpha})}. \tag{4.67}$$

Performing inverse Fourier transform and double inverse Laplace transforms of Eq. (4.67) yields the corresponding aging diffusion equation

$$_0\mathcal{D}_t^{\alpha} p(x, t_a, t) = \frac{1}{2} \frac{\partial^2}{\partial x^2} p(x, t_a, t) - {_0}\mathcal{D}_t^{\alpha} \frac{\partial^2}{\partial x^2} p(x, t_a, t)$$
$$- \omega(t_a, t) *_t \frac{t^{-\alpha}}{\Gamma(1 - \alpha)} \delta(x). \tag{4.68}$$

As expected, taking $\lambda = 0$ in Eq. (4.66) also leads to Eq. (4.68). There are also other forms of Eq. (4.66), e.g., adding the motionless part of Eq. (4.40) to the equation; for the longer time scale $t \gg 1$, then $s^{\alpha}k^2$ can be reasonably omitted, i.e., the term $_0\mathcal{D}_t^{\alpha} \frac{\partial^2}{\partial x^2} P(x, t_a, t)$ in Eq. (4.68) can be omitted.

4.6   Derivations of Aging Feynman-Kac Equation

Now, based on the ACTRW model, we derive the forward Feynman-Kac equation with (tempered) power law waiting time. We start from a simple case, i.e., the PDF of the step length is taken as $w(x) = \frac{1}{2}[\delta(x - a) + \delta(x + a)]$, which implies that the particles can only move to the left or right direction with the same probability; then we use Gaussian distribution and power law distribution, respectively, as the PDFs of jump length and waiting time. If letting $p = 0$, one obtains a generalization of the Montroll-Weiss equation for ACTRW, which agrees with the previous result [Barkai and Cheng (2003); Klafter and Sokolov (2011)]. Besides, we obtain the corresponding backward Feynman-Kac equations. Based on the derived equations, some applications are presented, such as the occupation time in half space T^+, the moments of T^+ and $(T^+)^2$. The behaviors of $\langle (T^+)^2 \rangle$ are different for strong and weak (none) aging. Furthermore, the asymptotic behaviors of the first passage time are analyzed for both strong and weak aging cases.

4.6.1 Forward Feynman-Kac Equation with Discrete Step Length PDF

We denote $G(x, A, t, t_a)$ as the joint PDF of x and A at time t with aging time t_a, where A is the functional. The difference between CTRW and ACTRW is their first step, i.e., the distribution of the waiting time of the first step is different, which plays an important role in the process of the derivation of the aging Feynman-Kac equation. Let $Q_1(x, A, t, t_a)$ be the joint PDF of x and A at time t with aging time t_a for the first step. $Q_0(x, A, t)$ is the joint PDF of x and A at the starting observation time t. We suppose that the walker's position at $t = t_a$ is $\bar{x}(t_a)$; as time is moving on (a new starting point), what we are interested in is its position at time t, i.e.,

$$x(t) = \bar{x}(t + t_a). \tag{4.69}$$

For simplicity, we first consider a particle walks on infinite one-dimensional lattices and the length of each lattice is a constant a. The particle is only allowed to jump to its nearest neighbors with the same probability to the left or right direction. Using the definition of ACTRW yields

$$Q_1(x, A, t, t_a) = \int_0^t \frac{1}{2}\omega(t_a, \tau)\Big(Q_0(x + a, A - \tau U(x + a), t - \tau) \\ + Q_0(x - a, A - \tau U(x - a), t - \tau)\Big)d\tau, \tag{4.70}$$

where $Q_0(x, A, t)$ is the initial distribution, and $\omega(t_a, \tau)$ is the forward waiting time PDF. Using Laplace transform w.r.t. A, i.e., $A \to p$, we have

$$Q_1(x, p, t, t_a) = \int_0^t \frac{1}{2}\omega(t_a, \tau) \cdot \Big(\exp(-p\tau U(x + a))Q_0(x + a, p, t - \tau) \\ + \exp(-p\tau U(x - a))Q_0(x - a, p, t - \tau)\Big)d\tau. \tag{4.71}$$

Taking Laplace transform $(t \to s)$ and using the shift property of Fourier transform $(x \to k)$, we have

$$\widetilde{\widehat{Q}}_1(k, p, s, t_a) = \cos(ka)\widehat{\omega}\Big(t_a, s + pU\Big(-i\frac{\partial}{\partial k}\Big)\Big)\widetilde{\widehat{Q}}_0(k, p, s), \tag{4.72}$$

where we use the relation $\mathcal{F}[g(x)w(x)] = g(-i\frac{\partial}{\partial k})\widetilde{w}(k)$ and denote $\widetilde{w}(k) = \mathcal{F}[w(x)]$. By the similar way above, for $n \geq 1$, there exists

$$Q_{N+1}(x, A, t, t_a) = \int_0^t \frac{\phi(\tau)}{2}\Big(Q_N(x + a, A - \tau U(x + a), t - \tau, t_a) \\ + Q_N(x - a, A - \tau U(x - a), t - \tau, t_a)\Big)d\tau, \tag{4.73}$$

where $\phi(\tau)$ is the PDF of waiting time between N-th and $(N+1)$-th jumps (assuming $N > 1$). Using double Laplace transforms ($A \to p$ and $t \to s$) and Fourier transform ($x \to k$) we can obtain

$$\widetilde{\widehat{Q}}_{N+1}(k,p,s,t_a) = \cos(ka)\widehat{\phi}\left(s + pU\left(-i\frac{\partial}{\partial k}\right)\right)\widetilde{\widehat{Q}}_N(k,p,s,t_a) \qquad (4.74)$$

with $N \geq 1$, which formally results in

$$\sum_{N=1}^{\infty} \widetilde{\widehat{Q}}_N(k,p,s,t_a) = \frac{\cos(ka)\widehat{\omega}(t_a, s + pU(-i\frac{\partial}{\partial k}))\widetilde{\widehat{Q}}_0(k,p,s)}{1 - \cos(ka)\widehat{\phi}(s + pU(-i\frac{\partial}{\partial k}))}. \qquad (4.75)$$

Then the joint PDF of a walk at time t with aging time t_a is given by

$$G(x,A,t,t_a) = \left(1 - \int_0^t \omega(t_a,\tau)d\tau\right)\delta(A - tU(x))\delta(x - x_0) \\ + \int_0^t \Phi(\tau)\sum_{N=1}^{\infty} Q_N(x, A - \tau U(x), t - \tau, t_a)d\tau, \qquad (4.76)$$

where x_0 is the initial position, i.e., $x_0 = x(t = 0)$ and $\Phi(t)$ is the survival probability. Performing Laplace transform from A to p leads to

$$G(x,p,t,t_a) = \left(1 - \int_0^t \omega(t_a,\tau)\right)\exp(-tpU(x))\delta(x - x_0) \\ + \int_0^t \Phi(\tau)\sum_{N=1}^{\infty} \exp(-\tau pU(x))Q_N(x,p,t-\tau,t_a)d\tau. \qquad (4.77)$$

In Laplace space, $t \to s$, there exists

$$\widehat{G}(x,p,s,t_a) = \frac{1 - \widehat{\omega}(t_a, s + pU(x))}{s + pU(x)}\delta(x - x_0) \\ + \widehat{\Phi}(s + pU(x))\sum_{N=1}^{\infty} \widehat{Q}_N(x,p,s,t_a). \qquad (4.78)$$

Taking Fourier transform of the above equation yields

$$\widetilde{\widehat{G}}(k,p,s,t_a) = \frac{1 - \widehat{\omega}(t_a, s + pU(-i\frac{\partial}{\partial k}))}{s + pU(-i\frac{\partial}{\partial k})}\exp(ikx_0) \\ + \frac{1 - \widehat{\phi}(s + pU(-i\frac{\partial}{\partial k}))}{s + pU(-i\frac{\partial}{\partial k})} \\ \times \frac{\cos(ka)\widehat{\omega}(t_a, s + pU(-i\frac{\partial}{\partial k}))\widetilde{\widehat{Q}}_0(k,p,s)}{1 - \cos(ka)\widehat{\phi}(s + pU(-i\frac{\partial}{\partial k}))}. \qquad (4.79)$$

Note that Eq. (4.79) is valid for all kinds of PDFs of waiting time. Omitting the singular part of Eq. (4.79), i.e., the unmoving part, and performing Laplace transform w.r.t. t_a, $t_a \to u$, result in

$$
\widetilde{\widehat{G}}(k,p,s,u) = \frac{1 - \widehat{\phi}(s + pU(-i\frac{\partial}{\partial k}))}{s + pU(-i\frac{\partial}{\partial k})}
$$
$$
\times \frac{\cos(ka)\widehat{\omega}(u, s + pU(-i\frac{\partial}{\partial k}))\widehat{Q}_0(k,p,s)}{1 - \cos(ka)\widehat{\phi}(s + pU(-i\frac{\partial}{\partial k}))}. \tag{4.80}
$$

Based on Eqs. (4.80) and (4.2), we study a special case of $\phi(t)$ and obtain its corresponding forward equation. Consider broad distribution of waiting times with index $\alpha < 1$. Taking the limit $k \to 0$, substituting $\widehat{\phi}(s)$ into Eq. (4.80), and expanding $\cos(ka)$ as series in k (i.e., $\cos(ka) \sim 1 - k^2 a^2/2$), we get

$$
s\widetilde{\widehat{G}}(k,p,s,u) = -pU\left(-i\frac{\partial}{\partial k}\right)\widetilde{\widehat{G}}(k,p,s,u)
$$
$$
- a^2 \frac{k^2(s + pU(-i\frac{\partial}{\partial k}))^{1-\alpha}}{2B_\alpha}\widetilde{\widehat{G}}(k,p,s,u) \tag{4.81}
$$
$$
+ \widehat{\omega}\left(u, s + pU\left(-i\frac{\partial}{\partial k}\right)\right)\widehat{Q}_0(k,p,s).
$$

Supposing that the initial distribution $Q_0(x, A, t) = \delta(t)\delta(x - x_0)\delta(A)$, using the above formulas, and performing the inverse transform, we get

$$
\frac{\partial}{\partial t}G(x,p,t,t_a) = \frac{a^2}{2B_\alpha}\frac{\partial^2}{\partial x^2}D_t^{1-\alpha}G(x,p,t,t_a)
$$
$$
- \delta(t) - pU(x)G(x,p,t,t_a) \tag{4.82}
$$
$$
+ \exp(-tpU(x))\omega(t_a,t)\delta(x - x_0),
$$

where $D_t^{1-\alpha}$ is the fractional substantial derivative defined in Eq. (2.13). Especially, if setting $p = 0$, then $G(x, p = 0, t, t_a) = \int_0^\infty G(x, A, t, t_a)dA$ reduces to the distribution of x; and Eq. (4.79) turns to the well known master equation of ACTRW model

$$
\widetilde{G}(k, p = 0, t, t_a) = \frac{1 - \widehat{\omega}(t_a, s)}{s} + \frac{1 - \widehat{\phi}(s)}{s}\frac{\cos(ka)\widehat{\omega}(t_a, s)}{1 - \cos(ka)\widehat{\phi}(s)} \tag{4.83}
$$

with $x_0 = 0$. Equation (4.83) is a generalization of the Montroll-Weiss equation for ACTRW. Omitting the motionless part of Eq. (4.83) yields [Barkai and Cheng (2003)]

$$
{}_0\mathcal{D}_t^\alpha G(x, p = 0, t, t_a) = \frac{a^2}{2B_\alpha}\frac{\partial^2}{\partial x^2}G(x, p = 0, t, t_a)
$$
$$
+ \frac{1}{B_\alpha}\omega(t_a, t) *_t \frac{t^{-\alpha}}{\Gamma(1 - \alpha)}\delta(x). \tag{4.84}
$$

In fact, Eq. (4.84) can also be obtained by taking $p = 0$ in Eq. (4.82). Furthermore, if $t_a = 0$, then $w(t_a, t) = \phi(t)$, and Eq. (4.79) reduces to the Feynman-Kac equation in frequency domain for the CTRW model [Carmi et al. (2010)].

4.6.2 Forward Feynman-Kac Equation with Continuous Step Length PDF

In the following, we consider another case, i.e., the displacement of each step is not a constant but a random variable following a symmetric PDF $w(x)$. It may be Gaussian distribution or symmetrical power law distribution. We consider the first step of the particle, i.e., the relation between $Q_0(x, A, t)$ and $Q_1(x, A, t, t_a)$. For ACTRW [Klafter and Sokolov (2011)], there exists

$$Q_1(x, A, t, t_a) = \int_0^t \int_{-\infty}^{\infty} w(t_a, \tau)w(z) \qquad (4.85)$$
$$\times Q_0\Big(x - z, A - \tau U(x - z), t - \tau\Big)dzd\tau.$$

For the general case, the random variable A may be negative. Therefore performing Fourier transform instead of Laplace transform [Dyke (2014)] of Eq. (4.85) w.r.t. A, leads to

$$Q_1(x, p, t, t_a) = \int_0^t \int_{-\infty}^{\infty} \exp(ip\tau U(x - z))w(t_a, \tau)w(z) \qquad (4.86)$$
$$\times Q_0(x - z, p, t - \tau)dzd\tau.$$

By Laplace transform w.r.t. t and Fourier transform w.r.t. x,

$$\widehat{\widetilde{Q}}_1(k, p, s, t_a) = \int_{-\infty}^{\infty} \int_{-\infty}^{\infty} \exp(ikx)\widehat{w}(t_a, s - ipU(x - z))w(z) \qquad (4.87)$$
$$\times \widehat{Q}_0(x - z, p, s)dzdx.$$

By variable substitution, i.e., $x - z = y$,

$$\widehat{\widetilde{Q}}_1(k, p, s, t_a) = \int_{-\infty}^{\infty} \int_{-\infty}^{\infty} \exp(iky + ikz)w(z)$$
$$\times \widehat{w}(t_a, s - ipU(y))\widehat{\widetilde{Q}}_0(y, p, s)dzdy$$
$$= \widetilde{w}(k)\widehat{w}\Big(t_a, s - ipU\Big(-i\frac{\partial}{\partial k}\Big)\Big)\widehat{\widetilde{Q}}_0(k, p, s).$$

For ACTRW, the waiting times of other steps, i.e., $N \geq 1$, following a common PDF $\phi(t)$, there exists

$$Q_{N+1}(x, A, t, t_a) = \int_0^t \int_{-\infty}^{\infty} \phi(\tau)w(z)$$
$$\times Q_N(x - z, A - \tau U(x - z), t - \tau, t_a)dzd\tau.$$

Using double Fourier transforms and Laplace transform, we can obtain

$$\widehat{\widetilde{Q}}_{N+1}(k,p,s,t_a) = \widetilde{w}(k)\widehat{\phi}\left(s - ipU\left(-i\frac{\partial}{\partial k}\right)\right)\widehat{\widetilde{Q}}_N(k,p,s,t_a).$$

From Eq. (4.77), we have

$$
\begin{aligned}
\widehat{\widetilde{G}}(k,p,s,t_a) =\ & \frac{1 - \widehat{\omega}(t_a, s - ipU(-i\frac{\partial}{\partial k}))}{s - ipU(-i\frac{\partial}{\partial k})}\exp(ikx_0) \\
& + \frac{1 - \widehat{\phi}(s - ipU(-i\frac{\partial}{\partial k}))}{s - ipU(-i\frac{\partial}{\partial k})} \\
& \times \frac{\widetilde{w}(k)\widehat{\omega}(t_a, s - ipU(-i\frac{\partial}{\partial k}))\widehat{\widetilde{Q}}_0(k,p,s)}{1 - \widetilde{f}(k)\widehat{\phi}(s - ipU(-i\frac{\partial}{\partial k}))}.
\end{aligned}
\tag{4.88}
$$

4.6.2.1 *Power Law Waiting Time*

We first consider the case of power law waiting time and Gaussian displacement. For simplicity, we ignore the unmoving part of Eq. (4.88). From Eq. (4.88), we have

$$\widehat{\widetilde{G}}(k,p,s,t_a) \sim \frac{B_\alpha}{(s - ipU(-i\frac{\partial}{\partial k}))^{1-\alpha}}\frac{\widehat{\omega}(t_a, s - ipU(-i\frac{\partial}{\partial k}))}{B_\alpha(s - ipU(-i\frac{\partial}{\partial k}))^\alpha + \frac{k^2}{2}}. \tag{4.89}$$

It can be noticed that Eq. (4.89) can be rearranged by the following formula

$$
\begin{aligned}
\left(\left(s - ipU\left(-i\frac{\partial}{\partial k}\right)\right)^\alpha + \frac{k^2}{2B_\alpha}\right)&\left(s - ipU\left(-i\frac{\partial}{\partial k}\right)\right)^{1-\alpha}\cdot\widehat{\widetilde{G}}(k,p,s,t_a) \\
& = \widehat{\omega}\left(t_a, s - ipU\left(-i\frac{\partial}{\partial k}\right)\right).
\end{aligned}
$$

Taking the inverse Laplace and Fourier transforms yields

$$
\begin{aligned}
\frac{\partial}{\partial t}G(x,p,t,t_a) =\ & \frac{1}{2B_\alpha}\frac{\partial^2}{\partial x^2}D_t^{1-\alpha}G(x,p,t,t_a) \\
& - \delta(t) + ipU(x)G(x,p,t,t_a) \\
& + \exp(iptU(x))\omega(t_a,t)\delta(x - x_0).
\end{aligned}
\tag{4.90}
$$

From the unmoving part of Eq. (4.88), we obtain the fractional Feynman-Kac equation for the ACTRW,

$$
\begin{aligned}
\frac{\partial}{\partial t}G(x,p,t,t_a) =\ & \frac{A_\beta}{B_\alpha}\nabla_x^\beta D_t^{1-\alpha}G(x,p,t,t_a) + ipU(x)G(x,p,t,t_a) \\
& - \delta(t) + \exp(iptU(x))\omega(t_a,t)\delta(x - x_0),
\end{aligned}
\tag{4.91}
$$

where ∇_x^β is the Riesz fractional operator [Metzler and Klafter (2000); Compte (1996)], being defined through

$$\mathcal{F}[\nabla_x^\beta w(x)] = -|k|^\beta \widetilde{w}(k). \tag{4.92}$$

4.6.2.2   *Tempered Power Law Waiting Time*

Next, we consider the aging effects of Feynman-Kac equation with exponentially tempered power law waiting time probability density [del Castillo-Negrete (2009); Sokolov *et al.* (2004); Meerschaert *et al.* (2008); Bruno *et al.* (2004)] defined in Eq. (4.1). From Eq. (4.88) without the unmoving part, using $w(k) \sim 1 - A_\beta |k|^\beta$, we have

$$\widehat{\widetilde{G}}(k,p,s,t_a) = \frac{(s + \lambda - ipU(-i\frac{\partial}{\partial k}))^\alpha - \lambda^\alpha}{s - ipU(-i\frac{\partial}{\partial k})}$$
$$\times \frac{\widehat{\omega}(t_a, s - ipU(-i\frac{\partial}{\partial k}))}{(s + \lambda - ipU(-i\frac{\partial}{\partial k}))^\alpha - \lambda^\alpha + A_\beta|k|^\beta}. \tag{4.93}$$

Then Eq. (4.93) can be given in another way

$$\frac{(s + \lambda - ipU(-i\frac{\partial}{\partial k}))^\alpha - \lambda^\alpha + A_\beta |k|^\beta)(s - ipU(-i\frac{\partial}{\partial k}))}{(s + \lambda - ipU(-i\frac{\partial}{\partial k}))^\alpha - \lambda^\alpha} \widehat{\widetilde{G}}(x,p,s,t_a)$$
$$= \widehat{\omega}\Big(t_a, s - ipU\Big(-i\frac{\partial}{\partial k}\Big)\Big). \tag{4.94}$$

Taking inverse Laplace and Fourier transforms, we obtain the final result

$$\frac{\partial}{\partial t} G(x,p,t,t_a) = A_\beta \nabla_x^\beta (e^{-t(\lambda - ipU(x))} t^{\alpha-1} E_{\alpha,\alpha}(-\lambda^\alpha t^\alpha)$$
$$\times \Big(\Big(\frac{\partial}{\partial t} - ipU(x)\Big) G(x,p,t,t_a) - \delta(t)\Big) + e^{iptU(x)}\omega(t_a,t)\delta(x) - \delta(t). \tag{4.95}$$

For tempered power law waiting time, $\omega(t_a,t)$ can be shown through simple integration [Deng *et al.* (2016)]. If $\beta > 2$, the operator ∇_x^β reduces to $\frac{\partial^2}{\partial x^2}$. Especially, setting $\lambda = 0$, Eq. (4.95) reduces to Eq. (4.91). Furthermore, supposing $p = 0$, Eq. (4.95) agrees with the aging diffusion equation with tempered power law waiting time and Gaussian step length distribution [Deng *et al.* (2016)].

4.6.3   *Backward Feynman-Kac Equation with Discrete Step Length PDF*

We further consider the distribution of A, which is useful in some applications, such as, calculating the first passage time [Redner (2001)] and solving the occupation time [Barkai (2006); Majumdar and Comtet (2002)]. One way is to integrate $G(x,p,t)$ over x from $-\infty$ to ∞, which is inconvenient in some cases. The other way is to let the process start at x_0; we can derive an equation of $G_{x_0}(A,t,t_a)$, i.e., the backward Feynman-Kac equation for ACTRW.

Supposing the length of each step is a, and the particles have the same probability of jumping to the left or right nearest point, from the definition of ACTRW we have the relation among $G_{x_0}(A, t, t_a)$, $G_{x_0-a}(A, t, t_a)$ and $G_{x_0+a}(A, t, t_a)$, i.e.,

$$
\begin{aligned}
G_{x_0}(A, t, t_a) = & \int_0^t \omega(t_a, \tau) \frac{1}{2} \Big(G_{x_0+a}(A - \tau U(x_0), t - \tau, t_a) \\
& + G_{x_0-a}(A - \tau U(x_0), t - \tau, t_a) \Big) d\tau \\
& + \int_t^\infty \omega(t_a, \tau) d\tau \delta(A - tU(x_0)).
\end{aligned}
\tag{4.96}
$$

Performing Laplace transform w.r.t. A, there exists

$$
\begin{aligned}
G_{x_0}(p, t, t_a) = & \int_0^t \omega(t_a, \tau) \frac{1}{2} \exp(-p\tau U(x_0)) \\
& \times \Big(G_{x_0+a}(p, t - \tau, t_a) + G_{x_0-a}(p, t - \tau, t_a) \Big) d\tau \\
& + \int_t^\infty \omega(t_a, \tau) d\tau \exp(-ptU(x_0)).
\end{aligned}
\tag{4.97}
$$

Using Laplace transform, $t \to s$, we have

$$
\begin{aligned}
\widehat{G}_{x_0}(p, s, t_a) = & \tfrac{1}{2} \widehat{\omega}(t_a, s + pU(x_0)) \Big(\widehat{G}_{x_0+a}(p, s, t_a) \\
& + \widehat{G}_{x_0-a}(p, s, t_a) \Big) + \frac{1 - \widehat{\omega}(t_a, s + pU(x_0))}{s + pU(x_0)}.
\end{aligned}
\tag{4.98}
$$

Taking Fourier transform, $x_0 \to k_0$, leads to

$$
\begin{aligned}
\widetilde{\widehat{G}}_{k_0}(p, s, t_a) = & \widehat{\omega}\Big(t_a, s + pU\Big(-i\frac{\partial}{\partial k_0} \Big) \Big) \cos(k_0 a) \widetilde{\widehat{G}}_{k_0}(p, s, t_a) \\
& + \frac{1 - \widehat{\omega}(t_a, s + pU(-i\frac{\partial}{\partial k_0}))}{s + pU(-i\frac{\partial}{\partial k_0})} \delta(k_0).
\end{aligned}
\tag{4.99}
$$

Equation (4.99) can be rewritten as

$$
\begin{aligned}
\Big(s + pU\Big(-i\frac{\partial}{\partial k_0} \Big) \Big) \widetilde{\widehat{G}}_{k_0}(p, s, t_a) = & \Big(s + pU\Big(-i\frac{\partial}{\partial k_0} \Big) \Big) \\
& \times \widehat{\omega}\Big(t_a, s + pU\Big(-i\frac{\partial}{\partial k_0} \Big) \Big) \cos(k_0 a) \widetilde{\widehat{G}}_{k_0}(p, s, t_a) \\
& + \Big(1 - \widehat{\omega}\Big(t_a, s + pU\Big(-i\frac{\partial}{\partial k_0} \Big) \Big) \Big) \delta(k_0).
\end{aligned}
\tag{4.100}
$$

Performing inverse Fourier transform of Eq. (4.100) w.r.t. k_0, there exists

$$
\begin{aligned}
(s + pU(x_0))\widehat{G}_{x_0}(p, s, t_a) - 1 = & (s + pU(x_0))\widehat{\omega}(t_a, s + pU(x_0)) \\
& \times \Big(1 + \frac{a^2}{2} \frac{\partial^2}{\partial x_0^2} \Big) \widehat{G}_{x_0}(p, s, t_a) - \widehat{\omega}(t_a, s + pU(x_0)).
\end{aligned}
\tag{4.101}
$$

Further taking inverse Laplace transform w.r.t. s, we get

$$\frac{\partial}{\partial t}G_{x_0}(p,t,t_a) = \frac{\partial}{\partial t}\Big(\exp(-ptU(x_0))\omega(t_a,t)\Big(1+\frac{a^2}{2}\frac{\partial^2}{\partial x_0^2}\Big)\Big)*_t G_{x_0}(p,t,t_a)$$

$$- pU(x_0)G_{x_0}(p,t,t_a) - \exp(-ptU(x_0))\omega(t_a,t)$$

$$+ pU(x_0)\Big(\exp(-ptU(x_0))\omega(t_a,t)\Big(1+\frac{a^2}{2}\frac{\partial^2}{\partial x_0^2}\Big)\Big)*_t G_{x_0}(p,t,t_a),$$

$$(4.102)$$

Especially, letting $t_a = 0$, $\widetilde{\widehat{G}}_{k_0}(p,s,t_a) \to \widetilde{\widehat{G}}_{k_0}(p,s)$ and Eq. (4.99) reduces to the Feynman-Kac equation for CTRW model,

$$\widetilde{\widehat{G}}_{k_0}(p,s) = \widehat{\phi}\Big(s + pU\Big(-i\tfrac{\partial}{\partial k_0}\Big)\Big)\cos(k_0 a)\widetilde{\widehat{G}}_{k_0}(p,s)$$

$$+ \frac{1-\widehat{\phi}(s+pU(-i\frac{\partial}{\partial k_0}))}{s+pU(-i\frac{\partial}{\partial k_0})}\delta(k_0).$$

Performing inverse transform on the above equation yields [Carmi *et al.* (2010)]

$$\frac{\partial}{\partial t}G_{x_0}(p,t) = \frac{a^2}{2b_\alpha}D_t^{1-\alpha}\frac{\partial^2}{\partial x_0^2}G_{x_0}(p,t) - pU(x_0)G_{x_0}(p,t).$$

4.6.4 *Backward Feynman-Kac Equation with Continuous Step Length PDF*

Here we derive the backward Feynman-Kac equation, and the functional is not necessarily positive. From the definition of ACTRW model, the particle waits for a time τ at the position x_0 with the waiting time PDF $\omega(t_a,\tau)$, then moves to $x_0 + y$, the PDF of y being $w(y)$. For simplicity, supposing $w(y)$ has the asymptotic behavior of power law, i.e., $w(y) \sim |y|^{-\beta-1}$, there exists

$$G_{x_0}(A,t,t_a) = \int_0^t \omega(t_a,t)\int_{-\infty}^{\infty} w(y)$$

$$\times\, G_{x_0+y}(A - \tau U(x_0), t-\tau, t_a)dyd\tau \qquad (4.103)$$

$$+\Big(1-\int_0^t \omega(t_a,\tau)d\tau\Big)\delta(A - tU(x_0)),$$

where the second term of the right hand side represents the probability of particles that remain immobile at the position x_0 over the observation time t. Performing Fourier transform w.r.t. A leads to

$$G_{x_0}(p,t,t_a) = \int_0^t \omega(t_a,t)\int_{-\infty}^{\infty}\exp(iptU(x_0))w(y)$$

$$\times\, G_{x_0+y}(p,t-\tau,t_a)dyd\tau \qquad (4.104)$$

$$+\Big(1-\int_0^t \omega(t_a,\tau)d\tau\Big)\exp(iptU(x_0)).$$

Then by Laplace and Fourier transforms, we have

$$\widehat{\widetilde{G}}_{k_0}(p,s,t_a) = \widehat{\omega}\left(t_a, s - ipU\left(-i\frac{\partial}{\partial k_0}\right)\right)\widetilde{w}(k_0)\widehat{\widetilde{G}}_{k_0}(p,s,t_a)$$
$$+\frac{1 - \widehat{\omega}(t_a, s - ipU(-i\frac{\partial}{\partial k_0}))}{s - ipU(-i\frac{\partial}{\partial k_0})}\delta(k_0). \tag{4.105}$$

Rearranging Eq. (4.105) yields

$$\left(s - ipU\left(-i\frac{\partial}{\partial k_0}\right)\right)\widehat{\widetilde{G}}_{k_0}(p,s,t_a) = \left(s - ipU\left(-i\frac{\partial}{\partial k_0}\right)\right)$$
$$\widehat{\omega}\left(t_a, s - ipU\left(-i\frac{\partial}{\partial k_0}\right)\right)f(k_0)\widehat{\widetilde{G}}_{k_0}(p,s,t_a) \tag{4.106}$$
$$+\left(1 - \widehat{\omega}\left(t_a, s - ipU\left(-i\frac{\partial}{\partial k_0}\right)\right)\right)\delta(k_0).$$

Performing inverse Laplace transform w.r.t. s, we obtain the final result

$$\frac{\partial}{\partial t}G_{x_0}(p,t,t_a)$$
$$= \frac{\partial}{\partial t}\left(\exp(iptU(x_0))\omega(t_a,t)\right) *_t \left(1 + A_\beta\nabla_{x_0}^\beta\right)G_{x_0}(p,t,t_a) \tag{4.107}$$
$$+ ipU(x_0)G_{x_0}(p,t,t_a) - ipU(x_0)\left(\exp(iptU(x_0))\omega(t_a,t)\right)$$
$$*_t \left(1 + A_\beta\nabla_{x_0}^\beta\right)G_{x_0}(p,t,t_a) - \exp(iptU(x_0))\omega(t_a,t),$$

where $\nabla_{x_0}^\beta$ is the Riesz operator. Compared with Eq. (4.102), Eq. (4.107) is a more general result.

4.7 Application

We further consider some applications of the distribution of the paths of particles performing aging anomalous diffusion. With the help of functional of anomalous diffusion path, it is convenient to obtain the occupation time in half space, fluctuations of occupation fraction, and the first passage time.

4.7.1 *Occupation Time in Half Space for ACTRW*

In the following, we consider the occupation time of particles in half space, which is an important topic in mathematics and physics. We introduce T^+, the occupation time in $x > 0$

$$T^+ = \int_0^t \theta(x(\tau))d\tau. \tag{4.108}$$

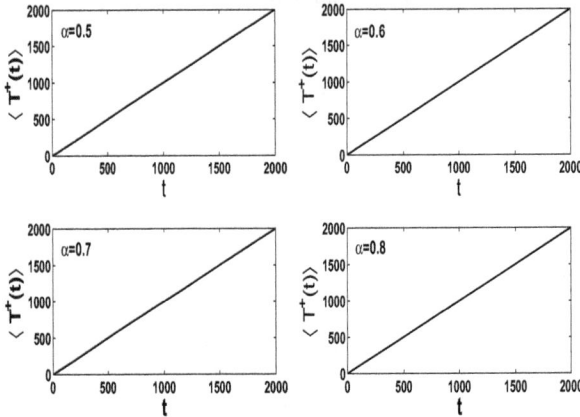

Fig. 4.15 The relation between $\langle T^+ \rangle$ and the observation time t for various α with $t > t_a$. The number of particles is 4000, $t_a = 10$, $t = 2000$, and $\alpha = 0.5, 0.6, 0.7,$ and 0.8. The real lines are obtained by averaging trajectories. It can be noticed that $\langle T^+ \rangle$ grows linearly with time t.

In order to derive the PDF of T^+, we consider the backward Feynman-Kac equation Eq. (4.101), i.e., power law waiting time and regular jump length, in Laplace space

$$\widehat{G}_{x_0}(p,s,t_a) = \begin{cases} \frac{K_a\widehat{\omega}(t_a,s)}{1-\widehat{\omega}(t_a,s)}\frac{\partial^2}{\partial x_0^2}\widehat{G}_{x_0}(p,s,t_a) + \frac{1}{s}, & x_0 < 0; \\ \frac{K_a\widehat{\omega}(t_a,s+p)}{1-\widehat{\omega}(t_a,s+p)}\frac{\partial^2}{\partial x_0^2}\widehat{G}_{x_0}(p,s,t_a) + \frac{1}{s+p}, & x_0 > 0, \end{cases} \tag{4.109}$$

where $K_a = a^2/2$ and $\widehat{\omega}(t_a,s)$ denotes the PDF of the forward waiting time. From Eq. (4.2), we obtain

$$\widehat{\omega}(t_a,s) = \frac{1}{\Gamma(\alpha)}\exp(st_a)\Gamma(\alpha,st_a). \tag{4.110}$$

Supposing $\widehat{G}_{x_0}(p,s,t_a) \to 0$ for $|x_0| \to \infty$, it's easy to solve the second order, ordinary differential equation about x_0,

$$\widehat{G}_{x_0}(p,s,t_a) = \begin{cases} C_0 \exp\left(x_0\sqrt{\frac{1-\widehat{\omega}(t_a,s)}{K_a\widehat{\omega}(t_a,s)}}\right) + \frac{1}{s}, & x_0 < 0; \\ C_1 \exp\left(-x_0\sqrt{\frac{1-\widehat{\omega}(t_a,s+p)}{K_a\widehat{\omega}(t_a,s+p)}}\right) + \frac{1}{s+p}, & x_0 > 0. \end{cases} \tag{4.111}$$

Here C_0 and C_1 are coefficients to be determined. It can be noticed the particles can never reach the right plane if the initial position $x_0 \to -\infty$, i.e., $G_{x_0}(T^+,t,t_a) = \delta(T^+)$; furthermore, performing Laplace transform we have $\widehat{G}_{x_0}(p,s,t_a) = \frac{1}{s}$, which agrees with Eq. (4.111). If $x_0 \to$

$+\infty$, the probability of the particles reaching the left plane is 0, namely $G_{x_0}(T^+, t, t_a) = \delta(T^+ - t)$ and $\widehat{G}_{x_0}(p, s, t_a) = \frac{1}{s+p}$, which is consistent with Eq. (4.111). Assuming that $\widehat{G}_{x_0}(p, s, t_a)$ and its first derivative about x_0 are continuous at $x_0 = 0$, we obtain

$$\begin{cases} C_0 + \frac{1}{s} = C_1 + \frac{1}{s+p}; \\ -C_0 \sqrt{\frac{1-\widehat{\omega}(t_a,s)}{K_a \widehat{\omega}(t_a,s)}} = C_1 \sqrt{\frac{1-\widehat{\omega}(t_a,s+p)}{K_a \widehat{\omega}(t_a,s+p)}} . \end{cases} \qquad (4.112)$$

From the above equation, there exists

$$\begin{cases} C_0 = -\frac{p}{s(s+p)} \dfrac{\sqrt{(1-\widehat{\omega}(t_a,s+p))\widehat{\omega}(t_a,s)}}{\sqrt{\widehat{\omega}(t_a,s+p)(1-\widehat{\omega}(t_a,s))}+\sqrt{\widehat{\omega}(t_a,s)(1-\widehat{\omega}(t_a,s+p))}}; \\ C_1 = \frac{p}{s(s+p)} \dfrac{\sqrt{(1-\widehat{\omega}(t_a,s))\widehat{\omega}(t_a,s+p)}}{\sqrt{\widehat{\omega}(t_a,s+p)(1-\widehat{\omega}(t_a,s))}+\sqrt{\widehat{\omega}(t_a,s)(1-\widehat{\omega}(t_a,s+p))}}. \end{cases} \qquad (4.113)$$

For simplicity, let the particle start at $x_0 = 0$. From Eq. (4.111) we have $\widehat{G}_0(p, t_a, s) = C_0 + 1/s$, i.e.,

$$\begin{aligned} \widehat{G}_0(p, s, t_a) &= \frac{1}{s(s+p)} \\ &\times \frac{(p+s)\sqrt{(1-\widehat{\omega}(t_a,s))\widehat{\omega}(t_a,s+p)} + s\sqrt{(1-\widehat{\omega}(t_a,s+p))\widehat{\omega}(t_a,s)}}{\sqrt{\widehat{\omega}(t_a,s+p)(1-\widehat{\omega}(t_a,s))} + \sqrt{\widehat{\omega}(t_a,s)(1-\widehat{\omega}(t_a,s+p))}} . \end{aligned}$$
$$(4.114)$$

Especially, if $t_a = 0$, $\omega(t_a = 0, t) = \phi(t)$, we have $\widehat{\omega}(t_a = 0, s) = \widehat{\phi}(s)$ and $\widehat{\omega}(t_a = 0, s + p) = \widehat{\phi}(s + p)$, and Eq. (4.114) reduces to

$$\widehat{G}_0(p, s) = \frac{s^{\alpha/2-1} + (s+p)^{\alpha/2-1}}{s^{\alpha/2} + (s+p)^{\alpha/2}}. \qquad (4.115)$$

Using the technique given by Godrèche and Luck [Godrèche and Luck (2001)], the PDF of $y = T^+/t$ can be shown by the Lamperti PDF [Carmi *et al.* (2010)], i.e.,

$$g(y) = \frac{\sin(\pi\alpha/2)}{\pi} \frac{y^{\alpha/2-1}(1-y)^{\alpha/2-1}}{y^\alpha + (1-y)^\alpha + 2y^{\alpha/2}(1-y)^{\alpha/2}\cos(\pi\alpha/2)}. \qquad (4.116)$$

For more details, see [Godrèche and Luck (2001); Lamperti (1958); Baldassarri *et al.* (1999)]. Equation (4.114) works well for comparable and large t and T^+, while it's difficult to invert Eq. (4.114) analytically. Furthermore, when T^+ is of the order t^0 or t, the rare fluctuations are recovered.

4.7.2 Fluctuation of Occupation Fraction

We further introduce $\eta(t) = \frac{T^+}{t}$, a quantity to illustrate the fraction of time that the particle spends in a given domain [Godrèche and Luck (2001); Thaler (2002); Bel and Barkai (2006)]. It's easy to obtain the moments of T^+ by using the relationship

$$\langle(\widehat{T}^+(s))^n\rangle = (-1)^n \frac{\partial^n}{\partial p^n} \widehat{G}_0(p, s, t_a)|_{p=0}. \tag{4.117}$$

From Eqs. (4.114) and (4.110), we get

$$\widehat{G}_0(p, s, t_a) = \frac{1}{s(s+p)} \cdot \frac{(p+s)g_1(p, s, t_a) + sg_2(p, s, t_a)}{g_1(p, s, t_a) + g_2(p, s, t_a)}, \tag{4.118}$$

where

$$g_1(p, s, t_a) = \sqrt{\Gamma(\alpha - e^{st_a}\Gamma(\alpha, st_a))e^{pt_a}\Gamma(\alpha, (s+p)t_a)}$$

and

$$g_2(p, s, t_a) = \sqrt{\Gamma(\alpha - e^{(s+p)t_a}\Gamma(\alpha, (s+p)t_a))\Gamma(\alpha, st_a)}.$$

Since it is difficult to take inverse transform on Eq. (4.118) analytically, we just consider the asymptotic behaviors of the moments, inlcuding the first and the second moments of T^+. Setting $n = 1$ and using the relation between Eqs. (4.117) and (4.118), yield

$$\langle\widehat{T}^+(s)\rangle = -\frac{\partial}{\partial p} \widehat{G}_0(p, s, t_a)|_{p=0} = \frac{1}{2s^2}, \tag{4.119}$$

by inverse Laplace transform, which leads to

$$\begin{cases} \langle T^+(t)\rangle = t/2; \\ \langle\eta(t)\rangle = 1/2. \end{cases} \tag{4.120}$$

That is to say, for both weak and strong aging systems, the aging time t_a makes no difference on $\langle T^+(t)\rangle$. In Fig. 4.15, we give the simulation results getting from trajectories for $t \gg t_a$; it can be noticed that $\langle T^+(t)\rangle$ increases linearly with time t, which agrees with our theoretical results Eq. (4.120). In Laplace space, the second moment of $T^+(t)$ is

$$\langle\widehat{T}^+(s)^2\rangle = \frac{(st_a)^\alpha - e^{st_a}\Gamma(\alpha, st_a)(4 + st_a)\Gamma(\alpha) + 4e^{2st_a}\Gamma(\alpha, st_a)^2}{4s^3\Gamma(\alpha, st_a)e^{st_a}(e^{st_a}\Gamma(\alpha, st_a) - \Gamma(\alpha))}. \tag{4.121}$$

We first consider weak aging system, i.e., $t_a \ll t$. For small y, the incomplete Gamma function has the following relation

$$\Gamma(\alpha, y) = \Gamma(\alpha)\left(1 - \exp(-y)\sum_{n=0}^{\infty} \frac{y^{n+\alpha}}{\Gamma(1+\alpha+n)}\right)$$

$$\sim \Gamma(\alpha)\left(1 - \exp(-y)\frac{y^\alpha}{\Gamma(1+\alpha)}\right). \tag{4.122}$$

Using Eqs. (4.121) and (4.122), for $t_a \ll t$, i.e., $t_a s \to 0$, there exists

$$\langle \widehat{T}^+(s)^2 \rangle \sim \frac{(1 - \alpha/4)}{s^3}. \tag{4.123}$$

Hence, the inverse Laplace transform of Eq. (4.123) w.r.t. s yields

$$\langle (T^+)^2(t) \rangle \sim \left(\frac{1}{2} - \frac{\alpha}{8} \right) t^2, \tag{4.124}$$

or $\langle \eta(t)^2 \rangle \sim (1/2 - \alpha/8)$. In Fig. 4.16, we can see that the theoretical results are consistent with the simulation ones. In addition, the fluctuation of the occupation fraction is

$$\langle \Delta_\eta^2(t) \rangle = \langle \eta^2(t) \rangle - (\langle \eta(t) \rangle)^2 \sim \frac{1}{4} - \frac{\alpha}{8}. \tag{4.125}$$

We further consider the strong aging system, i.e., $t \ll t_a$. For large y, $\Gamma(\alpha, y)$ behaves as

$$\Gamma(\alpha, y) \sim y^{\alpha-1} e^{-y}. \tag{4.126}$$

Substituting Eq. (4.126) into Eq. (4.121), there exists

$$\langle T^+(t)^2 \rangle = \frac{1}{2} t^2, \tag{4.127}$$

which is verified by Fig. 4.17. Furthermore, we get

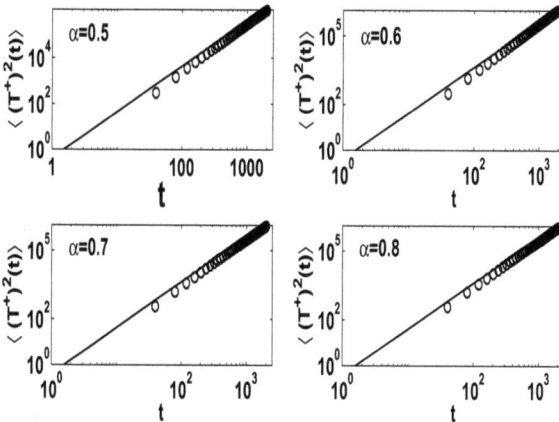

Fig. 4.16   Time evolution of the ensemble average of the occupation time for weak aging systems. The parameters are taken as $t_a = 10$, $t = 2000$, and $\alpha = 0.5$, 0.6, 0.7, and 0.8. The real lines are the analytical result shown in Eq. (4.124) and the symbols are obtained by averaging 4000 trajectories.

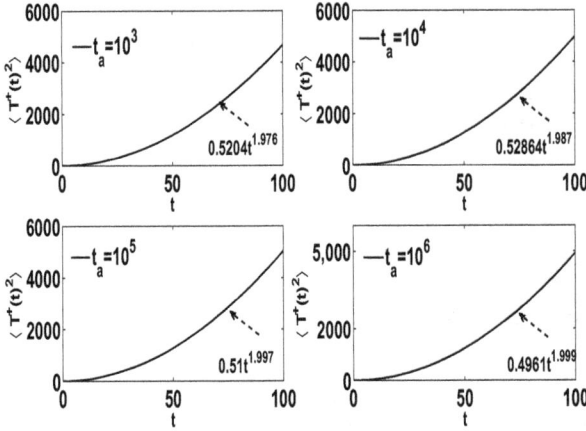

Fig. 4.17   Time evolution of the ensemble average of the occupation time for strong aging systems. The parameters are taken as $\alpha = 0.6$, $t = 100$, $N = 4000$ and $t_a = 10^3$, 10^4, 10^5, and 10^6. The solid lines are obtained by averaging 4000 trajectories. $\langle (T^+)^2(t) \rangle = 0.5204 \times t^{1.976}$, $0.5286 \times t^{1.987}$, $0.51 \times t^{1.997}$, and $0.4961 \times t^{1.999}$ are the fitting results for small t and large t_a, which agree with Eq. (4.127).

$$\langle \Delta_\eta^2(t) \rangle \sim \frac{1}{4}. \tag{4.128}$$

From Eqs. (4.128) and (4.125), it can be noted that the fluctuation of occupation fraction for strong aging system is larger than the one for weak aging system. This can be intuitively explained as follows. For $t \ll t_a$, i.e., the observation time t is small, a large number of particles do not finish their first steps and they still stay at the initial position, which indicates that long waiting time plays an important role. Furthermore, the coefficient of $\langle (T^+)^2(t) \rangle$ increases as the decrease of α for weak aging, which is confirmed by Fig. 4.18. However, for strong aging systems, the coefficient of $\langle (T^+)^2(t) \rangle$ is a constant.

4.7.3   *Distribution of First Passage Time*

First passage times [Krüsemann *et al.* (2014); Bel and Barkai (2005); Buonocore *et al.* (1987); Deng *et al.* (2017)] are central features of many families of stochastic processes, including Poisson process, Wiener process, gamma process, and Markov chains, to name but a few. The first passage time, also called the first hitting time, is considered as the time T_f, which in this subsection represents the time a particle starting at $x_0 = -b$ to hit $x = 0$

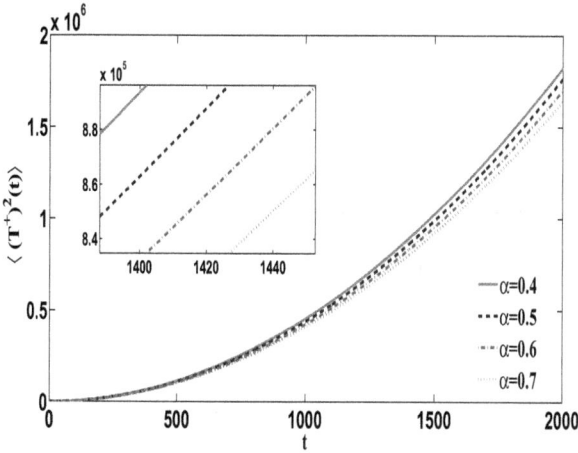

Fig. 4.18   Time evolution of the ensemble average of the occupation time for weak aging system. The parameters are taken as $t_a = 2$, $t = 2000$, and $\alpha = 0.4$, 0.5, 0.6, and 0.7. The symbolled lines are obtained by averaging 15000 trajectories.

for the first time with $b > 0$. Using the relation between the occupation time functional and the distribution of the first passage time [Kac (1951)]

$$P(T_f > t) = P\left(\max_{0 \leq \tau \leq t} x(\tau) < 0 \right) = \lim_{p \to \infty} G_{x_0}(p, t, t_a),$$

where $G_{x_0}(p, t, t_a)$ is the Laplace transform of $G_{x_0}(T^+, t, t_a)$ w.r.t. $T^+ = \int_0^t \theta(x(t)) dt$. Utilizing $P(\max_{0 < \tau < t} x(\tau) < 0) = P(T^+ = 0) = \lim_{T^+ \to 0} \int_0^{T^+} G_{x_0}(A, t, t_a) dA$ and the initial value theorem of Laplace transform [Dyke (2014)], there exists

$$P(T^+ = 0) = \lim_{p \to \infty} p\mathcal{L}\left[\int_0^{T^+} G_{x_0}(A, t, t_a) dA \right] = \lim_{p \to \infty} G_{x_0}(p, t, t_a).$$

Supposing $x_0 = -b$ and $p \to \infty$, from Eqs. (4.109) and (4.113), we obtain

$$\lim_{p \to \infty} \widehat{G}_{x_0}(p, s, t_a) = \lim_{p \to \infty} \left(\frac{1}{s} - \frac{p}{s(s+p)} \frac{g_3(p, s, t_a) e^{-b\sqrt{\frac{\Gamma(\alpha) - e^{st_a}\Gamma(\alpha, st_a)}{K_a e^{st_a}\Gamma(\alpha, st_a)}}}}{g_3(p, s, t_a) + g_4(p, s, t_a)} \right),$$

$$(4.129)$$

where

$$g_3(p, s, t_a) = \sqrt{\Gamma(\alpha) - e^{(s+p)t_a}\Gamma(\alpha, (s+p)t_a)\Gamma(\alpha, st_a)}$$

and

$$g_4(p, s, t_a) = \sqrt{e^{pt_a}\Gamma(\alpha, (s+p)t_a)\Gamma(\alpha) - e^{st_a}\Gamma(\alpha, st_a)\Gamma(\alpha, st_a)}.$$

Using Eq. (4.126), it yields

$$\lim_{p \to \infty} \widehat{G}_{x_0}(p, s, t_a) = \frac{1}{s} - \frac{1}{s}\exp\left(-b\sqrt{\frac{\Gamma(\alpha) - e^{st_a}\Gamma(\alpha, st_a)}{K_a e^{st_a}\Gamma(\alpha, st_a)}}\right). \quad (4.130)$$

According to the definition of the first passage time, we get its corresponding PDF

$$g(t_a, t) = \frac{\partial}{\partial t}(1 - P(T_f > t)) = -\frac{\partial}{\partial t}\lim_{p \to \infty} G_{-b}(p, t, t_a).$$

Performing Laplace transform w.r.t. t, leads to

$$\widehat{g}(t_a, s) = \exp\left(-b\sqrt{\frac{\Gamma(\alpha) - \exp(st_a)\Gamma(\alpha, st_a)}{K_a \exp(st_a)\Gamma(\alpha, st_a)}}\right). \quad (4.131)$$

This demonstrates that Eq. (4.131) is valid for all kinds of s and t_a. However, it seems difficult to invert Eq. (4.131) analytically. In the following, we consider the asymptotic behaviors of the density of the first passage time. For $st_a \to 0$, i.e., $t_a \ll t$, a simple calculation gives

$$\widehat{g}(t_a, s) = \exp\left(-b\sqrt{\frac{1 - e^{st_a}\left(1 - e^{-st_a}\frac{(st_a)^\alpha}{\Gamma(1+\alpha)}\right)}{K_a e^{st_a}\left(1 - e^{-st_a}\frac{(st_a)^\alpha}{\Gamma(1+\alpha)}\right)}}\right)$$

$$\sim \exp\left(-b\sqrt{\frac{1}{K_a\Gamma(1+\alpha)}}(st_a)^{\frac{\alpha}{2}}\right), \quad (4.132)$$

which yields

$$g(t_a, t) \sim \frac{(\Gamma(1+\alpha)K_a)^{\frac{1}{\alpha}}}{t_a b^{\frac{2}{\alpha}}}\ell_{\frac{\alpha}{2}}\left(\frac{(\Gamma(1+\alpha)K_a)^{\frac{1}{\alpha}}}{b^{\frac{2}{\alpha}}}\frac{t}{t_a}\right) \quad (4.133)$$

with $\ell_\alpha(t)$ being the one sided Lévy distribution [Klafter and Sokolov (2011)]. For large t/t_a, $g(t_a, t)$ decays as a power law

$$g(t_a, t) \sim t_a^{\frac{\alpha}{2}} t^{-1-\frac{\alpha}{2}}, \quad (4.134)$$

which is confirmed by Fig. 4.19. For small t_a and large t, $g(t_a, t) \sim t^{-1-\alpha/2}$ and $g(t_a, t)$ tends to 0 slowly, which is the same as the behaviors of none aging systems [Barkai (2001)]. When $st_a \to \infty$, corresponding to $t \ll t_a$, from Eq. (4.131), we have

$$\widehat{g}(t_a, s) \sim \exp\left(-b\sqrt{\frac{\Gamma(\alpha)}{K_a t_a^{\alpha-1}}}s^{\frac{1-\alpha}{2}}\right). \quad (4.135)$$

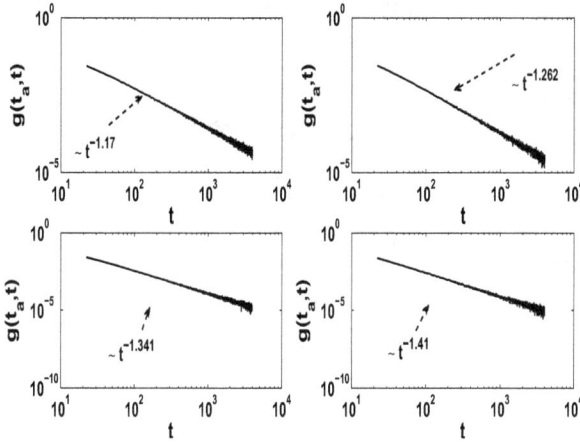

Fig. 4.19 Behaviors of first passage time density $g(t_a, t)$ generated from the trajectories of particles for constant jump length with $t_a = 10$, $t = 4000$, $b = 0.05$ and $\alpha = 0.4$, 0.5, 0.6, and 0.7, respectively. The solid lines are obtained by averaging 10^6 trajectories and the formulas are the fitting results.

Taking the inverse Laplace transform of Eq. (4.135), there exists

$$g(t_a, t) \sim \left(\left(\frac{\sqrt{K_a}}{b\sqrt{\Gamma(\alpha)}} \right)^{\frac{2}{1-\alpha}} \frac{1}{t_a} \right) \ell_{\frac{1-\alpha}{2}} \left(\left(\frac{\sqrt{K_a}}{b\sqrt{\Gamma(\alpha)}} \right)^{\frac{2}{1-\alpha}} \frac{t}{t_a} \right). \qquad (4.136)$$

Keeping in mind that t/t_a is a small number, we can not expand $\ell_\alpha(t)$ as $t^{-\alpha-1}$ to obtain its asymptotic form. By numerically inverting Eq. (4.135) and combining it with the form of Eq. (4.136), we predict that

$$g(t_a, t) \sim t_a^{\alpha-1} t^{-\alpha}. \qquad (4.137)$$

Equation (4.137) is confirmed by the simulation of trajectories of the particles; see Fig. 4.21. From Eqs. (4.134) and (4.137), we can find that the aging times play an important role for the first passage time. For the weak aging system, the scaling exponent of t is $-1 - \frac{\alpha}{2}$, while it changes to $-\alpha$ for strong aging system. If the power of t_a is α for the weak aging system, it becomes $\alpha - 1$ for strong aging system.

To summarize, the functionals of the trajectories of particles are very general statistical observables characterizing the motion of particles. Based on the ACTRW, we derive the forward and backward Feynman-Kac equations governing the distribution of the functionals of the trajectories of particles performing aging anomalous diffusion with (tempered) power law waiting time and/or jump length distributions. For deeply studying

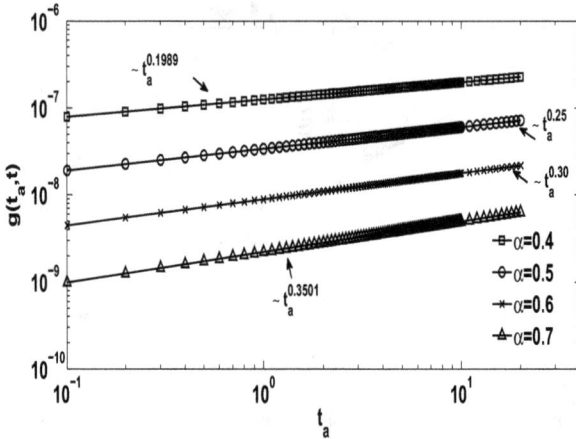

Fig. 4.20   The relation between first passage time density and t_a for weak aging system. The parameters are taken as $t_a = 20$, $t = 10^6$, $b = 0.05$ and $\alpha = 0.5$, 0.6, 0.7, and 0.8. The symbolled lines are obtained by the inversion of Eq. (4.132). The formulae are the fitting results for $\alpha = 0.5$, 0.6, 0.7, and 0.8, respectively.

the aging phenomena of anomalous diffusion, according to the built models, we more specifically calculate the statistical observables: fraction of the

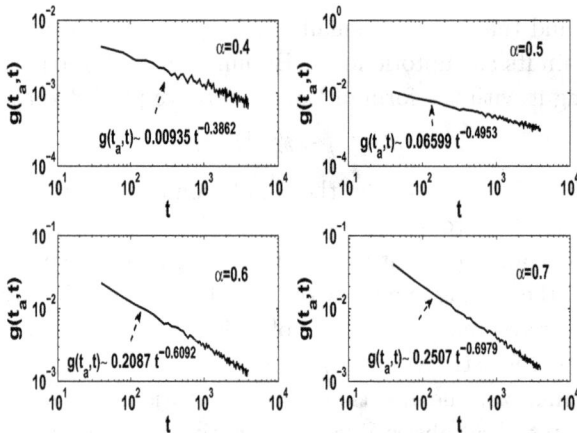

Fig. 4.21   First passage time density for strong aging system. The parameters are taken as $t_a = 10^5$, $t = 4000$, $b = 0.05$ and $\alpha = 0.4$, 0.5, 0.6, and 0.7. The solid lines are obtained by averaging 2×10^5 trajectories. The formulae in the lower-left corner of subplots are the fitting results for $\alpha = 0.4$, 0.5, 0.6, and 0.7, respectively.

occupation time and the first passage time. Besides, the fluctuation of the occupation fraction is also analyzed, discovering that the aging time has no influence on its first moment but greatly impacts the second moment. Another striking discovery is that the distribution of the first passage time $g(t_a, t) \sim t_a^{\frac{\alpha}{2}} t^{-1-\frac{\alpha}{2}}$ for weak (none) aging systems, while for strong aging systems, $g(t_a, t) \sim t_a^{\alpha-1} t^{-\alpha}$.

Chapter 5

Fokker-Planck and Feynman-Kac Equations with Multiple Internal States

This chapter is mainly based on the foundations of fractional Fokker-Planck equations and Feynman-Kac equations introduced in Chap. 2 and Chap. 3, respectively. As illustrated in Sec. 1.6, more and more phenomena that the ordinary CTRW model or Lévy walk can't depict very well are found, for example the particles move in the multi-phase viscous liquid consisting of different kinds of materials which will make the CTRW with different kinds of waiting time and jump length distributions (or the different kinds of walking time distributions and velocities for Lévy walks) when the particles stay in different phase. In this chapter, we first introduce the basic model and notations by noticing the fundamental concept of internal states introduced in Sec. 1.6. Then the fractional Fokker-Planck equations for the PDF of the CTRW with multiple internal states will also be obtained, and the corresponding MSDs will also be calculated. Then for CTRW with multiple internal states, the forward and backward Feynman-Kac equations, and the governing equations for the functionals of the internal states are also obtained. For Feynman-Kac equation, there're many applications introduced in Chap. 3, and in this chapter we simply consider the first passage time of an interval. And we also calculate the occupation time of one internal state based on the governing equations for the functionals of the internal states. Comparing the obtained fractional Fokker-Planck and Feynman-Kac equations for the CTRW process with multiple internal states with the ordinary CTRW model whose governing equation is illustrated in Eq. (2.33) in Sec. 2.2, there will exist an extra term in the equation.

Besides the Lévy walks with multiple internal states are also discussed in this chapter. Because the internal states and the corresponding transition matrix are abstract, we can construct many examples. One typical example given in this chapter is that the smart animals will not always return to the area they have already finished searching. And it turns out that for CTRW whose internal states obey different kinds of the transition matrices will always be influenced by the different ways of moving, while for Lévy walk, such influence will disappear when the walking is a superdiffusion except the case that transition matrix causes the walking roughly along one direction.

And there're still many potential applications for the processes with multiple internal states, such as the non-circle process whose trajectory, intuitively speaking, forms no circle, and such process is also important in the area of random walk.

5.1 Model and Notations

In this section, we mainly introduce the model and the notations we are going to use in the following derivations. In Sec. 1.6, we have already briefly introduced our model and the concepts of internal states, and also provided one example to help the comprehension of the internal states and the transition among different internal states determined by a given transition matrix. In this chapter we only consider the finite number of internal states denoted as N. In fact, it should be noted that the process of internal states is Markovian, and the transition matrix M with the dimension of $N \times N$ has been introduced in Sec. 1.6. Here we use the bras $\langle \cdot |$ and kets $| \cdot \rangle$ to denote the row and column vectors, respectively [Niemann *et al.* (2016); Xu and Deng (2018)]. According to the theory of Markov chains [Feller (1971)], the ergodic and periodic chain has an equilibrium distribution, denoted as $\langle \mathrm{eq_M} |$, such that $\langle \mathrm{eq_M} | M = \langle \mathrm{eq_M} |$. The construction of the transition matrix M (the factor of i-th column j-th row $m_{i,j}$ representing the probability of transition from i-th internal state to the j-th one) indicates that the sum of factors of each row is one, i.e., $M|\Sigma\rangle = |\Sigma\rangle$, where $|\Sigma\rangle$ represents the column vector of which every factor is one. It is obvious that $M^T|\mathrm{eq_M}\rangle = |\mathrm{eq_M}\rangle$ and $\langle \Sigma|M^T = \langle \Sigma|$. Besides according to the introduction in Sec. 1.6, we also need an initial distribution to start our process and denote it as $|\mathrm{init}\rangle$ (here we only give the column vector version of initial distribution). Because of the normalization of equilibrium and initial distributions, we note that $\langle \Sigma|\mathrm{init}\rangle = 1$ and $\langle \Sigma|\mathrm{eq_M}\rangle = 1$. The difference between ordinary CTRW and CTRW with N internal states is that the ordinary model has a single pair of waiting time and jump length distributions while the generalized model has N pairs. Here we denote $\phi_m(t) = \mathrm{diag}(\phi^{(1)}(t), \phi^{(2)}(t), \ldots, \phi^{(N)}(t))$ and

$\Lambda(\mathbf{x}) = \mathrm{diag}(\lambda^{(1)}(\mathbf{x}), \lambda^{(2)}(\mathbf{x}), \ldots, \lambda^{(N)}(\mathbf{x}))$ to represent the distribution matrices of waiting time and jump length respectively, where the superscript of each factor of the waiting time and jump length distributions represents the serial number of internal state.

In Sec. 1.6, we have also briefly introduced how the particle moves. Here we only give an overview of the CTRW with multiple internal states. First in order to start our process, we choose one of the internal states to decide which pair of waiting time and jump length distributions controls the wait and jump in the first step. For example if we choose the i-th internal state according to the initial distribution with the probability of the i-th factor of $|\text{init}\rangle$, then the waiting time and jump length of the first step are generated by the $\phi^{(i)}(t)$ and $\lambda^{(i)}(x)$, respectively. Then by the transition matrix M, the i-th row specifically, we may transit the internal state. After the transition, the particle determines one of the internal states and the corresponding waiting time and jump length distributions to determine the waiting time and jump length.

The transition matrix of the Markov process has a great influence on the final results. One of the most important properties of a Markov chain is reducibility or irreducibility. Here is the definition.

Definition 5.1. A set C consisting of internal states is called closed if no state outside C can be reached from any state in C. For an arbitrary internal states set C, the smallest closed set containing C is called the closure of C.

A Markov chain is irreducible if there is no closed set except for the set of all states.

We also note that a chain is irreducible, if and only if every state can be reached from every other state no matter how many steps it will take. Besides, sometimes we can judge the chain is reducible or not simply based on the form of transition matrix. For example if the transition matrix can be written as the form

$$M = \begin{pmatrix} P & 0 \\ U & V \end{pmatrix},$$

then we can conclude the chain with transition matrix M is reducible. Further we can conclude that the corresponding internal states of the block matrix P are closed, i.e., the other internal states cannot be reached by the internal states of which matrix P consists. In the next section, we will begin to obtain the fractional Fokker-Planck equations for the CTRW with multiple internal states and the MSD for such process. And for Lévy walk with multiple internal states, the internal state represents the pair of walking duration and velocity distributions, and it'll be detailedly discussed in Sec. 5.5.

5.2 Fractional Fokker-Planck Equations for CTRW with Multiple Internal States

After the introduction of the CTRW model with multiple internal states and notations in Sec. 5.1, in this section, we further derive the generalized fractional Fokker-Planck equations for the CTRW process with multiple internal states. Here we use the notation $g^{(i)}(\mathbf{x}, t), i = 1, 2, \ldots, N$ to represent the PDF of finding the particle at position $\mathbf{x}$ in d-dimensional space, time t and staying in the i-th internal state. Let $|G(\mathbf{x}, t)\rangle$ represent the column vector of gathering all $\{g^{(i)}(\mathbf{x}, t), i = 1, 2, \ldots, N\}$, that is

$$|G(\mathbf{x}, t)\rangle = \begin{pmatrix} g^{(1)}(\mathbf{x}, t) \\ g^{(2)}(\mathbf{x}, t) \\ \vdots \\ g^{(N)}(\mathbf{x}, t) \end{pmatrix}.$$

Similarly to the derivations of ordinary fractional Fokker-Planck equation for the CTRW model, we also first consider the PDF of the particle which just arrives at position $\mathbf{x}$, time t and i-th internal state, after n steps, and denoted as $q_n^{(i)}(\mathbf{x}, t)$. And $|Q(\mathbf{x}, t)\rangle$ represents the column vector consisting of $q_n^{(i)}(\mathbf{x}, t), i = 1, 2, \ldots, N$, that is

$$|Q(\mathbf{x}, t)\rangle = \begin{pmatrix} q^{(1)}(\mathbf{x}, t) \\ q^{(2)}(\mathbf{x}, t) \\ \vdots \\ q^{(N)}(\mathbf{x}, t) \end{pmatrix}.$$

Before the derivation, we first concern about the matrix of survival probability

$$W(t) = \mathrm{diag}(w^{(1)}(t), \ldots, w^{(N)}(t))$$

$$= \mathrm{diag}\left(\int_t^\infty \phi^{(1)}(\tau)d\tau, \ldots, \int_t^\infty \phi^{(N)}(\tau)d\tau \right) \tag{5.1}$$

$$= I - \int_0^t \phi_m(\tau)d\tau,$$

where I denotes the identity matrix. Taking Laplace transform w.r.t. t on Eq. (5.1), then we have

$$\widehat{W}(s) = \frac{1}{s}\left(I - \hat{\phi}_m(s)\right). \tag{5.2}$$

According to the transition of the process, there exists

$$q_n^{(i)}(\mathbf{x}, t) = \sum_{j=1}^{N} \int_0^t \int_{R^d} m_{ji} \kappa^{(j)}(\mathbf{x} - \mathbf{y}, t - \tau) q_{n-1}^{(j)}(\mathbf{y}, \tau) d\mathbf{y} d\tau, \qquad (5.3)$$

where $\kappa^{(j)}(\mathbf{x}, t)$ represents the joint PDF of waiting time and jump length when the particle stays in the j-th internal state. If space and time are independent as the assumption of CTRW model, then $\kappa^{(j)}(\mathbf{x}, t) = \lambda^{(j)}(\mathbf{x}) \phi^{(j)}(t)$. Equivalently, we can rewrite the above equation as the vector and matrix form

$$\big|Q_n(\mathbf{x}, t)\big\rangle = \int_0^t \int_{R^d} M^T H(\mathbf{x} - \mathbf{y}, t - \tau) \big|Q_{n-1}(\mathbf{y}, \tau)\big\rangle d\mathbf{y} d\tau, \qquad (5.4)$$

where the matrix $H(\mathbf{x}, t) = \mathrm{diag}\{\kappa^{(1)}(\mathbf{x}, t), \kappa^{(2)}(\mathbf{x}, t), \ldots, \kappa^{(N)}(\mathbf{x}, t)\}$. Taking Fourier and Laplace transforms w.r.t. $\mathbf{x}$ and t respectively, then we obtain

$$\big|\tilde{\hat{Q}}_n(\mathbf{k}, s)\big\rangle = M^T \tilde{\hat{H}}(\mathbf{k}, s) \big|\tilde{\hat{Q}}_{n-1}(\mathbf{k}, s)\big\rangle. \qquad (5.5)$$

Next we begin to build up the relationship between $\big|Q_n(\mathbf{x}, t)\big\rangle$ and $\big|G(\mathbf{x}, t)\big\rangle$. We also consider the factors of the vector first,

$$g^{(i)}(\mathbf{x}, t) = \int_0^t w^{(i)}(\tau) \sum_{n=0}^{\infty} q_n^{(i)}(\mathbf{x}, t - \tau) d\tau.$$

That is in the matrix form

$$\big|G(\mathbf{x}, t)\big\rangle = \int_0^t W(\tau) \sum_{n=0}^{\infty} \big|Q_n(\mathbf{x}, t - \tau)\big\rangle d\tau. \qquad (5.6)$$

We still apply Fourier and Laplace transforms of Eq. (5.6). There exists

$$\big|\tilde{\hat{G}}(\mathbf{k}, s)\big\rangle = \hat{W}(s) \sum_{n=0}^{\infty} \big|\tilde{\hat{Q}}_n(\mathbf{k}, s)\big\rangle. \qquad (5.7)$$

Combining Eq. (5.2), Eq. (5.5) and Eq. (5.7), we finally obtain

$$\big|\tilde{\hat{G}}(\mathbf{k}, s)\big\rangle = \frac{I - \hat{\phi}_m(s)}{s} \big[I - M^T \tilde{\hat{H}}(\mathbf{k}, s)\big]^{-1} \big|\mathrm{init}\big\rangle. \qquad (5.8)$$

In this section, we only consider the case that time and space are independent, that is $H(\mathbf{x}, t) = \Lambda(\mathbf{x}) \phi_m(t)$. The waiting time distribution will be chosen to behave as power laws asymptotically. That is $\hat{\phi}^{(i)}(s) \sim 1 - B_{\alpha_i} s^{\alpha_i}, 0 < \alpha_i < 1, i = 1, 2, \ldots, N$ in the Laplace space. Here we note that the asymptotical behavior only makes sense when time t is large enough (that is also to say s is sufficiently small). We rewrite the waiting time as the matrix form, i.e., $\hat{\phi}_m(s) \sim I - \hat{\phi}_m^*(s)$, where $\hat{\phi}_m^*(s) = \text{diag}(B_{\alpha_1} s^{\alpha_1}, \ldots, B_{\alpha_N} s^{\alpha_N})$. As for the jump length distribution, according to different kinds of processes we can choose the corresponding forms. Here we first consider all of the jump lengths to be Gaussian distribution, thus in the Fourier space $\tilde{\Lambda}(\mathbf{k}) \sim (1 - \sigma^2 |\mathbf{k}|^2) I$. Basing on Eq. (5.8) and utilizing the above waiting time and jump length distributions, we finally obtain the Fokker-Planck equations with N internal states after taking inverse Fourier and Laplace transforms

$$M^T \frac{\partial}{\partial t} |G(\mathbf{x}, t)\rangle$$
$$= (M^T - I) \text{diag}(B_{\alpha_1}^{-1}, \ldots, B_{\alpha_N}^{-1}) \text{diag}(D_t^{1-\alpha_1}, \ldots, D_t^{1-\alpha_N}) |G(\mathbf{x}, t)\rangle$$
$$+ M^T \text{diag}(K_{\alpha_1}, \ldots, K_{\alpha_N}) \text{diag}(D_t^{1-\alpha_1}, \ldots, D_t^{1-\alpha_N}) \Delta |G(\mathbf{x}, t)\rangle, \tag{5.9}$$

where $D_t^{1-\alpha_i}, i = 1, \ldots, N$ are the Riemann-Liouville derivatives defined as [Samko *et al.* (1993)]

$$D_t^{1-\alpha_i} g(x, t) = \frac{1}{\Gamma(\alpha)} \frac{\partial}{\partial t} \int_0^t \frac{g(x, \tau)}{(t - \tau)^{1-\alpha_i}} d\tau;$$

the Laplace operator Δ is defined as $\mathcal{F}\{\Delta |G(\mathbf{x}, t)\rangle\} = -|\mathbf{k}|^2 \mathcal{F}\{|G(\mathbf{x}, t)\rangle\}$; the factors $K_{\alpha_i} = \sigma^2 / B_{\alpha_i}$ of the diagonal matrix represent diffusion coefficients with the dimension $\text{cm}^2/\text{sec}^{\alpha_i}$. Here we note that if the process has a single internal state, i.e., $N = 1$ and the dimension of space is one, then $M^T = I = 1$. And from the form of Eq. (5.9) we can easily obtain the first term of the right hand side of the equation will vanish and the ordinary fractional Fokker-Planck equation will be recovered.

In the following part of this section, we mainly discuss MSD of the process in one spacial dimension. Before our calculation, it should be noted that the PDF $g(x, t)$ which represents finding the particle at position x at time t has the following relationship with the column vector $|G(x, t)\rangle$,

$$g(x, t) = \langle \Sigma | G(x, t)\rangle.$$

Without loss of generality, here we assume $0 < \alpha_1 \leqslant \alpha_2 \leqslant \ldots \leqslant \alpha_N < 1$. By substituting the asymptotic forms of waiting time and jump length distributions into the Eq. (5.8) we can obtain

$$|\tilde{\hat{G}}(k, s)\rangle = \frac{1}{s} [I - \hat{\phi}_m(s)] \tilde{\hat{R}}^{-1}(k, s) |\text{init}\rangle, \tag{5.10}$$

where $\tilde{\hat{R}}(k, s) := I - M^T \hat{\phi}_m(s) \tilde{\Lambda}(k) \sim I - M^T + \sigma^2 k^2 M^T + M^T \hat{\phi}_m^*(s)$. Whether the matrix $\tilde{\hat{R}}(k, s)$ is reducible or not is determined by the transition matrix M. First we consider the irreducible transition matrix M.

Then the asymptotic expression of the inverse matrix of $\tilde{\hat{R}}(k, s)$ has the form

$$\tilde{\hat{R}}^{-1}(k, s) \sim \frac{|\text{eq}_M\rangle\langle\Sigma|}{\langle\Sigma|\hat{\phi}_m^*(s)|\text{eq}_M\rangle + \sigma^2 k^2 \langle\Sigma|\text{eq}_M\rangle}. \tag{5.11}$$

By utilizing Eq. (5.11) and Eq. (5.10), we finally obtain the asymptotical behavior of $g(x, t)$ in the Fourier-Laplace space,

$$\tilde{\hat{g}}(k, s) = \langle\Sigma|\tilde{\hat{G}}(k, s)\rangle \sim \frac{1}{s}\frac{\langle\Sigma|\hat{\phi}_m^*(s)|\text{eq}_M\rangle}{\langle\Sigma|\hat{\phi}_m^*(s)|\text{eq}_M\rangle + \sigma^2 k^2}. \tag{5.12}$$

Equivalently, we can rewrite the above equation as the form

$$\tilde{\hat{g}}(k, s) \sim \frac{1}{s}\frac{\varepsilon_1 B_{\alpha_1} s^{\alpha_1} + \cdots + \varepsilon_N B_{\alpha_N} s^{\alpha_N}}{\varepsilon_1 B_{\alpha_1} s^{\alpha_1} + \cdots + \varepsilon_N B_{\alpha_N} s^{\alpha_N} + \sigma^2 k^2}. \tag{5.13}$$

From Eq. (5.13) we can finally obtain the MSD of the process, that is

$$\langle x^2(t)\rangle = \mathcal{L}^{-1}\left\{ -\frac{\partial^2}{\partial k^2}\tilde{\hat{g}}(k, s)\Big|_{k=0} \right\}$$

$$\sim \mathcal{L}^{-1}\left\{ \frac{2\sigma^2}{s(\varepsilon_1 B_{\alpha_1} s^{\alpha_1} + \ldots + \varepsilon_N B_{\alpha_N} s^{\alpha_N})} \right\}. \tag{5.14}$$

If we take inverse Laplace transform, the MSD of the process will asymptotically behave as $\langle x^2(t)\rangle \sim \mathcal{L}^{-1}\left\{\frac{2K_{\alpha_1}}{\varepsilon_1 s^{1+\alpha_1}}\right\} \sim \frac{2K_{\alpha_1}}{\varepsilon_1 \Gamma(1+\alpha_1)}t^{\alpha_1}$ when time t is sufficiently long, i.e., s small enough. This result indicates that the MSD of process with irreducible transition matrix behaves asymptotically as t^α and the α is the smallest one among all exponents of the power law waiting time distributions. Furthermore, for the irreducible Markov chain the equilibrium distribution $\langle\text{eq}_M|$ isn't influenced by the initial distribution, that is also to say the initial distribution has no influence on the final MSD or PDF. And this result is also in accordance with the theory of Markov chain [Feller (1971)].

On the other hand, we consider the reducible transition matrix. In this case, the initial distribution always makes a great influence on the final results, and the reducible transition matrix is very complicated. Thus we first consider the transition matrix M with the form of $\text{diag}(M_1, \ldots M_j)$, where the block matrices $M_i, i = 1, \ldots, j$ are irreducible with the dimension of $n_i \times n_i$, and $n_1 + n_2 + \ldots + n_j = N$. From such form of transition matrix, one can find that this process actually consists of several different independent Markov chains with irreducible transition matrices

$M_1, M_2, \ldots, M_j$ and these different chains are connected with each other by the initial distribution. In order to be consistent with the structure of M, since $\hat{\phi}_m^*(s)$ and I are diagonal matrices, we can also rewrite $\hat{\phi}_m^*(s)$ and I as the block matrices form $\mathrm{diag}(\hat{\phi}_{m_1}^*(s), \ldots, \hat{\phi}_{m_j}^*(s))$ and $\mathrm{diag}(I_1, \ldots, I_j)$, respectively. Similarly the matrix $\tilde{R}(k,s)$ can also be written as the form of $\mathrm{diag}(\tilde{R}_1(k,s), \ldots, \tilde{R}_j(k,s))$, where $\tilde{R}_i(k,s) = I_i - M_i^T + M_i^T \hat{\phi}_{m_i}^*(s) + \sigma^2 k^2 M_i^T$. To make the presentations clear, we also write the vectors as the form $|\mathrm{init}\rangle = (|\mathrm{init}\rangle_1, \ldots, |\mathrm{init}\rangle_j)^T$, $|\mathrm{eqM}\rangle = (|\mathrm{eqM}\rangle_1, \ldots, |\mathrm{eqM}\rangle_j)^T$, and $\langle \Sigma| = (\langle \Sigma|_1, \ldots, \langle \Sigma|_j)$. For the convenience of statement, the subscripts of the factors of the block matrices are redefined as well. Let $\{B_{\alpha_{ir}} s^{\alpha_{ir}}\}$, $\{\lambda_{i,r}\}$, and $\{\varepsilon_{i,r}\}$ etc, $r = 1, 2, \ldots, n_i$ be the factors of $\hat{\phi}_{m_i}^*(s)$, $|\mathrm{init}\rangle_i$, and $|\mathrm{eqM}\rangle_i$ etc, respectively. According to the inverse of a block diagonal matrix, we have

$$\tilde{R}^{-1}(k,s) = \mathrm{diag}(\tilde{R}_1^{-1}(k,s), \ldots, \tilde{R}_j^{-1}(k,s)),$$

where

$$\tilde{R}_i^{-1}(k,s) \sim \frac{|\mathrm{eqM}\rangle_i \langle \Sigma|_i}{\langle \Sigma|_i \Phi_i^*(s) |\mathrm{eqM}\rangle_i + \sigma^2 k^2 \langle \Sigma|_i |\mathrm{eqM}\rangle_i}$$

with $i = 1, \ldots, j$, simply according to the inverse matrix for the irreducible case shown in Eq. (5.11). Thus the PDF $g(x,t)$ in the Fourier-Laplace space has the form

$$\tilde{g}(k,s) \sim \sum_{\substack{i=1 \\ i \neq i_1, i_2, \cdots, i_j}}^{j} \frac{1}{s} \frac{\langle \Sigma|_i \Phi_i^*(s) |\mathrm{eqM}\rangle_i \langle \Sigma|_i |\mathrm{init}\rangle_i}{\langle \Sigma|_i \Phi_i^*(s) |\mathrm{eqM}\rangle_i + \sigma^2 k^2 \langle \Sigma|_i |\mathrm{eqM}\rangle_i}, \qquad (5.15)$$

where $i_m = m$ if $|\mathrm{init}\rangle_m = 0$ otherwise $i_m = 0$, $m = 1, 2, \cdots, j$. If we consider a very special case, the transition matrix is an identity matrix and $i_m = 0$ for $m = 1, 2, \cdots, N$. Then the PDF of the process in the Fourier-Laplace space will behave as

$$\tilde{g}(k,s) \sim \sum_{i=1}^{N} \frac{1}{s} \frac{\lambda_i s^{\alpha_i}}{s^{\alpha_i} + K_{\alpha_i} k^2} \sim \frac{1}{s} - k^2 \left[\sum_{i=1}^{N} K_{\alpha_i} \lambda_i s^{-\alpha_i - 1} \right].$$

According to the PDF in the Fourier-Laplace space as obtained in Eq. (5.15), we can also calculate MSD with the same method shown in Eq. (5.14). That is

$$\langle x^2(t) \rangle \sim \mathcal{L}^{-1} \left\{ \sum_{\substack{i=1 \\ i \neq i_1, i_2, \cdots, i_j}}^{j} \frac{2\sigma^2 \langle \Sigma|_i |\mathrm{init}\rangle_i \langle \Sigma|_i |\mathrm{eqM}\rangle_i}{s \langle \Sigma|_i \Phi_i^*(s) |\mathrm{eqM}\rangle_i} \right\}.$$

According to the above result, the MSD behaves asymptotically as $\langle x^2(t) \rangle \sim t^{\alpha^*}$ for large time t with $\alpha^* = \max\limits_{\substack{1 \leqslant i \leqslant j \\ i \neq i_1, i_2, \cdots, i_j}} \left\{ \min\limits_{1 \leqslant r \leqslant n_i} \{\alpha_{ir}\} \right\}$. Besides from the results of PDF and MSD for the reducible case, we can also conclude that the initial distribution of the process will influence the final results.

Then we consider the last case that the transition matrix cannot be written as the form of block diagonal matrix. This case is too complicated to analyze in detail. So for this case, we only have a brief look at it. For a given stochastic process, i.e., the initial distribution and the transition matrix are already known, then the equilibrium distribution may not be single. And for this case, the equilibrium distribution can be calculated by taking the limit

$$\langle \text{eq}_\text{M}| = \lim_{n \to \infty} \langle \text{init}| M^n.$$

Some factors of equilibrium distribution must be zero for the reducible transition matrix, otherwise the equilibrium distribution whose all factors are positive indicates the internal state can transit to any other state from the given initial distribution, or just transits within some of the internal stats and connects with each other with the initial distribution. That is also to say the stochastic process is irreducible or can be written as the diagonal block matrix form. Then we can neglect the internal states whose corresponding factors of equilibrium distribution are zeros. The rest internal states can form a new irreducible Markov chain or the transition matrix with the form of diagonal blocks. And the equilibrium distribution is regarded as the new initial distribution to start the stochastic process. We note that for the different initial distributions the new stochastic process may not be the same. In order to make our point clearer, we make an example.

Example 5.1. Here we consider a Markov process whose transition matrix is

$$M = \begin{pmatrix} 1/2 & 0 & 1/2 & 0 \\ 1/2 & 0 & 0 & 1/2 \\ 1/4 & 0 & 3/4 & 0 \\ 0 & 0 & 0 & 1 \end{pmatrix}.$$

First we take the initial distribution as $\langle \text{init}| = (1/3, 1/3, 1/3, 0)$. Then the equilibrium distribution for this case is $\langle \text{eq}_\text{M}| = (5/18, 0, 5/9, 1/6)$. From the equilibrium distribution we can observe that the factor of second internal state is zero. Thus we can consider the Markov process with the transition matrix

$$M_2 = \begin{pmatrix} 1/2 & 1/2 & 0 \\ 1/4 & 3/4 & 0 \\ 0 & 0 & 1 \end{pmatrix}$$

by deleting the second column and row. And the new initial distribution will be $(5/18, 5/9, 1/6)$, i.e., the equilibrium distribution after neglecting the second factor. For the second case we can consider the initial distribution as $\langle \text{init} | = (1/2, 0, 1/2, 0)$ and the corresponding equilibrium distribution will be $\langle \text{eq}_\text{M} | = (1/3, 0, 2/3, 0)$. Then we can obtain the following transition matrix for the new process after deleting the second and fourth columns and rows. That is

$$M_3 = \begin{pmatrix} 1/2 & 1/2 \\ 1/4 & 3/4 \end{pmatrix}.$$

And the new initial distribution for this case is $(1/3, 2/3)$. And the transition matrix M_2 and M_3 are the diagonal block and irreducible matrices respectively. From this example we can also find that for a reducible Markov chain, the initial distribution has a great influence on the process even though we don't change the transition matrix.

5.3 Equations Governing Distribution of Functionals of Paths and Internal States of Process

In Chap. 3, we have introduced both the derivations and applications of Feynman-Kac equation. In this section, we turn to focus on the Feynman-Kac equations for the CTRW model with multiple internal states. In fact, the Feynman-Kac equation can also be considered as the equation that governs the PDF of the functional of the paths. While for the CTRW with multiple internal states, we can conclude that there's also another discrete random variable except for the position of the particle, that is the internal state. And sometimes we simply want to obtain the PDF of the functional of the internal state, such as the occupation time of each state. Motivated by this, we will further derive the equation governing the PDF of the functional of the internal states and calculate the distribution of the occupation time of some internal state in Sec. 5.4.

Therefore, in this section we mainly discuss two different functionals of particles and the governing equations for these two different kinds of functionals. One is still defined as $A = \int_0^t U(x(\tau))d\tau$, where $x(t)$ is the trajectory of a particle and $U(x)$ is a prescribed function, and such functional is also defined in Eq. (3.6). For the process with multiple internal states, except for the trajectory of the particle, the internal state is also the stochastic random variable as we stated before and we can also define the corresponding functional as $A_s = \int_0^t U(i(\tau))d\tau$, where $i(\tau)$ represents the particle stays in the i-th state at time τ, naturally its value belonging to $\{1, 2, \ldots, N\}$. And in this section, we still consider the distributions of jump length and waiting time as Gaussian and power law respectively, which have been shown in Sec. 5.2, so that the asymptotic behaviours of the distributions of waiting time and jump length have the forms $\hat{\phi}_m(s) \sim I - \text{diag}(B_{\alpha_1}s^{\alpha_1}, \ldots, B_{\alpha_N}s^{\alpha_N}), 0 < \alpha_1, \ldots, \alpha_N < 1$ and $\tilde{\Lambda}(k) \sim (1 - \sigma^2 k^2)I$, respectively.

Similar to the derivations of Fokker-Planck equations for the CTRW with multiple internal states in Sec. 5.2, we also use the notation $g^{(i)}(x, A, t), i = 1, 2, \ldots, N$ to represent the joint PDF of the particle staying at position x with the functional A and in the i-th internal state at time t. Then we take $|G(x, A, t)\rangle$ to represent the column vector gathering all of $g^{(i)}(x, A, t)$

$$|G(x, A, t)\rangle = \begin{pmatrix} g^{(1)}(x, A, t) \\ g^{(2)}(x, A, t) \\ \vdots \\ g^{(N)}(x, A, t) \end{pmatrix}.$$

First we consider the governing equation of $|G(x, A, t)\rangle$ and the Laplace transform w.r.t. A denoted as $|G(x, \rho, t)\rangle$, i.e.,

$$|G(x, \rho, t)\rangle = \int_0^\infty e^{-\rho A}|G(x, A, t)\rangle dA.$$

The governing equation for PDF $g(x, A, t)$ of the ordinary case (single one internal state) has been discussed in [Carmi *et al.* (2010)]. In order to obtain the Feynman-Kac equations, we also need to introduce the PDF denoted as $q_n^{(i)}(x, A, t)$ to represent the density of finding particle just arriving at the position x with the functional A at time t in the i-th internal state after n steps. Then we have

$$g^{(i)}(x, A, t) = \int_0^t w^{(i)}(\tau) \sum_{n=0}^\infty q_n^{(i)}(x, A - \tau U(x), t - \tau)d\tau.$$

That is

$$|G(x, A, t)\rangle = \int_0^t W(\tau) \sum_{n=0}^\infty |Q_n(x, A - \tau U(x), t - \tau)\rangle d\tau. \tag{5.16}$$

Besides, we can also construct the equations for $q_n^{(i)}(x, A, t)$

$$q_n^{(i)}(x, A, t) = \sum_{j=1}^{N} \int_0^t \int_{-\infty}^{\infty} m_{ji}\kappa^j(y, \tau)q_{n-1}^{(j)}(x - y, A - \tau U(x - y), t - \tau)dyd\tau.$$

We can also rewrite the above equation as the form of matrix,

$$\big|Q_{n+1}(x, A, t)\big\rangle = \int_0^t \int_{-\infty}^{\infty} M^T H(y, \tau)\big|Q_n(x - y, A - \tau U(x - y), t - \tau)\big\rangle dyd\tau,$$
(5.17)

where $H(x, t) = \Lambda(x)\phi_m(t)$. First we take Laplace transform w.r.t. A,

$$\big|Q_{n+1}(x, \rho, t)\big\rangle$$
$$= \int_0^t \int_{-\infty}^{\infty} M^T \Lambda(y)\phi_m(\tau) \exp\big[-\rho\tau U(x - y)\big]\big|Q_n(x - y, \rho, t - \tau)\big\rangle dyd\tau.$$

Then applying Laplace transform w.r.t. t,

$$\big|\hat{Q}_{n+1}(x, \rho, s)\big\rangle = \int_{-\infty}^{\infty} M^T \Lambda(y)\hat{\phi}_m(s + \rho U(x - y))\big|\hat{Q}_n(x - y, \rho, s)\big\rangle dy.$$

Finally taking Fourier transform $x \to k$, we have

$$\big|\tilde{\hat{Q}}_{n+1}(k, \rho, s)\big\rangle = M^T \tilde{\Lambda}(k)\hat{\phi}_m\left(s + \rho U\left(-i\frac{\partial}{\partial k}\right)\right)\big|\tilde{\hat{Q}}_n(k, \rho, s)\big\rangle.$$

And if taking the initial condition as $\big|G(x, A, t)\big\rangle = \delta(A)\delta(x)\delta(t)\big|\text{init}\big\rangle$, we obtain

$$\sum_{n=0}^{\infty} \big|\tilde{\hat{Q}}_n(k, \rho, s)\big\rangle = \left[I - M^T \tilde{\Lambda}(k)\hat{\phi}_m\left(s + \rho U\left(-i\frac{\partial}{\partial k}\right)\right)\right]^{-1}\big|\text{init}\big\rangle. \quad (5.18)$$

Similarly, we can also take Laplace transforms w.r.t. A, t and Fourier transform w.r.t. x on Eq. (5.16). Then by utilizing Eq. (5.18), we obtain

$$\big|\tilde{\hat{G}}(k, \rho, s)\big\rangle$$
$$= \hat{W}\left(s + \rho U\left(-i\frac{\partial}{\partial k}\right)\right)\sum_{n=0}^{\infty}\big|\tilde{\hat{Q}}_n(k, \rho, s)\big\rangle$$
$$= \hat{W}\left(s + \rho U\left(-i\frac{\partial}{\partial k}\right)\right)\left[I - M^T \tilde{\Lambda}(k)\hat{\phi}_m\left(s + \rho U\left(-i\frac{\partial}{\partial k}\right)\right)\right]^{-1}\big|\text{init}\big\rangle$$
$$= \frac{\hat{\phi}_m^*(s + \rho U(-i\frac{\partial}{\partial k}))}{s + \rho U(-i\frac{\partial}{\partial k})}\left[I - M^T \tilde{\Lambda}(k)\hat{\phi}_m\left(s + \rho U\left(-i\frac{\partial}{\partial k}\right)\right)\right]^{-1}\big|\text{init}\big\rangle,$$
(5.19)

where $\hat{\phi}_m^*(s) = I - \hat{\phi}_m(s)$, i.e., $\hat{\phi}_m^*(s) = \mathrm{diag}(B_{\alpha_1}s^{\alpha_1},\ldots,B_{\alpha_N}s^{\alpha_N})$. According to Eq. (5.19), we obtain

$$M^T\frac{\partial}{\partial t}\big|G(x,\rho,t)\big\rangle$$
$$= (M^T - I)\mathrm{diag}\big(B_{\alpha_1}^{-1},\ldots,B_{\alpha_N}^{-1}\big)\mathrm{diag}(\mathcal{D}_t^{1-\alpha_1},\ldots,\mathcal{D}_t^{1-\alpha_N})\big|G(x,\rho,t)\big\rangle$$
$$+ M^T\mathrm{diag}(K_{\alpha_1},\ldots,K_{\alpha_N})\frac{\partial^2}{\partial x^2}\mathrm{diag}(\mathcal{D}_t^{1-\alpha_1},\ldots,\mathcal{D}_t^{1-\alpha_N})\big|G(x,\rho,t)\big\rangle$$
$$- \rho U(x)M^T\big|G(x,\rho,t)\big\rangle,$$

$$(5.20)$$

where $\mathcal{D}_t^{1-\alpha_i}$, $i = 1,\ldots,N$, is fractional substantial derivative defined as [Friedrich *et al.* (2006)]

$$\mathcal{D}_t^{1-\alpha_i}g^{(i)}(x,\rho,t) = \frac{1}{\Gamma(\alpha_i)}\Big[\frac{\partial}{\partial t} + \rho U(x)\Big]\int_0^t \frac{e^{-(t-\tau)\rho U(x)}}{(t-\tau)^{1-\alpha}}g^{(i)}(x,\rho,\tau)d\tau.$$

$$(5.21)$$

Thus we obtain the forward Feynman-Kac equations for the stochastic process with multiple internal states. Additionally we consider the backward Feynman-Kac equations. According to CTRW model with multiple internal states, we have

$$g_{x_0}^{(i)}(A,t) = \sum_{j=1}^{N}\int_0^t\int_{-\infty}^{\infty}\phi^{(i)}(\tau)\lambda^{(i)}(y)m_{ij}g_{x_0-y}^{(j)}(A - \tau U(x_0),t-\tau)dyd\tau$$
$$+ w^{(i)}(t)\delta(A - tU(x_0)).$$

That is

$$\big|G_{x_0}(A,t)\big\rangle = \int_0^t\int_{-\infty}^{\infty}\phi_m(\tau)\Lambda(y)M\big|G_{x_0-y}(A - \tau U(x_0),t-\tau)\big\rangle dyd\tau$$
$$+ W(t)\delta(A - tU(x_0))\big|\Sigma\big\rangle.$$

Then we also perform Laplace transform w.r.t. A and t, Fourier transform w.r.t. x_0,

$$\big|\hat{\tilde{G}}_{k_0}(\rho,s)\big\rangle = \hat{\phi}_m\Big(s + \rho U\Big(-i\frac{\partial}{\partial k_0}\Big)\Big)\tilde{\Lambda}(k_0)M\big|\hat{\tilde{G}}_{k_0}(\rho,s)\big\rangle$$
$$+ \frac{\hat{\phi}_m^*(s + \rho U(-i\frac{\partial}{\partial k_0}))}{s + \rho U(-i\frac{\partial}{\partial k_0})}\delta(k_0)\big|\Sigma\big\rangle$$
$$\sim M\big|\hat{\tilde{G}}_{k_0}(\rho,s)\big\rangle - \hat{\phi}_m^*\Big(s + \rho U\Big(-i\frac{\partial}{\partial k_0}\Big)\Big)M\big|\hat{\tilde{G}}_{k_0}(\rho,s)\big\rangle$$
$$- \Lambda^*(k_0)M\big|\hat{\tilde{G}}_{k_0}(\rho,s)\big\rangle + \frac{\hat{\phi}_m^*(s + \rho U(-i\frac{\partial}{\partial k_0}))}{s + \rho U(-i\frac{\partial}{\partial k_0})}\delta(k_0)\big|\Sigma\big\rangle.$$

By utilizing the matrices of waiting time and jump length distributions and taking inverse Fourier and Laplace transforms w.r.t. k_0 and s, respectively, we finally obtain the corresponding backward Feynman-Kac equations

$$M \frac{\partial}{\partial t} |G_{x_0}(\rho, t)\rangle$$
$$= \text{diag}(B_{\alpha_1}^{-1}, \ldots, B_{\alpha_N}^{-1}) \text{diag}(\mathcal{D}_t^{1-\alpha_1}, \ldots, \mathcal{D}_t^{1-\alpha_N})(M - I)|G_{x_0}(\rho, t)\rangle$$
$$+ \text{diag}(K_{\alpha_1}, \ldots, K_{\alpha_N}) \text{diag}(\mathcal{D}_t^{1-\alpha_1}, \ldots, \mathcal{D}_t^{1-\alpha_N}) \frac{\partial^2}{\partial x_0^2} M |G_{x_0}(\rho, t)\rangle$$
$$- \rho U(x_0) M |G_{x_0}(\rho, t)\rangle.$$

$$(5.22)$$

Here we note that if one is only interested in the PDF of the process starting at position x_0 with functional A at time t, we just calculate the $g_{x_0}(A, t) = \sum_{i=1}^N \lambda_{x_0}^{(i)} g_{x_0}^{(i)}(A, t)$.

For the final part of this section, we will consider the governing equations of the column vector $|G_s(A_s, t)\rangle$ whose factors are $g_s^{(i)}(A_s, t)$, $i = 1, \ldots, N$. That is

$$|G_s(A_s, t)\rangle := \begin{pmatrix} g_s^{(1)}(A_s, t) \\ \vdots \\ g_s^{(N)}(A_s, t) \end{pmatrix}.$$

According to the movement of CTRW model with multiple internal states, we can construct the equation

$$|G_s(A_s, t)\rangle = \int_0^t W(\tau) \sum_{n=0}^{\infty} \begin{pmatrix} q_{s,n}^{(1)}(A_s - \tau U(1), t - \tau) \\ \vdots \\ q_{s,n}^{(N)}(A_s - \tau U(N), t - \tau) \end{pmatrix} d\tau. \quad (5.23)$$

Besides, we have

$$q_{s,n+1}^{(i)}(A_s, t) = \int_0^{\infty} m_{1i} q_{s,n}^{(1)}(A_s - U(1)\tau, t - \tau) \phi^{(1)}(\tau) d\tau$$
$$+ \ldots + \int_0^{\infty} m_{Ni} q_{s,n}^{(N)}(A_s - U(N)\tau, t - \tau) \phi^{(N)}(\tau) d\tau,$$

that is

$$\begin{pmatrix} q_{s,n+1}^{(1)}(A_s, t) \\ \vdots \\ q_{s,n+1}^{(N)}(A_s, t) \end{pmatrix} = \int_0^t M^T \phi_m(\tau) \begin{pmatrix} q_{s,n}^{(1)}(A_s - \tau U(1), t - \tau) \\ \vdots \\ q_{s,n}^{(N)}(A_s - \tau U(N), t - \tau) \end{pmatrix} d\tau.$$

Applying Laplace transform of the above equation w.r.t. A_s, we obtain

$$|Q_{s,n+1}(\rho_s, t)\rangle = \int_0^t M^T \phi_m(\tau) \begin{pmatrix} \exp(-\rho_s \tau U(1)) & & \\ & \ddots & \\ & & \exp(-\rho_s \tau U(N)) \end{pmatrix}$$

$$\times \begin{pmatrix} q_{s,n}^{(1)}(\rho_s, t - \tau) \\ \vdots \\ q_{s,n}^{(N)}(\rho_s, t - \tau) \end{pmatrix} d\tau.$$

After performing Laplace transform w.r.t. t on the above equation we obtain

$$|\hat{Q}_{s,n+1}(\rho_s, s)\rangle$$

$$= M^T \begin{pmatrix} \hat{\phi}^{(1)}(s + \rho_s U(1)) & & \\ & \ddots & \\ & & \hat{\phi}^{(N)}(s + \rho_s U(N)) \end{pmatrix} |\hat{Q}_{s,n}(\rho_s, s)\rangle.$$

Considering the initial condition $|Q_{s,0}(A_s, t)\rangle = \delta(A_s)\delta(t)|\text{init}\rangle$, we have

$$\sum_{n=0}^{\infty} |\hat{Q}_{s,n}(\rho_s, s)\rangle =$$

$$\left[I - M^T \begin{pmatrix} \hat{\phi}^{(1)}(s + \rho_s U(1)) & & \\ & \ddots & \\ & & \hat{\phi}^{(N)}(s + \rho_s U(N)) \end{pmatrix} \right]^{-1} |\text{init}\rangle. \qquad (5.24)$$

Then from Eq. (5.23) and Eq. (5.24), we have

$$|\hat{G}_s(\rho_s, s)\rangle$$

$$= \begin{pmatrix} \hat{w}^{(1)}(s + \rho_s U(1)) & & \\ & \ddots & \\ & & \hat{w}^{(N)}(s + \rho_s U(N)) \end{pmatrix} \sum_{n=0}^{\infty} |\hat{Q}_{s,n}(\rho_s, s)\rangle$$

$$= \begin{pmatrix} \frac{[1 - \hat{\phi}^{(1)}(s + \rho_s U(1))]}{s + \rho_s U(1)} & & \\ & \ddots & \\ & & \frac{[1 - \hat{\phi}^{(N)}(s + \rho_s U(N))]}{s + \rho_s U(N)} \end{pmatrix}$$

$$\times \left[I - M^T \begin{pmatrix} \hat{\phi}^{(1)}(s + \rho_s U(1)) & & \\ & \ddots & \\ & & \hat{\phi}^{(N)}(s + \rho_s U(N)) \end{pmatrix} \right]^{-1} |\text{init}\rangle.$$

$$(5.25)$$

Finally,

$$M^T s |\hat{G}_s(\rho_s, s)\rangle - |\text{init}\rangle$$

$$= (M^T - I) \begin{pmatrix} B_{\alpha_1}^{-1}(s + \rho_s U(1))^{1-\alpha_1} & & \\ & \ddots & \\ & & B_{\alpha_N}^{-1}(s + \rho_s U(N))^{1-\alpha_N} \end{pmatrix}$$

$$\times |\hat{G}_s(\rho_s, s)\rangle - M^T \rho_s \begin{pmatrix} U(1) & & \\ & \ddots & \\ & & U(N) \end{pmatrix} |\hat{G}_s(\rho_s, s)\rangle.$$

Here, we define a new operator

$$\mathfrak{D}_t^{1-\alpha_i} g_s^{(i)}(\rho_s, t)$$

$$= \frac{1}{\Gamma(\alpha_i)} \left[\frac{\partial}{\partial t} + \rho_s U(i) \right] \int_0^t \frac{\exp(-(t-\tau)\rho_s U(i))}{(t-\tau)^{1-\alpha_i}} g_s^{(i)}(\rho_s, \tau) d\tau.$$

Then we have the equation

$$M^T \frac{\partial}{\partial t} |G_s(\rho_s, t)\rangle$$

$$= (M^T - I)\text{diag}\left(B_{\alpha_1}^{-1}, \ldots, B_{\alpha_N}^{-1}\right)\text{diag}(\mathfrak{D}_t^{1-\alpha_1}, \ldots, \mathfrak{D}_t^{1-\alpha_N}) \qquad (5.26)$$

$$\times |G_s(\rho_s, t)\rangle - \rho_s M^T \text{diag}\left(U(1), \ldots U(N)\right) |G_s(\rho_s, t)\rangle.$$

Equations (5.20), (5.22) and (5.26) have various applications. In Sec. 5.4 we will give some specific applications to the equations we obtained.

5.4   Some Applications of Feynman-Kac Equations and Governing Equations of Functionals of Internal States

After the derivations of the forward and backward Feynman-Kac equations Eqs. (5.20), (5.22) and the equations of the PDF of functionals of internal state states Eq. (5.26), now we begin to give some specific applications to these equations respectively. In this section we simply discuss the applications of Eq. (5.22) and Eq. (5.26).

First we use Eq. (5.22) to calculate the distribution of the first passage time denoted as t_f, which represents the time the particle starting at x_0 ($< B$) and first reaching $x = B$. Here we define the functional as $A_f = \int_0^t U(x(\tau))d\tau$, where

$$U(x) = \begin{cases} 0 & x < B \\ 1 & x > B. \end{cases}$$

Here we consider a simple but representative example, and the transition matrix of the process with two internal states is chosen as $M = \begin{pmatrix} 0 & 1 \\ 1 & 0 \end{pmatrix}$, which represents the alternating of these two internal states from one to the other; the coefficients $K_{\alpha_1} = K_{\alpha_2} = B_{\alpha_1} = B_{\alpha_2} = 1$, for the sake of simplifying the calculations. From the backward Feynman-Kac equations Eq. (5.22), when $x_0 < B$ there exists

$$\begin{cases} s\hat{g}_{x_0}^{(2)}(\rho, s) - 1 = & -\hat{g}_{x_0}^{(1)}(\rho, s)s^{1-\alpha_1} + \hat{g}_{x_0}^{(2)}(\rho, s)s^{1-\alpha_1} + s^{1-\alpha_1}\frac{\partial^2}{\partial x_0^2}\hat{g}_{x_0}^{(2)}(\rho, s), \\ s\hat{g}_{x_0}^{(1)}(\rho, s) - 1 = & \hat{g}_{x_0}^{(1)}(\rho, s)s^{1-\alpha_2} - \hat{g}_{x_0}^{(2)}(\rho, s)s^{1-\alpha_2} + s^{1-\alpha_2}\frac{\partial^2}{\partial x_0^2}\hat{g}_{x_0}^{(1)}(\rho, s). \end{cases}$$

$$(5.27)$$

While for $x_0 > B$,

$$\begin{cases} s\hat{g}_{x_0}^{(2)}(\rho, s) - 1 = - \hat{g}_{x_0}^{(1)}(\rho, s)(s + \rho)^{1-\alpha_1} + \hat{g}_{x_0}^{(2)}(\rho, s)(s + \rho)^{1-\alpha_1} \\ \qquad\qquad + (s + \rho)^{1-\alpha_1}\frac{\partial^2}{\partial x_0^2}\hat{g}_{x_0}^{(2)}(\rho, s) - \rho\hat{g}_{x_0}^{(2)}(\rho, s), \\ s\hat{g}_{x_0}^{(1)}(\rho, s) - 1 = \hat{g}_{x_0}^{(1)}(\rho, s)(s + \rho)^{1-\alpha_2} - \hat{g}_{x_0}^{(2)}(\rho, s)(s + \rho)^{1-\alpha_2} \\ \qquad\qquad + (s + \rho)^{1-\alpha_2}\frac{\partial^2}{\partial x_0^2}\hat{g}_{x_0}^{(1)}(\rho, s) - \rho\hat{g}_{x_0}^{(1)}(\rho, s). \end{cases}$$

$$(5.28)$$

By solving these differential equations, from Eq. (5.27) we obtain

$$\hat{g}_{x_0}^{(1)}(\rho, s) = \frac{1}{s} + e^{\sqrt{\frac{a_0 + b_0}{2}}x_0}C_1,$$

and

$$\hat{g}_{x_0}^{(2)}(\rho, s) = \frac{1}{s} + \frac{1}{2}(2 + a_0 + b_0 - 2s^{\alpha_2})e^{\sqrt{\frac{a_0 + b_0}{2}}x_0}C_1,$$

where C_1 is a constant to be determined, $a_0 = -2 + s^{\alpha_1} + s^{\alpha_2}$ and $b_0 = (4 + s^{2\alpha_1} + s^{2\alpha_2} - 2s^{\alpha_1 + \alpha_2})^{\frac{1}{2}}$. Therefore, for the first case $x_0 < B$, we have

$$\hat{g}_{x_0}(\rho, s) = \lambda_{x_0}^{(1)}\hat{g}_{x_0}^{(1)}(\rho, s) + \lambda_{x_0}^{(2)}\hat{g}_{x_0}^{(2)}(\rho, s),$$

where $\lambda_{x_0}^{(1)}$ and $\lambda_{x_0}^{(2)}$ are the factors of the initial distribution of particles starting at x_0, i.e., $\langle \text{init}| = (\lambda_{x_0}^{(1)}, \lambda_{x_0}^{(2)})$. Thus

$$\hat{g}_{x_0}(\rho, s)$$
$$= \frac{1}{s} + \frac{1}{2}\left[2 + a_0 + b_0 - a_0\lambda_{x_0}^{(1)} - b_0\lambda_{x_0}^{(1)} + 2\left(-1 + \lambda_{x_0}^{(1)}\right)s^{\alpha_2}\right]e^{\sqrt{\frac{a_0 + b_0}{2}}x_0}C_1.$$

$$(5.29)$$

While for the second condition $x_0 > B$, from Eq. (5.28) we can obtain

$$\hat{g}_{x_0}^{(1)}(\rho, s) = \frac{1}{\rho + s} + e^{-\sqrt{\frac{a_1 + b_1}{2}}x_0}C_2$$

and

$$\hat{g}_{x_0}^{(2)}(\rho, s) = \frac{1}{\rho + s} + \frac{1}{2}[2 + a_1 + b_1 - 2(\rho + s)^{\alpha_2}]e^{-\sqrt{\frac{a_1 + b_1}{2}}x_0}C_2,$$

where C_2 is also a constant to be determined, $a_1 = -2 + (\rho + s)^{\alpha_1} + (\rho + s)^{\alpha_2}$ and $b_1 = \left(4 + (\rho + s)^{2\alpha_1} + (\rho + s)^{2\alpha_2} - 2(\rho + s)^{\alpha_1 + \alpha_2}\right)^{\frac{1}{2}}$. Then there exists

$$
\begin{aligned}
\hat{g}_{x_0}(\rho, s) = &\frac{1}{\rho + s} + \frac{1}{2}\Big[2 + a_1 + b_1 - a_1\lambda_{x_0}^{(1)} - b_1\lambda_{x_0}^{(1)} \\
&+ 2\left(-1 + \lambda_{x_0}^{(1)}\right)(\rho + s)^{\alpha_2}\Big]e^{-\sqrt{\frac{a_1 + b_1}{2}}x_0}C_2.
\end{aligned}
\tag{5.30}
$$

In order to determine the constants C_1 and C_2, we require the continuities of $\hat{g}_{x_0}(\rho, s)$ and its first derivative at $x_0 = 0$. Then we let $x_0 = 0$ and eventually obtain

$$\hat{g}_0(\rho, s) = \frac{1}{s}\left[1 - \frac{\sqrt{a_1 + b_1}\rho}{(\sqrt{a_0 + b_0} + \sqrt{a_1 + b_1})(\rho + s)}e^{-\sqrt{\frac{a_0 + b_0}{2}}B}\right].$$

Thus

$$\lim_{\rho \to \infty} \hat{g}_0(\rho, s) = \frac{1}{s}\left[1 - \exp\left(-\sqrt{\frac{a_0 + b_0}{2}}B\right)\right].$$

Then we can obtain the PDF of the first passage time

$$f(t) = \mathcal{L}^{-1}\left\{\exp\left(-\sqrt{\frac{a_0 + b_0}{2}}B\right)\right\}.$$

When time t is big enough and $\alpha_1 > \alpha_2$, then $a_0 \sim -2 + s^{\alpha_2}$ and $b_0 \sim \sqrt{4 + s^{2\alpha_2}} \sim 2 + \frac{1}{4}s^{2\alpha_2}$. Thus we conclude that $\exp\left(-\sqrt{\frac{a_0 + b_0}{2}}B\right) \sim \exp\left(-\sqrt{\frac{s^{\alpha_2}}{2}}B\right) \sim 1 - \frac{B}{\sqrt{2}}s^{\frac{\alpha_2}{2}}$, i.e.,

$$f(t) \sim \frac{B}{\sqrt{2}\,|\,\Gamma(-\alpha_2/2)\,|}t^{-\alpha_2/2 - 1}.$$

Specially, if taking $\alpha_1 = \alpha_2 = \alpha$ we can easily obtain $a_0 + b_0 \sim 2s^{\alpha}$. Thus we conclude

$$f(t) \sim \frac{B}{|\,\Gamma(-\alpha/2)\,|}t^{-\alpha/2 - 1}.$$

If $\alpha_1 = \alpha_2$, then the process actually has only one internal state. So the result of process with single internal state is recovered. The figure of the distribution of the first passage time is shown in Fig. 5.1. From the figure we can conclude that the process with different waiting time distributions can make a difference on the distribution.

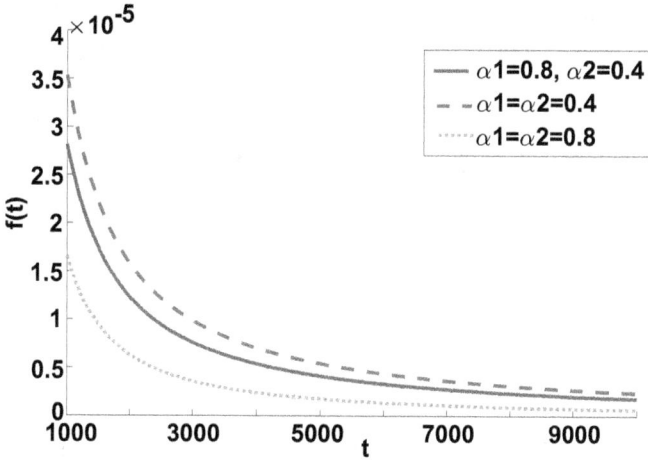

Fig. 5.1 The distribution of the first passage time. Parameter B is chosen to be 1. Several different pairs of α_1 and α_2 ($\alpha_1 = \alpha_2 = 0.4$ shown as segment in the top, $\alpha_1 = 0.8, \alpha_2 = 0.4$ shown as real line in the middle, $\alpha_1 = \alpha_2 = 0.8$ shown as dash line in the bottom) are chosen to make a comparison.

Next we consider the application of Eq. (5.26). A direct application of this equation is to calculate the average occupation time of each internal state and the distribution of the fraction of the occupation time, i.e., the distribution of $t^{(i)}/t$, where $t^{(i)}$ represents the occupation time of i-th internal state. And we denote the distribution as $l_{t^{(i)}/t}(x)$. Without loss of generality, we only consider the occupation of the first internal state. Thus the function is

$$U[i(\tau)] = \begin{cases} 1 & i(\tau) = 1 \\ 0 & \text{else.} \end{cases}$$

From Eq. (5.25), there exists

$$|\hat{G}_s(\rho_s, s)\rangle = \operatorname{diag}\left(B_{\alpha_1}(s + \rho_s)^{\alpha_1 - 1}, \ldots, B_{\alpha_2} s^{\alpha_2 - 1}\right) R'^{-1} |\text{init}\rangle, \quad (5.31)$$

where

$$R' = I - M^T \operatorname{diag}(\hat{\phi}^{(1)}(s + \rho_s), \hat{\phi}^{(2)}(s), \ldots, \hat{\phi}^{(N)}(s))$$
$$\sim I - M^T + M^T \operatorname{diag}\left(B_{\alpha_1}(s + \rho_s)^{\alpha_1}, \ldots, B_{\alpha_N} s^{\alpha_N}\right).$$

If the transition matrix M is irreducible, similar to the discussion in Sec. 5.2 we can also obtain the asymptotic behaviour of the inverse matrix of R',

$$R'^{-1} \sim \frac{|\text{eq}_M\rangle\langle\Sigma|}{\varepsilon_1 B_{\alpha_1}(\rho_s + s)^{\alpha_1} + \varepsilon_2 B_{\alpha_2} s^{\alpha_2} + \ldots + \varepsilon_N B_{\alpha_N} s^{\alpha_N}}.$$

Then by utilizing the above result and Eq. (5.31), we can get the asymptotic form of $|\hat{G}_s(\rho_s, s)\rangle$. Further we can obtain the expression of $g_s(A_s, t)$ which represents the PDF of the functional A_s of internal state at time t. The Laplace transforms of $g_s(A_s, t)$ w.r.t. A_s and t is denoted as $\hat{g}_s(\rho_s, s)$. According to the relation $\hat{g}_s(\rho_s, s) = \langle\Sigma|\hat{G}_s(\rho_s, s)\rangle$, there exists

$$\hat{g}_s(\rho_s, s) \sim \frac{\varepsilon_1 B_{\alpha_1}(\rho_s + s)^{\alpha_1 - 1} + \varepsilon_2 B_{\alpha_2} s^{\alpha_2 - 1} + \ldots + \varepsilon_N B_{\alpha_N} s^{\alpha_N - 1}}{\varepsilon_1 B_{\alpha_1}(\rho_s + s)^{\alpha_1} + \varepsilon_2 B_{\alpha_2} s^{\alpha_2} + \ldots + \varepsilon_N B_{\alpha_N} s^{\alpha_N}}. \tag{5.32}$$

From Eq. (5.32) we can obtain the average occupation time of the first internal state, i.e., the average of the functional $\langle A_s \rangle$,

$$\langle A_s \rangle = -\frac{\partial}{\partial \rho_s} g_s(\rho_s, t)\Big|_{\rho_s = 0} \tag{5.33}$$

$$\sim \mathcal{L}^{-1}\left\{\frac{B_{\alpha_1}\varepsilon_1 s^{\alpha_1 - 2}}{B_{\alpha_1}\varepsilon_1 s^{\alpha_1} + B_{\alpha_2}\varepsilon_2 s^{\alpha_2} + \ldots + B_{\alpha_N}\varepsilon_N s^{\alpha_N}}\right\}.$$

From the result in the above equation, we can obtain three different conclusions based on the relation between α_1 and $\alpha_{\min}^{2,N} := \min\{\alpha_2, \ldots, \alpha_N\}$:

(1) If $\alpha_1 < \alpha_{\min}^{2,N}$, then

$$\langle A_s \rangle \sim \mathcal{L}^{-1}\left\{\frac{1}{s^2}\right\} = t.$$

(2) If $\alpha_1 > \alpha_{\min}^{2,N}$, without loss of generality, we assume that $\alpha_2 = \ldots = \alpha_m = \alpha_{\min}^{2,N}$ with $2 \leq m \leq N$. And we obtain

$$\langle A_s \rangle \sim \mathcal{L}^{-1}\left\{\frac{B_{\alpha_1}\varepsilon_1}{B_{\alpha_2}\varepsilon_2 + \ldots + B_{\alpha_m}\varepsilon_m}\frac{1}{s^{\alpha_{\min}^{2,N} - \alpha_1 + 2}}\right\}$$

$$\sim \frac{B_{\alpha_1}\varepsilon_1}{B_{\alpha_2}\varepsilon_2 + \ldots + B_{\alpha_m}\varepsilon_m}\frac{1}{\Gamma(2 - \alpha_{\min}^{2,N} + \alpha_1)}t^{1 - \alpha_{\min}^{2,N} + \alpha_1}.$$

(3) If $\alpha_1 = \alpha_{\min}^{2,N}$. Here we still assume $\alpha_2 = \ldots = \alpha_m = \alpha_{\min}^{2,N} = \alpha_1, 2 \leq m \leq N$. Then we have

$$\langle A_s \rangle \sim \mathcal{L}^{-1}\left\{\frac{\varepsilon_1 B_{\alpha_1}}{\varepsilon_1 B_{\alpha_1} + \ldots + \varepsilon_m B_{\alpha_m}}\frac{1}{s^2}\right\} = \frac{\varepsilon_1 B_{\alpha_1}}{\varepsilon_1 B_{\alpha_1} + \ldots + \varepsilon_m B_{\alpha_m}}t.$$

Further if we take $B_{\alpha_1} = \ldots = B_{\alpha_m}$,

$$\langle A_s \rangle \sim \frac{\varepsilon_1}{\varepsilon_1 + \ldots + \varepsilon_m}t.$$

It can be noted that the coefficient $\varepsilon_1/(\varepsilon_1 + \ldots + \varepsilon_m)$ in the above equation represents the weight of the first internal state among all the first m states.

From the results shown in the first and second cases, we take the process with two alternating internal states for instance. If $\alpha_1 < \alpha_2$, then the average occupation time of the second internal state behaves as $t^{1-\alpha_2+\alpha_1}$, while the occupation time of the first internal state asymptotically behaves as t. However this result doesn't mean that the particles just stay at the first internal state for the whole time t.

For the third case discussed above, it's necessary to calculate the PDF of the fraction of the occupation time of the first internal state denoted as $t^{(1)}$ over the whole time t, and denote this PDF as $l_{t^{(1)}/t}(x)$. Intuitively we may simply think $l_{t^{(1)}/t}(x)$ goes like $\delta(x - \varepsilon_1/(\varepsilon_1 + \ldots + \varepsilon_m))$. However, it behaves like arcsine law as shown in the following.

Before we begin our further discussions, we introduce some results of inversion of the scaling form of a double Laplace transforms. Here we consider a PDF denoted as $f_Y(y,t)$ of a stochastic process. After applying double Laplace transforms w.r.t. t and y on $f_Y(y,t)$, denoted as $\hat{f}_Y(u,s)$, there exists the property

$$\hat{f}_Y(u,s) = \frac{1}{s} g\left(\frac{u}{s}\right)$$

in the regime $s, u \to 0$, where g is a function. Then the following properties hold [Godrèche and Luck (2001)]:

(1) The random variable $X_t = Y_t/t$ possesses a limiting distribution when $t \to \infty$, i.e.,

$$f_X(x) = \lim_{t \to \infty} f_{Y/t}(x = y/t, t).$$

(2) There exists a relation between the distribution $f_X(x)$ and the function g, as shown in the following

$$f_X(x) = -\frac{1}{\pi x} \lim_{\epsilon \to 0} \operatorname{Im} g\left(-\frac{1}{x + i\epsilon}\right). \qquad (5.34)$$

In fact, from Eq. (5.32) there exists

$$\hat{g}_s(\rho_s, s) \sim \frac{1}{s} \frac{(\varepsilon_2 + \ldots + \varepsilon_m) + \varepsilon_1(\frac{\rho_s}{s} + 1)^{\alpha_1 - 1}}{(\varepsilon_2 + \ldots + \varepsilon_m) + \varepsilon_1(\frac{\rho_s}{s} + 1)^{\alpha_1}}.$$

According to the inversion of the scaling form of a double Laplace transforms as shown in Eq. (5.34),

$$\lim_{t \to \infty} l_{t^{(1)}/t}(x) = \frac{\sin(\pi \alpha_1)}{\pi} \left[\varepsilon_1(\varepsilon_2 + \ldots + \varepsilon_m)(1 - x)^{\alpha_1 - 1} x^{\alpha_1 - 1}\right]$$

$$\left[(\varepsilon_2 + \ldots + \varepsilon_m)^2 x^{2\alpha_1} + \varepsilon_1^2 (1 - x)^{2\alpha_1} \qquad (5.35)\right.$$

$$\left. + 2\varepsilon_1(\varepsilon_2 + \ldots + \varepsilon_m) \cos(\alpha_1 \pi) x^{\alpha_1} (1 - x)^{\alpha_1}\right]^{-1},$$

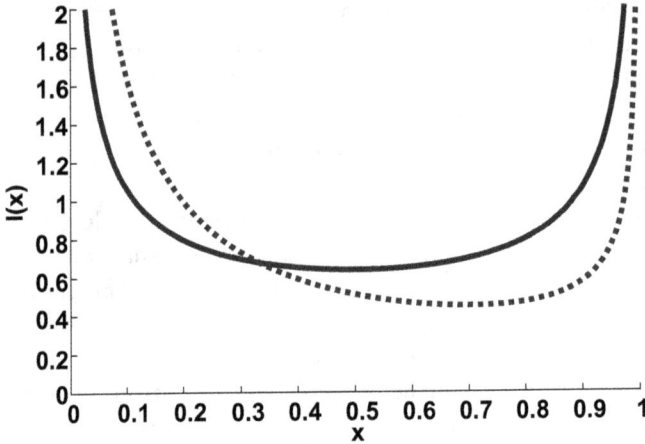

Fig. 5.2 The distribution of $t^{(1)}/t$. All of the waiting time distributions are chosen to behave asymptotically as $t^{-1-\alpha}$, where $\alpha = 0.5$. The real line and dash line are the theoretical results of the processes with 2 and 3 internal states, respectively. Besides the $\varepsilon_1 = \varepsilon_2 = 1/2$ and $\varepsilon_1 = \varepsilon_2 = \varepsilon_3 = 1/3$ for the processes with 2 and 3 internal states (the corresponding transit matrix can also be constructed) respectively.

The above result can be known as the generalized arcsine law. The differences between the classical arcsine law and the generalized one are shown in Fig. 5.2. Besides, if we choose $\alpha_1 = \alpha_2$ and $B_{\alpha_1} = B_{\alpha_2}$, which indicates the distributions of the waiting time are the same, from Eq. (5.35), we obtain

$$\lim_{t \to \infty} l_{t^{(1)}/t}(x) = \frac{[\sin(\pi\alpha)/\pi]x^{\alpha-1}(1-x)^{\alpha-1}}{x^{2\alpha} + 2x^\alpha(1-x)^\alpha \cos(\pi\alpha) + (1-x)^{2\alpha}}.$$

Further choosing $\alpha = 1/2$, we obtain the classical arcsine law known as

$$\lim_{t \to \infty} l_{t^{(1)}/t}(x) = \frac{1}{\pi\sqrt{x(1-x)}}.$$

In the final part of this section, we consider the average occupation time of the first internal state and $l_{t^{(1)}/t}(x)$ for the reducible process. Here we only consider the transition matrix with the form of $M = \mathrm{diag}(M_1, \ldots, M_j)$. And we can obtain the inverse matrix of R'

$$R'^{-1} = \mathrm{diag}\left(R_1'^{-1}, \ldots, R_j'^{-1}\right),$$

where

$$R_1'^{-1} = \frac{|eq_M\rangle_1 \langle\Sigma|_1}{\varepsilon_{11}B_{\alpha_{11}}(\rho_s + s)^{\alpha_{11}} + \varepsilon_{12}B_{\alpha_{12}}s^{\alpha_{12}} + \ldots + \varepsilon_{1n_1}B_{\alpha_{1n_1}}s^{\alpha_{1n_1}}}.$$

and

$$R_i'^{-1} = \frac{|\mathrm{eq_M}\rangle_i \langle \Sigma|_i}{\varepsilon_{i1} B_{\alpha_{i1}} s^{\alpha_{i1}} + \varepsilon_{i2} B_{\alpha_{i2}} s^{\alpha_{i2}} + \ldots + \varepsilon_{in_i} B_{\alpha_{in_i}} s^{\alpha_{in_i}}}, i = 2, \ldots, j.$$

Thus we have

$$\hat{g}_s(\rho_s, s) \sim \frac{1}{s} \left[\langle \Sigma|_1 |\mathrm{init}\rangle_1 \frac{\varepsilon_{11} B_{\alpha_{11}} (\rho_s + s)^{\alpha_{11}-1} s + \ldots + \varepsilon_{1n_1} B_{\alpha_{1n_1}} s^{\alpha_{1n_1}}}{\varepsilon_{11} B_{\alpha_{11}} (\rho_s + s)^{\alpha_{11}} + \ldots + \varepsilon_{1n_1} B_{\alpha_{1n_1}} s^{\alpha_{1n_1}}} \right. $$

$$\left. + \left(1 - \langle \Sigma|_1 |\mathrm{init}\rangle_1 \right) \right].$$

$$(5.36)$$

Similar with the method used in Eq. (5.33), we have

$$\langle A_s \rangle \sim \mathcal{L}^{-1} \left\{ \frac{\langle \Sigma|_1 |\mathrm{init}\rangle_1 B_{\alpha_{11}} \varepsilon_{11} s^{\alpha_{11}-2}}{B_{\alpha_{11}} \varepsilon_{11} s^{\alpha_{11}} + B_{\alpha_{12}} \varepsilon_{12} s^{\alpha_{12}} + \ldots + B_{\alpha_{1n_1}} \varepsilon_{1n_1} s^{\alpha_{1n_1}}} \right\}.$$

We can also have three different cases:

(1) If $\alpha_{11} < \alpha_{\min}^{12,1n_1}$, where $\alpha_{\min}^{12,1n_1} = \min\{\alpha_{12}, \ldots, \alpha_{1n_1}\}$. Then

$$\langle A_s \rangle \sim \langle \Sigma|_1 |\mathrm{init}\rangle_1 \mathcal{L}^{-1} \left\{ \frac{1}{s^2} \right\} = \langle \Sigma|_1 |\mathrm{init}\rangle_1 t.$$

(2) If $\alpha_{11} > \alpha_{\min}^{12,1n_1}$, without loss of generality, we still assume $\alpha_{12} = \ldots = \alpha_{1m} = \alpha_{\min}^{12,1n_1}$ with $2 \le m \le n_1$. Thus

$$\langle A_s \rangle \sim t^{1-\alpha_{\min}^{12,1n_1}+\alpha_{11}}.$$

(3) If $\alpha_{12} = \ldots = \alpha_{1m} = \alpha_{\min}^{12,1n_1} = \alpha_{11}$, then

$$\langle A_s \rangle \sim \frac{\langle \Sigma|_1 |\mathrm{init}\rangle_1 \varepsilon_{11} B_{\alpha_{11}}}{\varepsilon_{11} B_{\alpha_{11}} + \ldots + \varepsilon_{1m} B_{\alpha_{1m}}} t.$$

Besides, for this case we can also calculate the distribution of fraction of the occupation time of first internal state $t^{(1)}$ over the total time t, which still is the generalized arcsine law,

$$\lim_{t \to \infty} l_{t^{(1)}/t}(x) = \frac{\sin(\alpha_{11}\pi)}{\pi} \left[\varepsilon_{11}(\varepsilon_{12} + \ldots + \varepsilon_{1i})(\lambda_{11} + \ldots + \lambda_{1n_1}) \right.$$

$$(1-x)^{\alpha_{11}-1} x^{\alpha_{11}-1} \left] \left[(\varepsilon_{12} + \ldots + \varepsilon_{1i})^2 x^{2\alpha_{11}} \right. \right.$$

$$+ \varepsilon_{11}^2 (1-x)^{2\alpha_{11}} + 2\varepsilon_{11}(\varepsilon_{12} + \ldots + \varepsilon_{1i})$$

$$\times \cos(\alpha_{11}\pi)(1-x)^{\alpha_{11}} x^{\alpha_{11}} \right]^{-1}.$$

5.5   Lévy Walk with Multiple Internal States

After introducing the theories of CTRW model with multiple internal states, in this section we will turn to discuss the Lévy walk with multiple internal states. For Lévy walk, realizing there's no jump or waiting, the internal states for this process can be considered as the corresponding pairs of walking time and velocity densities. And there still exists initial distribution and transition matrix as introduced in Sec. 5.1. The only difference is the concept of internal states.

     Lévy walk with multiple internal states can also be considered as a generalization of Lévy walk [Xu and Deng (2018)]. The distributions of walking duration of each movement and velocities are now denoted as $\phi^{(i)}(\tau)$ and $h^{(i)}(\mathbf{v})$, $i = 1, \cdots, N$. Here we first derive the equations of $q^{(i)}(\mathbf{r}, t)$, the PDF that the particle may change its velocity at position $\mathbf{r}$ and i-th internal state at time t. Assuming that the particle starts at origin, then

$$q^{(i)}(\mathbf{r}, t) = \sum_{j=1}^{N} \int_0^t d\tau \int d\mathbf{v} m_{ji} \phi^{(j)}(\tau) h^{(j)}(\mathbf{v}) q^{(j)}(\mathbf{r} - \mathbf{v}\tau, t - \tau) + \xi_i \delta(\mathbf{r})\delta(t).$$

(5.37)

And we can obtain

$$\big|Q(\mathbf{r}, t)\big\rangle = \int_0^t d\tau \int d\mathbf{v} M^T \phi_m(\tau) H(\mathbf{v}) \big|Q(\mathbf{r} - \mathbf{v}\tau, t - \tau)\big\rangle + \delta(\mathbf{r})\delta(t)\big|\text{init}\big\rangle$$

(5.38)

by denoting $\big|Q(\mathbf{r}, t)\big\rangle = (Q^{(1)}(\mathbf{r}, t), \dots, Q^{(N)}(\mathbf{r}, t))^T$, $\phi_m(\tau) = \text{diag}(\phi^{(1)}(\tau),$ $\dots, \phi^{(N)}(\tau))$, $H(\mathbf{v}) = \text{diag}(h^{(1)}(\mathbf{v}), \dots, h^{(N)}(\mathbf{v}))$. Next we still use the notation $\big|G(\mathbf{r}, t)\big\rangle$ to represent the column vector consisting of $g^{(i)}(\mathbf{r}, t)$ which denotes the probability of the particle staying at position $\mathbf{r}$ with i-th internal state at time t. And we have

$$\big|G(\mathbf{r}, t)\big\rangle = \int_0^t d\tau \int d\mathbf{v} W(\tau) H(\mathbf{v}) \big|Q(\mathbf{r} - \mathbf{v}\tau, \tau)\big\rangle,$$

(5.39)

where

$$W(\tau) := \text{diag}(W^{(1)}(\tau), \dots, W^{(N)}(\tau)) = I - \int_0^\tau \phi_m(t') dt'.$$

(5.40)

Then after Fourier and Laplace transforms w.r.t. x and t respectively, there exists

$$\big|\tilde{\hat{G}}(\mathbf{k}, s)\big\rangle = \int d\mathbf{v} H(\mathbf{v}) \hat{W}(s + i\mathbf{k}\mathbf{v}) \left[I - \int M^T \hat{\phi}_m(s + i\mathbf{k}\mathbf{v}) H(\mathbf{v}) d\mathbf{v} \right]^{-1} \big|\text{init}\big\rangle.$$

(5.41)

Additionally, if the Lévy walk has a single internal state, then according to Eq. (5.41), we have

$$\tilde{G}(\mathbf{k}, s) = \frac{\int d\mathbf{v} h(\mathbf{v}) \hat{W}(s + i\mathbf{k}\mathbf{v})}{1 - \int d\mathbf{v} \hat{\phi}(s + i\mathbf{k}\mathbf{v}) h(\mathbf{v})} \tag{5.42}$$

indicating Eq. (1.76) is recovered.

Based on $\int H(\mathbf{v}) d\mathbf{v} = I$, there exists

$$I - \int M^T \hat{\phi}_m(s + i\mathbf{k}\mathbf{v}) H(\mathbf{v}) d\mathbf{v}$$

$$= I - M^T \int \left[I - (s + i\mathbf{k}\mathbf{v}) \hat{W}(s + i\mathbf{k}\mathbf{v}) \right] H(\mathbf{v}) d\mathbf{v} \tag{5.43}$$

$$= I - M^T + M^T \langle (s + i\mathbf{k}\mathbf{v}) \hat{W}(s + i\mathbf{k}\mathbf{v}) \rangle_v,$$

where $\langle \cdots \rangle_v = \int \cdots H(\mathbf{v}) d\mathbf{v}$. Then by taking $s = 0$ and $\mathbf{k} = 0$ and utilizing Eq. (5.43), we get

$$I - M^T \langle \hat{\phi}_m(s + i\mathbf{k}\mathbf{v}) \rangle_v = I - M^T. \tag{5.44}$$

According to the similar methods used in Eq. (5.11), we can also obtain the asymptotic inverse matrix when $\mathbf{k}$ and s are small

$$\left[I - M^T \langle \hat{\phi}_m(s + i\mathbf{k}\mathbf{v}) \rangle_v \right]^{-1} \sim \frac{|\mathrm{eq_M}\rangle\langle\Sigma|}{\langle\Sigma| \langle (s + i\mathbf{k}\mathbf{v}) \hat{W}(s + i\mathbf{k}\mathbf{v}) \rangle_v |\mathrm{eq_M}\rangle}. \tag{5.45}$$

Thus the following equation is obtained

$$\tilde{G}(\mathbf{k}, s) = \langle \Sigma | \tilde{G}(\mathbf{k}, s) \rangle \sim \frac{\langle\Sigma| \langle \hat{W}(s + i\mathbf{k}\mathbf{v}) \rangle_v |\mathrm{eq_M}\rangle}{\langle\Sigma| \langle (s + i\mathbf{k}\mathbf{v}) \hat{W}(s + i\mathbf{k}\mathbf{v}) \rangle_v |\mathrm{eq_M}\rangle}. \tag{5.46}$$

5.6 More Applications for CTRW and Lévy Walk with Multiple Internal States

Now we complete the introduction of the CTRW and Lévy walk with multiple internal states. In this section we mainly discuss more applications of the process with multiple internal states. One of significant applications is the non-immediate repeat processes, which can be widely observed in the natural world. For example, the random walk of intelligent animals, such as elephants, they can always remember the way they have passed through more or less. Then based on their memories, they decide the next paths randomly except returning to the areas that they have just been through. In this section, for CTRW with multiple internal states we construct the transition matrix and the corresponding jump length and waiting time distributions (while for Lévy walk case, we need to construct the corresponding distributions of velocities and walking time) to model the animal that can only remember the previous step and doesn't return to the area it comes from. From Fig. 5.3, we can see after each step the particle moves, the area

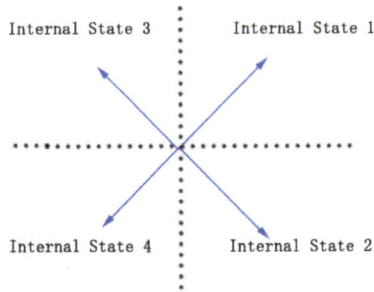

Fig. 5.3   The description of the four internal states of the process. After each movement, based on the current trajectory the area will be divided into four parts. And each part represents the corresponding internal state.

can be divided into four different parts and each part represents an internal state. We choose the four internal states denoted as $++, +-, -+, --$ to represent the particle will move to the quadrant of positive x and y, positive x and negative y, negative x and positive y, negative x and y, respectively at its next step.

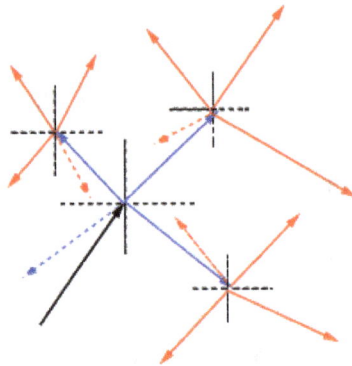

Fig. 5.4   The movements of the particle that can't immediately return to the area it comes from. The dash lines represent the zone that the particle will not move towards for the next step.

Here we first consider the particle simply cannot return to the area it comes from at its next jump as illustrated in Fig. 5.4. That is if the particle stays at the ++ (first internal state) and it won't make a double negative step, i.e., −− (fourth internal state) in the next step. Similarly, we can consider the transitions of other internal states. So that we can obtain the transition matrix of the internal states for this case as

$$M_1 = \begin{pmatrix} 1/3 & 1/3 & 1/3 & 0 \\ 1/3 & 1/3 & 0 & 1/3 \\ 1/3 & 0 & 1/3 & 1/3 \\ 0 & 1/3 & 1/3 & 1/3 \end{pmatrix}.$$

For the CTRW with multiple internal states, we need to construct the corresponding diagonal matrix of the jump length distributions, that is,

$$\Lambda(x,y) = \text{diag}\left(\gamma^+(x)\gamma^+(y), \gamma^+(x)\gamma^-(y), \gamma^-(x)\gamma^+(y), \gamma^-(x)\gamma^-(y)\right)$$

where

$$\gamma^+(l) = \begin{cases} \sqrt{\frac{2}{\pi\sigma^2}} \exp\left(-\frac{l^2}{2\sigma^2}\right) & l \geq 0 \\ 0 & l < 0 \end{cases},$$

and

$$\gamma^-(l) = \begin{cases} 0 & l \geq 0 \\ \sqrt{\frac{2}{\pi\sigma^2}} \exp\left(-\frac{l^2}{2\sigma^2}\right) & l < 0 \end{cases}.$$

Then performing Fourier transform w.r.t. l on $\gamma^+(l)$ and $\gamma^-(l)$, there exists

$$\tilde{\gamma}^\pm(k) = \exp\left(-\frac{\sigma^2}{2}k^2\right)\left(1 \pm \text{erf}\left(\frac{\sqrt{2}}{2}i\sigma k\right)\right).$$

In the following, we simply take $\sigma = 1$ for the sake of simplifying the calculations. According to the asymptotic behavior of error function [Abramowitz and Stegun (1984)], we have

$$\text{erf}\left(\frac{\sqrt{2}}{2}ik\right) \sim \sqrt{\frac{2}{\pi}}ik.$$

Thus we can obtain the jump length distribution matrix in the Fourier space,

$$\tilde{\Lambda}(k_1, k_2) = \text{diag}\left(\exp\left(-\frac{k_1^2 + k_2^2}{2}\right)\left(1 + \sqrt{\frac{2}{\pi}}ik_1\right)\left(1 + \sqrt{\frac{2}{\pi}}ik_2\right),\right.$$

$$\exp\left(-\frac{k_1^2 + k_2^2}{2}\right)\left(1 + \sqrt{\frac{2}{\pi}}ik_1\right)\left(1 - \sqrt{\frac{2}{\pi}}ik_2\right),$$

$$\exp\left(-\frac{k_1^2 + k_2^2}{2}\right)\left(1 - \sqrt{\frac{2}{\pi}}ik_1\right)\left(1 + \sqrt{\frac{2}{\pi}}ik_2\right),$$

$$\left.\exp\left(-\frac{k_1^2 + k_2^2}{2}\right)\left(1 - \sqrt{\frac{2}{\pi}}ik_1\right)\left(1 - \sqrt{\frac{2}{\pi}}ik_2\right)\right)$$

$$\sim \text{diag}\left(1 + \sqrt{\frac{2}{\pi}}i(k_1 + k_2) - \frac{1}{2}(k_1^2 + k_2^2) - \frac{2}{\pi}k_1 k_2,\right.$$

$$1 + \sqrt{\frac{2}{\pi}}i(k_1 - k_2) - \frac{1}{2}(k_1^2 + k_2^2) + \frac{2}{\pi}k_1 k_2,$$

$$1 + \sqrt{\frac{2}{\pi}}i(-k_1 + k_2) - \frac{1}{2}(k_1^2 + k_2^2) + \frac{2}{\pi}k_1 k_2,$$

$$\left.1 - \sqrt{\frac{2}{\pi}}i(k_1 + k_2) - \frac{1}{2}(k_1^2 + k_2^2) - \frac{2}{\pi}k_1 k_2\right).$$

$$(5.47)$$

Next we choose the PDF matrix of waiting time,

$$\hat{\phi}_m(s) \sim I - s^\alpha I.$$

And the initial distribution is chosen to be $\langle \text{init}| = (1/4, 1/4, 1/4, 1/4)$. Then by utilizing Eq. (5.10) and Eq. (5.47) we can eventually obtain

$$\tilde{g}(k_1, k_2, s) = \langle \Sigma|\hat{\tilde{G}}(k_1, k_2, s)\rangle \sim \frac{1}{s}\frac{2\pi s^\alpha}{(2+\pi)k_1^2 + (2+\pi)k_2^2 + 2\pi s^\alpha}. \quad (5.48)$$

From Eq. (5.48) we can also obtain the fractional Fokker-Planck equation

$$\frac{\partial}{\partial t}g(x, y, t) = \frac{2+\pi}{2\pi}\frac{\partial^2}{\partial x^2}D_t^{1-\alpha}g(x, y, t) + \frac{2+\pi}{2\pi}\frac{\partial^2}{\partial y^2}D_t^{1-\alpha}g(x, y, t).$$

From the equation above we can conclude that the process will be faster than the ordinary case (the repeatable case). And we can also calculate the MSD along x axis of this case as

$$\langle x^2(t)\rangle = \mathcal{L}^{-1}\left\{-\frac{\partial^2}{\partial k_1^2}\tilde{g}(k_1, k_2 = 0, s)\Big|_{k_1=0}\right\}$$

$$= \mathcal{L}^{-1}\left\{\frac{2+\pi}{\pi}s^{-1-\alpha}\right\} \sim \frac{2+\pi}{\pi\Gamma(1+\alpha)}t^\alpha.$$

$$(5.49)$$

We note that the PDF of the repeatable random walk in two dimensions in Fourier-Laplace space has the form

$$\tilde{g}(k_1, k_2, s) \sim \frac{1}{s} \frac{2s^\alpha}{k_1^2 + k_2^2 + 2s^\alpha}. \tag{5.50}$$

Comparing the results between Eq. (5.48) and Eq. (5.50), we can easily find they are different. Further we also obtain the asymptotic behavior of MSD along x axis for sufficiently long time t,

$$\langle x^2(t) \rangle \sim \frac{1}{\Gamma(1+\alpha)} t^\alpha. \tag{5.51}$$

By comparing the coefficients of the MSD shown in Eq. (5.49) and Eq. (5.51) obtained by the models of non-immediately-repeated and repeatable processes respectively, we can conclude the previous one is bigger and accelerate the process while the diffusion exponent doesn't change.

Then we consider the case that particle won't return to the area it comes from, besides one of the other three areas can't be reached either. As shown in Fig. 5.5, we have two different subcases (in fact, there're three

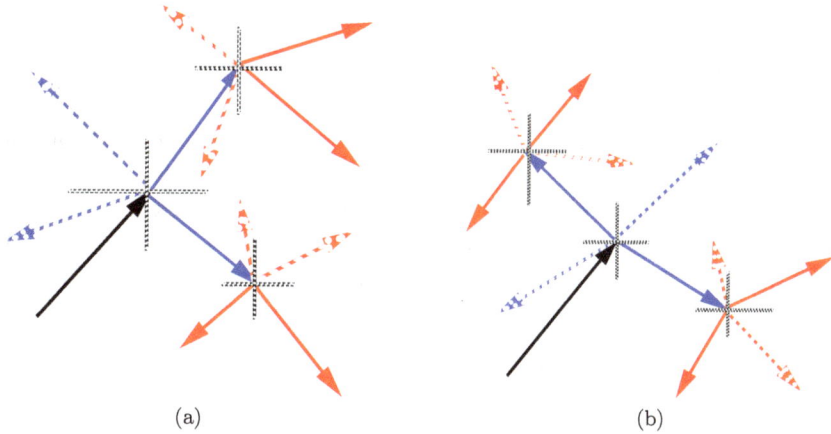

(a)   (b)

Fig. 5.5   The movements of the particle not going back to the area it comes from and one of the other three areas. The dash arrows represent the area the particle will not move into for the next step.

different cases, however, two of them are equivalent). For the subcase (a) shown in Fig. 5.5, similar with the method of obtaining the transition matrix M_1, we can construct the corresponding transition matrix as

$$M_2 = \begin{pmatrix} 0 & 1/2 & 1/2 & 0 \\ 1/2 & 0 & 0 & 1/2 \\ 1/2 & 0 & 0 & 1/2 \\ 0 & 1/2 & 1/2 & 0 \end{pmatrix}.$$

While for the subcase (b) shown in Fig. 5.5, the corresponding transition matrix is

$$M_3 = \begin{pmatrix} 1/2 & 1/2 & 0 & 0 \\ 0 & 1/2 & 0 & 1/2 \\ 1/2 & 0 & 1/2 & 0 \\ 0 & 0 & 1/2 & 1/2 \end{pmatrix}.$$

By using the same methods, we can obtain the PDFs of subcases (a) and (b) in the Fourier-Laplace space, which are the same as

$$\tilde{\hat{g}}(k_1, k_2, s) \sim \frac{1}{s} \frac{2s^\alpha}{k_1^2 + k_2^2 + 2s^\alpha}. \tag{5.52}$$

Comparing the results in Eq. (5.52) with Eq. (5.50), we conclude that the asymptotic behaviors of subcases (a) and (b) for the long time are the same as repeatable case. Besides the governing equations for the subcases (a) and (b) are also the same, that is

$$\frac{\partial}{\partial t} g(x, y, t) = \frac{1}{2} \frac{\partial^2}{\partial x^2} D_t^{1-\alpha} g(x, y, t) + \frac{1}{2} \frac{\partial^2}{\partial y^2} D_t^{1-\alpha} g(x, y, t).$$

And MSD along x axis is $\langle x^2(t) \rangle \sim \frac{1}{\Gamma(1+\alpha)} t^\alpha$. Of course, they are still the same.

For the last case, we consider the particle will not go back to the area it passes through at the previous step, and only choose one area out of the other three. Thus we have two subcases as illustrated in Fig. 5.6. For the

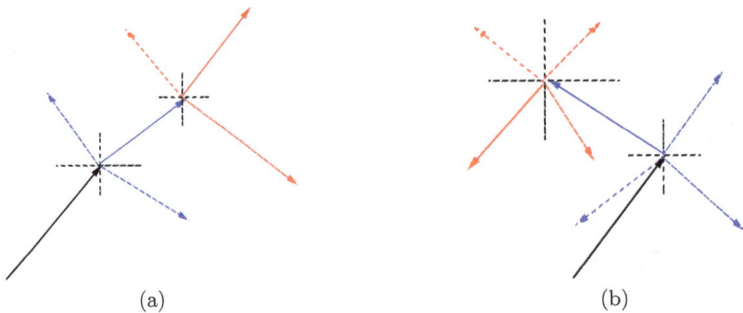

(a)   (b)

Fig. 5.6 The movements of the particle not going back to the area it comes from and two of the other three areas. The dash arrows still represent the areas the particle will not move into for the next step.

subcase (a) in Fig. 5.6, we can also construct the corresponding transition matrix

$$M_4 = \begin{pmatrix} 1 & 0 & 0 & 0 \\ 0 & 1 & 0 & 0 \\ 0 & 0 & 1 & 0 \\ 0 & 0 & 0 & 1 \end{pmatrix},$$

which represents the particle mainly moves towards one direction. And the corresponding PDF in Fourier-Laplace space is

$$\tilde{\hat{g}}(k_1, k_2, s)$$
$$\sim \frac{1}{s} \frac{s^\alpha (8\pi k_1^4 + 8\pi k_2^4 + 16\pi k_1^2 s^\alpha + 16\pi k_2^2 s^\alpha + 16(-4+\pi)k_1^2 k_2^2 + 8\pi^2 s^{3\alpha})}{32k_1^4 + 32k_2^4 + 32k_1^2 \pi s^{2\alpha} + 32k_2^2 \pi s^{2\alpha} - 64k_1^2 k_2^2 + 8\pi^2 s^{4\alpha}}.$$

The form of the above equation is too complex to take inverse transforms and obtain the equation for the $q(x, y, t)$ although the process is very simple to describe. From the above equation, we can also obtain the MSD along x axis $\langle x^2(t) \rangle \sim \frac{4}{\Gamma(1+2\alpha)} t^{2\alpha}$. For this subcase we can see the exponent of MSD has changed, which means the type of the diffusion may change. Specifically, when $0 < \alpha < 1/2$, $\alpha = 1/2$, and $1/2 < \alpha < 1$ the process behaves as subdiffusion, normal diffusion, and superdiffusion, respectively. Next we consider the subcase (b), which has the transition matrix

$$M_5 = \begin{pmatrix} 0 & 0 & 1 & 0 \\ 1 & 0 & 0 & 0 \\ 0 & 0 & 0 & 1 \\ 0 & 1 & 0 & 0 \end{pmatrix}.$$

From the trajectory of subcase (b) shown in Fig. 5.6, we can see it moves like a vortex. And its PDF in the Fourier-Laplace space is

$$\tilde{\hat{g}}(k_1, k_2, s) \sim \frac{1}{s} \frac{2\pi s^\alpha}{(\pi - 2)k_1^2 + (\pi - 2)k_2^2 + 2\pi s^\alpha}.$$

The fractional Fokker-Planck equation for this subcase is

$$\frac{\partial}{\partial t} g(x, y, t) = \frac{\pi - 2}{2\pi} \frac{\partial^2}{\partial x^2} D_t^{1-\alpha} g(x, y, t) + \frac{\pi - 2}{2\pi} \frac{\partial^2}{\partial y^2} D_t^{1-\alpha} g(x, y, t).$$

The corresponding MSD is $\langle x^2(t) \rangle \sim \frac{\pi-2}{\Gamma(1+\alpha)\pi} t^\alpha$, which means the particle moves slower than the repeatable case by comparing the coefficients.

From the above calculations and analyses, it can be noted that the constraints of the reachable regions does not always make the diffusion go faster. According to the limitation, the process can move faster, the same or even slower.

Then we consider the applications of Lévy walk with multiple internal states. Here we still utilize the transition matrices $M_1, \cdots, M_5$, and construct the diagonal matrix of the velocity $\mathbf{v} = (v_x, v_y)$ for non-immediately-repeating Lévy walk

$$
H_{\text{uniform}}(\mathbf{v}) = \text{diag}\Big\{ \frac{2}{\pi v_0} \delta\big(\sqrt{v_x^2 + v_y^2} - v_0\big) \kappa(v_x)\kappa(v_y)
$$

$$
\frac{2}{\pi v_0} \delta\big(\sqrt{v_x^2 + v_y^2} - v_0\big) \kappa(v_x)\kappa(-v_y)
$$

$$
\frac{2}{\pi v_0} \delta\big(\sqrt{v_x^2 + v_y^2} - v_0\big) \kappa(-v_x)\kappa(v_y)
$$

$$
\frac{2}{\pi v_0} \delta\big(\sqrt{v_x^2 + v_y^2} - v_0\big) \kappa(-v_x)\kappa(-v_y) \Big\},
$$

where

$$
\kappa(x) = \begin{cases} 1 & x > 0; \\ 0 & x \le 0. \end{cases}
$$

And here we also consider the case that all of the walking duration distributions are the same. That is $\phi^{(1)}(\tau) = \cdots = \phi^{(4)}(\tau) = \alpha/(\tau_0(1 + \tau/\tau_0)^{1+\alpha})$, where τ_0, $\alpha > 0$. Here we consider $1 > \alpha > 0$, $2 > \alpha > 1$ and $\alpha > 2$. Then after Laplace transform, one has

$$
\hat{\phi}_m(s) \sim \left[1 - \frac{\tau_0}{\alpha - 1} s - \tau_0^\alpha \Gamma(1 - \alpha) s^\alpha + \frac{\tau_0^2}{(\alpha - 2)(\alpha - 1)} s^2 \right] I. \tag{5.53}
$$

Noticing that $\big\langle M^T \hat{\phi}_m(s + i k \mathbf{v})\big\rangle_v = M^T \big\langle \hat{\phi}_m(s + i k \mathbf{v})\big\rangle_v$, we can consider each internal state specifically, that is for the first internal state

$$
\int_{-\infty}^{\infty} dv_x \int_{-\infty}^{\infty} dv_y h^{(1)}(v_x, v_y) \hat{\phi}^{(1)}(s + ik_x v_x + ik_y v_y)
$$

$$
= \int_{-\infty}^{\infty} dv_x \int_{-\infty}^{\infty} dv_y \frac{2}{\pi v_0} \delta\big(\sqrt{v_x^2 + v_y^2} - v_0\big) \kappa(v_x)\kappa(v_y) \hat{\phi}^{(1)}(s + ik_x v_x + ik_y v_y)
$$

$$
= \int_{0}^{\infty} d\rho \int_{0}^{\frac{\pi}{2}} d\theta \frac{2\rho}{\pi v_0} \delta(\rho - v_0) \hat{\phi}^{(1)}(s + ik_x \rho \cos\theta + ik_y \rho \sin\theta)
$$

$$
= \frac{2}{\pi} \int_{0}^{\frac{\pi}{2}} 1 - \frac{\tau_0}{\alpha - 1}(s + ik_x v_0 \cos\theta + ik_y v_0 \sin\theta) - \tau_0^\alpha \Gamma(1 - \alpha)(s + ik_x v_0
$$

$$
\times \cos\theta + ik_y v_0 \sin\theta)^\alpha + \frac{\tau_0^2}{(\alpha - 2)(\alpha - 1)}(s + ik_x v_0 \cos\theta + ik_y v_0 \sin\theta)^2 d\theta.
$$

$$
\tag{5.54}
$$

Similarly, we can obtain the other internal states by simply changing the boundary of the integral according to the corresponding internal

Table 5.1   Asymptotic behaviours of MSDs for different transition matrices and regions of α.

Cases	$0 < \alpha < 1$	$1 < \alpha < 2$	$2 < \alpha$
Ordinary	$\frac{(1-\alpha)v_0^2}{2}t^2$	$\frac{v_0^2\tau_0^{\alpha-1}(\alpha-1)}{(3-\alpha)(2-\alpha)}t^{3-\alpha}$	$\frac{v_0^2\tau_0}{\alpha-2}t$
$M1$	$\frac{(1-\alpha)v_0^2}{2}t^2$	$\frac{v_0^2\tau_0^{\alpha-1}(\alpha-1)}{(3-\alpha)(2-\alpha)}t^{3-\alpha}$	$\frac{\tau_0 v_0^2[-8-\pi^2+\alpha(4+\pi^2)]}{(\alpha-2)(\alpha-1)\pi^2}t$
$M2$	$\frac{(1-\alpha)v_0^2}{2}t^2$	$\frac{v_0^2\tau_0^{\alpha-1}(\alpha-1)}{(3-\alpha)(2-\alpha)}t^{3-\alpha}$	$\frac{v_0^2\tau_0}{\alpha-2}t$
$M3$	$\frac{(1-\alpha)v_0^2}{2}t^2$	$\frac{v_0^2\tau_0^{\alpha-1}(\alpha-1)}{(3-\alpha)(2-\alpha)}t^{3-\alpha}$	$\frac{v_0^2\tau_0}{\alpha-2}t$
$M4$	$\frac{v_0^2}{2}t^2$	$\frac{\pi^2+\alpha(8-\pi^2)}{2\pi^2}v_0^2 t^2$	$\frac{4}{\pi^2}v_0^2 t^2$
$M5$	$\frac{(1-\alpha)v_0^2}{2}t^2$	$\frac{v_0^2\tau_0^{\alpha-1}(\alpha-1)}{(3-\alpha)(2-\alpha)}t^{3-\alpha}$	$\frac{\tau_0 v_0^2[8-\pi^2+\alpha(-4+\pi^2)]}{(\alpha-2)(\alpha-1)\pi^2}t$

state. Besides we have $\left\langle \hat{W}(s+i\mathbf{kv})\right\rangle_v$ for each internal state with the same method. Then after some calculations [Xu and Deng (2018)], we can obtain the MSD along one direction (here we consider along X-axis) which is enough to measure how fast the diffusion is, although the process moves in two-dimensional space.

From the results shown in Tabel 5.1, one can conclude that non-immediately-repeating has no influence on the MSD of Lévy walk for $0 < \alpha < 1$ and $1 < \alpha < 2$ for the transition matrices M_1, M_2, M_3 and M_5. From here we can see Lévy walk is a stable process. Besides, we can also conclude that M_4 has a major influence on the MSD. And this result is reasonable because the Lévy walk with transition matrix M_4 cannot change its internal states once determined by the initial distribution and mainly moves towards one direction.

Further we can also construct different transition matrices and the corresponding matrices of distributions of jump length (or velocity) and waiting time to deal with some other problems, such as the self avoiding random walk and intelligent animals who can remember multiple steps they have taken. Some other applications can be dug out in the future.

Chapter 6

Fractional Reaction Diffusion Equations and Corresponding Feynman-Kac Equations

In this chapter, we focus on the fractional reaction diffusion equations, which is the generalization of the classical reaction diffusion equations in the anomalous diffusion scenario. Furthermore, the Feynman-Kac equations of the reaction diffusion processes are also investigated.

6.1 Fractional Reaction Diffusion Equations

Chemical reaction leads to the transformation of one type of particles to another type. Recall the governing equation for the PDF $\rho(x,t)$ of the reaction diffusion process. Assume that the reaction rate law is $\dot{\rho} = r(\rho)\rho$. For normal diffusion, the well-known reaction diffusion equation is

$$\frac{\partial \rho(x,t)}{\partial t} = K_1 \frac{\partial^2 \rho(x,t)}{\partial x^2} + r(\rho)\rho \tag{6.1}$$

with K_1 being the diffusion constant. Equation (6.1) macroscopically governs the PDF of the underlying reaction diffusion process. For anomalous diffusion, the situation becomes complicated, and particularly the resulted equation can not be obtained by simply adding the reaction term into the non-Markovian transport equation. In the following subsections, we will illustrate this idea in detail. Furthermore, one can refer to the seminal work [Mendez *et al.* (2010)] for the general derivation of mesoscopic reaction-transport equations. For the multispecies system with anomalous diffusion and reaction dynamics, refer to [Langlands *et al.* (2008); Fedotov *et al.* (2010); Froemberg *et al.* (2011)]; and for the front propagation problems, see [Yadav *et al.* (2007); Iomin and Sokolov (2012); Iomin and Méndez (2013)].

6.1.1   *Reaction-Anomalous Diffusion Equations*

In [Sokolov *et al.* (2006)], the authors derive the reaction subdiffusion equation for the simplest reaction scheme, namely, monomolecular conversion $A \to B$, with the reaction rate κ. Inspired by the generalized master equation for CTRW [Chechkin *et al.* (2005)], the authors start from a discrete scheme and consider particles occupying sites of a one-dimensional lattice. A balance condition for the mean number A_i of particles A on the site i reads

$$\frac{dA_i(t)}{dt} = I_i^+(t) - I_i^-(t) - \kappa A_i(t), \tag{6.2}$$

where $I_i^-(t)$ is the loss per unit time due to the departure of particles from the site i (loss flux), $I_i^+(t)$ is the gain flux, and κA_i is the loss caused by the chemical conversion $A \to B$. Suppose the particles only jump to the nearest neighbours with equal probabilities. Thus, one obtains

$$\frac{dA_i(t)}{dt} = \frac{1}{2}I_{i-1}^-(t) + \frac{1}{2}I_{i+1}^-(t) - I_i^-(t) - \kappa A_i(t). \tag{6.3}$$

According to [Sokolov *et al.* (2006)], the loss flux at time t is connected with the gain flux at the site in the past: the particles which leave the site i at time t either were at i from the very beginning (and survived without being converted into B), or arrived at i at some previous time $t' < t$ and survived during the time interval $[t', t]$. Denote $\phi(t)$ as the waiting time distribution. Thus,

$$I_i^-(t) = \phi(t)e^{-\kappa t}A_i(0) + \int_0^t \phi(t - t')e^{-\kappa(t-t')}I_i^+(t')\,dt'. \tag{6.4}$$

Using Eq. (6.2) one gets

$$I_i^-(t) = \phi_s(t)A_i(0) + \int_0^t \phi_s(t - t')\left[\frac{dA_i(t')}{dt'} + \kappa A_i(t') + I_i^-(t')\right]dt', \tag{6.5}$$

where $\phi_s(t) = \phi(t)e^{-\kappa t}$ is the nonproper waiting time density provided that the particle survived. Correspondingly, in the Laplace domain, $\hat{\phi}_s(u) = \hat{\phi}(u + \kappa)$. Thus, the Laplace transform of Eq. (6.4) leads to

$$\widehat{I_i^-}(u) = \widehat{\Phi}_\kappa(u)\widehat{A}_i(u), \tag{6.6}$$

with $\widehat{\Phi}_\kappa(u)$ given by

$$\widehat{\Phi}_\kappa(u) = \frac{(u + \kappa)\hat{\phi}(u + \kappa)}{1 - \hat{\phi}(u + \kappa)}. \tag{6.7}$$

Back to the time domain, one has

$$I_i^-(t) = \int_0^t \Phi_\kappa(t - t') A_i(t') \, dt', \qquad (6.8)$$

where $\Phi_\kappa(t)$ is the inverse Laplace transform of $\hat{\Phi}_\kappa(u)$. Combining Eq. (6.3) and Eq. (6.8), one obtains

$$\frac{dA_i(t)}{dt} = \int_0^t \Phi_\kappa(t - t') \left[\frac{A_{i-1}(t')}{2} + \frac{A_{i+1}(t')}{2} - A_i(t') \right] dt' - \kappa A_i(t). \qquad (6.9)$$

Transition to a continuum in space $x = ai$ gives [Sokolov *et al.* (2006)]

$$\frac{\partial A(x,t)}{\partial t} = \frac{a^2}{2} \int_0^t \Phi_\kappa(t - t') \Delta A(x,t') \, dt' - \kappa A(x,t)$$

$$= \frac{a^2}{2} \int_0^t \Phi_0(t - t') e^{-\kappa(l - t')} \Delta A(x,t') \, dt' - \kappa A(x,t), \qquad (6.10)$$

where the reaction rate explicitly affects the transport term. Particularly, when the exponential waiting time distribution $\phi(t)$ is assumed, the kernel $\Phi_0(t)$ has the form of δ function and thus the classical reaction diffusion equation is recovered. However, in the case of the power law waiting time distribution, one has $\hat{\phi}(u) \simeq 1 - (\tau u)^\alpha \Gamma(1 - \alpha)$ where τ is the appropriate time scale. Thus, the Laplace transform of the integral kernel reads

$$\hat{\Phi}_\kappa(u) = \frac{1}{\tau^\alpha \Gamma(1 - \alpha)} (u + \kappa)^{1 - \alpha}. \qquad (6.11)$$

Correspondingly, the integral operator $T_t(1 - \alpha, \kappa) f(t) = \tau^\alpha \Gamma(1 - \alpha) \int_0^t \Phi_\kappa(t - t') f(t') \, dt'$ is expressed as

$$T_t(1 - \alpha, \kappa) f = \frac{d}{dt} \int_0^t \frac{e^{-\kappa(t - t')}}{(t - t')^{1 - \alpha}} f(t') \, dt' + \kappa \int_0^t \frac{e^{-\kappa(t - t')}}{(t - t')^{1 - \alpha}} f(t') \, dt'. \qquad (6.12)$$

Finally, the equation for the A concentration reads

$$\frac{\partial A(x,t)}{\partial t} = K_\alpha T_t(1 - \alpha, \kappa) \Delta A(x,t) - \kappa A(x,t), \qquad (6.13)$$

where the generalized diffusion coefficient is $K_\alpha = a^2 \left[2\tau^\alpha \Gamma(1 - \alpha) \right]^{-1}$. As we can see from the above equation, the reaction affects the diffusion part of the equation and they are not simply additive.

In [Schmidt *et al.* (2007)], the authors further consider the simple monomolecular conversion $A \to B$ and present a different derivation of the reaction-anomalous diffusion equations for a CTRW transport mechanism. A particles are assumed to be converted into B at a constant conversion rate κ independent of their position. Thus, the survival probability of a single A particle in the time interval $[t', t]$ is $e^{-\kappa(t-t')}$. Then, based on the CTRW framework, the authors [Schmidt *et al.* (2007)] write down the equation for the PDF of the positions x of the particles, which have just made a jump at time t

$$\eta_A(x,t) = \int_{-\infty}^{\infty} \int_0^t \eta_A(x',t') e^{-\kappa(t-t')} \psi(x-x', t-t') \, dx' dt' + A(x,0)\delta(t). \tag{6.14}$$

Here $\psi(x,t)$ is the jump PDF given by the probability density in space and time to make a jump of length x at time t after the last jump. For the decoupled waiting time and jump length distributions, $\psi(x,t)$ is assumed to be $\psi(x,t) = w(x)\phi(t)$. The concentration $A(x,t)$ is then connected with $\eta_A(x,t)$ via

$$A(x,t) = \int_0^t dt' \eta_A(x,t') e^{-\kappa(t-r')} W(t-t'), \tag{6.15}$$

where $W(t)$ is the probability of the waiting time longer than t and it is given by

$$W(t) = 1 - \int_0^t dt' \phi(t'). \tag{6.16}$$

Conducting Fourier-Laplace transforms of Eqs. (6.14) and (6.15) leads to

$$\hat{\tilde{A}}(k,u) = \frac{[1 - \tilde{\phi}(u+\kappa)]\hat{A}(k,0)}{(u+\kappa)[1 - \hat{\tilde{\psi}}(k, u+\kappa)]}. \tag{6.17}$$

Once specifying the jump length and waiting time distributions, one can return to the space and time domain in the continuum limit of large scales and long times. Assume the notation

$$K_{\mu,\alpha} = \sigma^\mu \left[\tau^\alpha \Gamma(1-\alpha) \right]^{-1}. \tag{6.18}$$

Therefore, for a Gaussian jump length distribution and an exponential waiting time distribution, one has

$$\frac{\partial A(x,t)}{\partial t} = K_{2,1} \Delta A(x,t) - \kappa A(x,t) \tag{6.19}$$

and for Lévy flights and exponential waiting time,

$$\frac{\partial A(x,t)}{\partial t} = K_{\mu,1} \Delta^{\mu/2} A(x,t) - \kappa A(x,t). \tag{6.20}$$

Under these circumstances, the transport term and the reaction term are additive. If one considers the Gaussian jump length and the board power law waiting time with $0 < \alpha < 1$, one obtains [Schmidt *et al.* (2007)]

$$\frac{\partial A(x,t)}{\partial t} = K_{2,\alpha} \mathcal{T}_t (1-\alpha, \kappa) \Delta A(x,t) - \kappa A(x,t) \tag{6.21}$$

with the transport operator $\mathcal{T}_t(1-\alpha,\kappa)$, which is time dependent and

$$\mathcal{T}_t(1-\alpha,\kappa)f(t)$$
$$= \frac{1}{\Gamma(\alpha)}\left(\frac{\mathrm{d}}{\mathrm{d}t}\int_0^t \frac{e^{-\kappa(t-t')}}{(t-t')^{1-\alpha}}f(t')\,\mathrm{d}t' + \kappa\int_0^t \frac{e^{-\kappa(t-t')}}{(t-t')^{1-\alpha}}f(t')\,\mathrm{d}t'\right). \tag{6.22}$$

So the reaction parameter κ enters the transport term and they are not simply additive. Note that the transport operator $\mathcal{T}_t(1-\alpha,\kappa)$ also has the form [Henry *et al.* (2006)]

$$\mathcal{T}_t(1-\alpha,\kappa)f(t) = \exp(-\kappa t)\mathcal{D}_t^{1-\alpha}\{\exp(\kappa t)f(t)\}, \tag{6.23}$$

where $\mathcal{D}_t^{1-\alpha}$ is the Riemann-Liouville fractional derivative. One can also easily formulate the equation for power law waiting times and Lévy jumps as

$$\frac{\partial A(x,t)}{\partial t} = K_{\mu,\alpha}\mathcal{T}_t(1-\alpha,\kappa)\Delta^{\mu/2}A(x,t) - \kappa A(x,t), \tag{6.24}$$

where $\Delta^{\mu/2}$ denotes the symmetrized (Riesz-Weyl) spatial fractional derivative.

6.1.2 Non-Markovian Transport with Nonlinear Reactions

In [Fedotov (2010)], the author considers how to incorporate the nonlinear reaction term into non-Markovian transport which is described by CTRW. Let $\rho(x,t)$ be the density of the particles at point x and time t. Assume the chemical reaction follows the mass action law and the reaction term is of the form $r(\rho)\rho$. In order to formulate the integral balance equations for the density $\rho(x,t)$, the author [Fedotov (2010)] also introduces the auxiliary density $j(x,t)$ which describes the number of particles arriving at point x exactly at time t. Assume $\rho_0(x)$ is the initial distribution and $\lambda(x)$ is the jump length distribution. Then one gets the balance equations for $\rho(x,t)$ and $j(x,t)$ according to [Fedotov (2010)]

$$\rho(x,t) = \rho_0(x)e^{\int_0^t r(\rho(x,u))du}W(t)$$
$$+ \int_0^t j(x,\tau)e^{\int_\tau^t r(\rho(x,u))du}W(t-\tau)d\tau, \tag{6.25}$$

and

$$j(x,t) = \int_{\mathbb{R}} \rho_0(x-z)e^{\int_0^t r(\rho(x,u))du}\lambda(z)\phi(t)dz$$
$$+ \int_0^t\int_{\mathbb{R}} j(x-z,\tau)e^{\int_\tau^t r(\rho(x-z,u))du}\lambda(z)\phi(t-\tau)dzd\tau. \tag{6.26}$$

The next step is to derive the evolution equation for the density $\rho(x, t)$. Instead of the standard technique of Fourier-Laplace transforms, the author [Fedotov (2010)] first differentiates the density $\rho(x, t)$ given by Eq. (6.25) w.r.t. time

$$\frac{\partial \rho}{\partial t} = j(x, t) + r(\rho)\rho - \rho_0(x) e^{\int_0^t r(\rho(x,u)) du} \phi(t)$$
$$- \int_0^t j(x, \tau) e^{\int_\tau^t r(\rho,u)) du} \phi(t - \tau) d\tau. \tag{6.27}$$

The last two terms can be interpreted as the density of particles that leave the point x exactly at time t

$$i(x, t) = \rho_0(x) e^{\int_0^t r(\rho(x,u)) du} \phi(t) + \int_0^t j(x, \tau) e^{\int_\tau^t r(\rho(x,u)) du} \phi(t-\tau) d\tau. \tag{6.28}$$

Then Eq. (6.27) can be rewritten as

$$\frac{\partial \rho}{\partial t} = \int_{\mathbb{R}} i(x - z, t)\lambda(z) dz - i(x, t) + r(\rho)\rho. \tag{6.29}$$

Furthermore, the author [Fedotov (2010)] expresses the density $i(x, t)$ in terms of $\rho(x, t)$ as

$$i(x, t) = \int_0^t K(t - \tau)\rho(x, \tau) e^{\int_\tau^t r(\rho(x,u)) du} d\tau, \tag{6.30}$$

where $K(t)$ is the standard memory kernel defined by its Laplace transform

$$\hat{K}(s) = \frac{\hat{\phi}(s)}{\hat{W}(s)} = \frac{s\hat{\phi}(s)}{1 - \hat{\phi}(s)}. \tag{6.31}$$

Finally the nonlinear master equation for the density $\rho(x, t)$ is [Fedotov (2010)]

$$\frac{\partial \rho}{\partial t} = \int_0^t K(t - \tau) \left[\int_{\mathbb{R}} \rho(x - z, \tau) e^{\int_\tau^t r(\rho(x-z,u)) du} \lambda(z) dz \right.$$
$$\left. - \rho(x, \tau) e^{\int_\tau^t r(\rho(x,u)) du} \right] d\tau + r(\rho)\rho. \tag{6.32}$$

Therefore, the reaction term influences the (non-Markovian) transport term and they are coupled via some integral operators. Assume the dispersal kernel $\lambda(z)$ is even and it has finite second moment $\sigma^2 = \int_{\mathbb{R}} z^2 \lambda(z) dz$. Then the corresponding master equation is [Fedotov (2010)]

$$\frac{\partial \rho}{\partial t} = \frac{\sigma^2}{2} \frac{\partial^2}{\partial x^2} \int_0^t K(t - \tau)\rho(x, \tau) e^{\int_\tau^t r(\rho(x,u)) du} d\tau + r(\rho)\rho. \tag{6.33}$$

6.2   Feynman-Kac Equations for Reaction and Diffusion Processes

In this section, we would like to provide a theoretical framework for deriving the forward and backward Feynman-Kac equations for the distributions of functionals when the particles undergo both diffusion and reaction processes. Using the derived equations, we can compute the distributions of some experimentally measurable statistics, for example, the occupation time in half space as well as the first passage time, and analyze their properties when the physical system exhibits both anomalous diffusion and spontaneous evanescence. Functionals of the particle's trajectory have practical applications in rather diverse fields, including probability theory, the Kardar-Parisi-Zhang varieties, finance, economics, even in characterizing the stochastic behaviour of daily temperature records [Majumdar (2005)]. Moreover, the functional can be defined in the form

$$A = \int_0^t U(x(\tau), \tau) d\tau. \tag{6.34}$$

Note that the function U explicitly depends on time, which may be used to model the first passage time to a moving boundary [Tuckwell and Wan (1984)]. Here $x(t)$ is the particle's trajectory and $U(x,t)$ is some specified function.

As for the chemical reactions, both nonlinear reaction rate $r(\rho(x,t))$ and linear reaction rates $r(t)$ and $r(x)$ are considered, i.e., the equations governing reaction process are, respectively,

$$\dot{\rho} = r(\rho)\rho, \tag{6.35}$$

$$\dot{\rho} = r(t)\rho, \tag{6.36}$$

or

$$\dot{\rho} = r(x)\rho. \tag{6.37}$$

Under these conditions, the framework of deriving the Feynman-Kac equations provided in [Turgeman *et al.* (2009)] for the diffusion process does not work for the reaction diffusion process. So, new theoretical frameworks are presented [Hou and Deng (2018)] to derive the forward and backward Feynman-Kac equations for the distributions of the more general functionals of the path of a particle undergoing both diffusion and chemical reaction with linear/nonlinear rates; the new framework also applies to the diffusion process and the functional $A = \int_0^t U(x(\tau)) d\tau$. Based on the very general forms of the derived forward and backward Feynman-Kac equations, we more specifically provide the equations for normal/anomalous diffusions and reactions with linear/nonlinear rate. And the derived equations are used to calculate the occupation time in half space, the first passage time to a fixed boundary, and the occupation time in half space with absorbing or reflecting boundary conditions.

6.2.1   Forward Feynman-Kac Equations for Nonlinear Reaction Rate $r(\rho(x,t))$

We define a stochastic process $\{x(t), A(t)\}$ describing the time-varying positions and functional values for the reaction diffusion processes. Denote $\rho(x, A, t)$ as the joint PDF of finding the particle at position x with the functional value A at time t. We extend the method in [Mendez *et al.* (2010); Fedotov (2010)] to formulate the balance equations for the density $\rho(x, A, t)$ and the auxiliary density $j(x, A, t)$. The latter one is the joint PDF of particles exactly arriving at position x with the functional value A at time t. The balance equation for $\rho(x, A, t)$ can be written as

$$
\begin{aligned}
\rho(x, A, t) = {} & \rho_0(x)\delta\left(A - \int_0^t U(x, t')dt'\right) e^{\int_0^t r(\rho(x,u))du}W(t) \\
& + \int_0^t j\left(x, A - \int_\tau^t U(x, t')dt', \tau\right) e^{\int_\tau^t r(\rho(x,u))du}W(t - \tau)d\tau,
\end{aligned}
\tag{6.38}
$$

where $W(t)$ is the survival probability function. The first term on the right hand side of Eq. (6.38) represents the particles that stay at their initial position x up to time t. $\rho_0(x)$ is the initially spatial distribution and all particles start at $t = 0$ without aging effect. The corresponding functional value of these motionless particles can be calculated as $A = \int_0^t U(x, t')dt'$, since they stay at position x during the time interval $(0, t)$. Hence the Dirac delta function $\delta(\cdot)$ is introduced. The particle density changes according to the nonlinear reaction rate $r(\rho(x, t))$ during the time interval $(0, t)$ with $\rho(x, t) = \int_{-\infty}^{+\infty} \rho(x, A, t)dA$. Consequently, the factors $e^{\int_0^t r(\rho(x,u))du}$ and $W(t)$ are multiplied. The second term on the right hand side of Eq. (6.38) describes the particles that arrive at some position x at some previous time $\tau < t$, react with the nonlinear rate $r(\rho(x, t))$ during the time interval $[\tau, t]$, and don't move during the same interval. Thus, the factors $e^{\int_\tau^t r(\rho(x,u))du}$ and $W(t-\tau)$ are multiplied in the integrand. Note that $j\left(x, A - \int_\tau^t U(x, t')dt', \tau\right)$ means that the particles arrive at the position x at the time τ with the functional value $A - \int_\tau^t U(x, t')dt'$. Recall that $A = \int_0^t U(x(t'), t')dt'$. Thus, the difference of the functional values between the time intervals $[0, \tau]$ and $[0, t]$ is $\int_\tau^t U(x, t')dt'$, since no jump takes place during $[\tau, t]$. In view of arbitrary τ, the integral regarding τ is naturally introduced from 0 to t. If the usual functional $A = \int_0^t U(x(t'))dt'$ is used, one just needs to replace $\int_0^t U(x, t')dt'$ and $\int_\tau^t U(x, t')dt'$ in Eq. (6.38) with $tU(x)$ and $(t - \tau)U(x)$, respectively. Moreover, the balance equation for $j(x, A, t)$ can be written in the following form

$$j(x, A, t) =$$

$$\int_\mathbb{R} \rho_0(x - z)\delta \left(A - \int_0^t U(x - z, t')dt' \right) e^{\int_0^t r(\rho(x-z,u))du} w(z)\phi(t)dz +$$

$$\int_0^t \int_\mathbb{R} j \left(x - z, A - \int_\tau^t U(x - z, t')dt', \tau \right) e^{\int_\tau^t r(\rho(x-z,u))du} w(z)\phi(t - \tau)dzd\tau.$$

$$(6.39)$$

Similarly, the first term on the right hand side of Eq. (6.39) expresses the particles that stay at the initial position $x - z$, react with the rate $r(\rho(x - z, u))$ and jump to x instantaneously at time t. The functional is calculated as $A = \int_0^t U(x - z, t')dt'$ and the reaction factor $e^{\int_0^t r(\rho(x-z,u))du}$ is multiplied. And the second term demonstrates the particles that arrive at the position $x - z$ at some previous time $\tau < t$, react up to time t (leading to the reaction factor $e^{\int_\tau^t r(\rho(x-z,u))du}$) and jump to x exactly at time t. Similarly, the difference of the functional values between the time interval $[0, \tau]$ and $[0, t]$ is $\int_\tau^t U(x - z, t')dt'$. Since z is arbitrary in $\mathbb{R}$ and τ is varying from 0 to t, the integrals are respectively introduced regarding z and τ. In order to derive the forward Feynman-Kac equation, we need to make the Fourier transforms $A \to p$, denoted by

$$\mathcal{F}\{f(A)\} = \int_{-\infty}^{+\infty} e^{ipA} f(A)dA, \qquad (6.40)$$

of both Eqs. (6.38) and (6.39), since A could be negative. Then we obtain

$$\rho(x, p, t) = \rho_0(x)e^{ip \int_0^t U(x,t')dt'} e^{\int_0^t r(\rho(x,u))du} W(t)$$

$$+ \int_0^t j(x, p, \tau)e^{ip \int_\tau^t U(x,t')dt'} e^{\int_\tau^t r(\rho(x,u))du} W(t - \tau)d\tau, \qquad (6.41)$$

and

$$j(x, p, t) = \int_\mathbb{R} \rho_0(x - z)e^{ip \int_0^t U(x-z,t')dt'} e^{\int_0^t r(\rho(x-z,u))du} w(z)\phi(t)dz$$

$$+ \int_0^t \int_\mathbb{R} j(x - z, p, \tau)e^{ip \int_\tau^t U(x-z,t')dt'} e^{\int_\tau^t r(\rho(x-z,u))du} w(z)\phi(t - \tau)dzd\tau.$$

$$(6.42)$$

Since Eqs. (6.41) and (6.42) have nonlinear terms, the standard Fourier-Laplace transform techniques used in [Turgeman *et al.* (2009); Wu *et al.* (2016); Carmi *et al.* (2010); Carmi and Barkai (2011)] cannot be applied directly. Instead, we differentiate the density $\rho(x, p, t)$ given by Eq. (6.41) w.r.t. t, which leads to

$$\frac{\partial \rho(x,p,t)}{\partial t} = [ipU(x,t) + r(\rho(x,t))]\rho(x,p,t) + j(x,p,t) - h(x,p,t), \quad (6.43)$$

where

$$h(x,p,t) = \rho_0(x)e^{ip\int_0^t U(x,t')dt'}e^{\int_0^t r(\rho(x,u))du}\phi(t)$$
$$+ \int_0^t j(x,p,\tau)e^{ip\int_\tau^t U(x,t')dt'}e^{\int_\tau^t r(\rho(x,u))du}\phi(t-\tau)d\tau. \quad (6.44)$$

It follows from Eqs. (6.42) and (6.44) that

$$j(x,p,t) = \int_{\mathbb{R}} h(x-z,p,t)w(z)dz. \quad (6.45)$$

In order to express $h(x,p,t)$ in terms of $\rho(x,p,t)$, we divide Eqs. (6.41) and (6.44) by the same factor $e^{ip\int_0^t U(x,t')dt'}e^{\int_0^t r(\rho(x,u))du}$ and then take the Laplace transform $t \to s$ on both equations. Consequently, we have

$$\mathcal{L}\{\rho(x,p,t)e^{-ip\int_0^t U(x,t')dt'}e^{-\int_0^t r(\rho(x,u))du}\}$$
$$= [\rho_0(x) + \mathcal{L}\{j(x,p,t)e^{-ip\int_0^t U(x,t')dt'}e^{-\int_0^t r(\rho(x,u))du}\}]\hat{W}(s), \quad (6.46)$$

and

$$\mathcal{L}\{h(x,p,t)e^{-ip\int_0^t U(x,t')dt'}e^{-\int_0^t r(\rho(x,u))du}\}$$
$$= [\rho_0(x) + \mathcal{L}\{j(x,p,t)e^{-ip\int_0^t U(x,t')dt'}e^{-\int_0^t r(\rho(x,u))du}\}]\hat{\phi}(s). \quad (6.47)$$

From Eqs. (6.46) and (6.47), we obtain

$$\mathcal{L}\{h(x,p,t)e^{-ip\int_0^t U(x,t')dt'}e^{-\int_0^t r(\rho(x,u))du}\}$$
$$= \frac{\hat{\phi}(s)}{\hat{W}(s)}\mathcal{L}\{\rho(x,p,t)e^{-ip\int_0^t U(x,t')dt'}e^{-\int_0^t r(\rho(x,u))du}\}. \quad (6.48)$$

Performing inverse Laplace transform of Eq. (6.48) gives

$$h(x,p,t) = \int_0^t K(t-\tau)\rho(x,p,\tau)e^{ip\int_\tau^t U(x,t')dt'}e^{\int_\tau^t r(\rho(x,u))du}d\tau, \quad (6.49)$$

where $K(t)$ is the memory kernel defined by its Laplace transform [Fedotov (2010)]

$$\hat{K}(s) = \frac{\hat{\phi}(s)}{\hat{W}(s)} = \frac{s\hat{\phi}(s)}{1-\hat{\phi}(s)}. \quad (6.50)$$

Substituting Eq. (6.49) into Eqs. (6.43) and (6.45) gives the general forward Feynman-Kac equation with nonlinear reaction rate $r(\rho(x,t))$, i.e.,

$$\frac{\partial \rho(x,p,t)}{\partial t} = -\int_0^t K(t-\tau)\rho(x,p,\tau)e^{ip\int_\tau^t U(x,t')dt'}e^{\int_\tau^t r(\rho(x,u))du}d\tau$$

$$+ \int_{\mathbb{R}}\int_0^t K(t-\tau)\rho(x-z,p,\tau)e^{ip\int_\tau^t U(x-z,t')dt'}e^{\int_\tau^t r(\rho(x-z,u))du}w(z)d\tau dz$$

$$+ [ipU(x,t) + r(\rho(x,t))]\rho(x,p,t),$$

(6.51)

where $K(t)$ is the memory kernel defined by its Laplace transform in Eq. (6.50). This generalized Feynman-Kac equation has few restrictions on the forms of the waiting time and jump length densities, which allows us to conveniently model standard transport, anomalous diffusion or tempered dynamics together with nonlinear reaction terms. Obviously, plugging $r(\rho(x,t)) \equiv 0$ into Eq. (6.51) gives the forward Feynman-Kac equation for the diffusion process without reaction. To avoid confusion, we denote this equation in terms of $n(x,p,t)$, namely,

$$\frac{\partial n(x,p,t)}{\partial t} = \int_{\mathbb{R}}\int_0^t K(t-\tau)n(x-z,p,\tau)e^{ip\int_\tau^t U(x-z,t')dt'}w(z)d\tau dz$$

$$- \int_0^t K(t-\tau)n(x,p,\tau)e^{ip\int_\tau^t U(x,t')dt'}d\tau + ipU(x,t)n(x,p,t).$$

(6.52)

Assume that the symmetric jump length distribution $w(z)$ satisfies

$$\int_{\mathbb{R}} z^2 w(z)dz = \sigma^2 < +\infty. \tag{6.53}$$

Expanding the first term on the right hand side of Eq. (6.51) for small z and truncating the Taylor series at the second order give

$$\frac{\partial \rho(x,p,t)}{\partial t} = \frac{\sigma^2}{2}\frac{\partial^2}{\partial x^2}\int_0^t K(t-\tau)\rho(x,p,\tau)e^{ip\int_\tau^t U(x,t')dt'}e^{\int_\tau^t r(\rho(x,u))du}d\tau$$

$$+ [ipU(x,t) + r(\rho(x,t))]\rho(x,p,t).$$

(6.54)

Now we consider several special cases of Eqs. (6.51) and (6.54). To start with, if $p = 0$, then $\rho(x,0,t) = \int_{-\infty}^{+\infty} e^{ipA}\rho(x,A,t)dA|_{p=0} = \rho(x,t)$. Thus, when $p = 0$, Eqs. (6.51) and (6.54) reduce to Eqs. (14) and (22) in [Fedotov (2010)] respectively, which describe the nonlinear master equations for reaction diffusion processes. Furthermore, we can specify the waiting time distributions and write down the corresponding Feynman-Kac equations in the continuum limit of large scales and long time as follows:

(1) Exponential waiting time distribution and Gaussian jump length distribution. Suppose $\phi(t) = \frac{1}{\tau}\exp(-\frac{t}{\tau})$. Thus, $\hat{\phi}(s) = \int_0^{+\infty} e^{-st}\phi(t)dt = \frac{1}{\tau s+1}$ and $\hat{K}(s) = \frac{s\hat{\phi}(s)}{1-\hat{\phi}(s)} = \frac{1}{\tau}$. Inserting $K(t) = \frac{1}{\tau}\delta(t)$ into Eq. (6.54) gives

$$\frac{\partial\rho(x,p,t)}{\partial t} = \frac{\sigma^2}{2\tau}\frac{\partial^2\rho(x,p,t)}{\partial x^2} + ipU(x,t)\rho(x,p,t) \qquad (6.55)$$
$$+ r(\rho(x,t))\rho(x,p,t),$$

which is the standard Feynman-Kac equation for the normal diffusion (Markovian) process with the reaction rate $r(\rho(x,t))$.

(2) Power law waiting time distribution and Gaussian jump length distribution. When $\phi(t) = \alpha\tau^\alpha t^{-1-\alpha}\mathbf{1}_{[\tau,+\infty)}(t)$ with $0 < \alpha < 1$, we can calculate $\hat{\phi}(s)$, which has the asymptotic form $\hat{\phi}(s) \sim 1 - \tau^\alpha\Gamma(1-\alpha)s^\alpha$ for $s \to 0$. Meanwhile, according to Eq. (6.50), $\hat{K}(s) \approx \frac{s^{1-\alpha}}{\tau^\alpha\Gamma(1-\alpha)}$. Thus, Eq. (6.54) reduces to

$$\frac{\partial\rho(x,p,t)}{\partial t} = \left[ipU(x,t) + r(\rho(x,t))\right]\rho(x,p,t)$$
$$+ \frac{\sigma^2}{2\tau^\alpha\Gamma(1-\alpha)}\frac{\partial^2}{\partial x^2}\left[e^{ip\int_0^t U(x,t')dt'}e^{\int_0^t r(\rho(x,u))du}\mathcal{D}_t^{1-\alpha}\right. \qquad (6.56)$$
$$\left.\left\{\rho(x,p,t)e^{-ip\int_0^t U(x,t')dt'}e^{-\int_0^t r(\rho(x,u))du}\right\}\right],$$

where $\mathcal{D}_t^{1-\alpha}$ is the Riemann-Liouville fractional derivative.

(3) Tempered power law waiting time distribution and Gaussian jump length distribution. Assume $\phi(t) = C_\tau^{-1}e^{-\lambda t}t^{-1-\alpha}\mathbf{1}_{[\tau,+\infty)}(t)$, where $0 < \alpha < 1$ and $C_\tau = \int_\tau^{+\infty} e^{-\lambda t}t^{-1-\alpha}dt$. Then it can be calculated that $\hat{\phi}(s)$ has the asymptotic form $\hat{\phi}(s) \sim 1 - \frac{\Gamma(1-\alpha)}{\alpha C_\tau}(s+\lambda)^\alpha + \frac{\Gamma(1-\alpha)}{\alpha C_\tau}\lambda^\alpha$ for $s \to 0$. Similarly, $\hat{K}(s) \approx \frac{\alpha C_\tau}{\Gamma(1-\alpha)}\frac{s}{(s+\lambda)^\alpha-\lambda^\alpha}$. Under this condition, Eq. (6.54) can be rewritten as

$$\frac{\partial\rho(x,p,t)}{\partial t} = [ipU(x,t) + r(\rho(x,t))]\rho(x,p,t)$$
$$+ \frac{\sigma^2\alpha C_\tau}{2\Gamma(1-\alpha)}\frac{\partial^2}{\partial x^2}\left[e^{ip\int_0^t U(x,t')dt'}e^{\int_0^t r(\rho(x,u))du}\right. \qquad (6.57)$$
$$\left.\frac{\partial}{\partial t}\int_0^t H(t-\tau)\rho(x,p,\tau)e^{-ip\int_0^\tau U(x,t')dt'}e^{-\int_0^\tau r(\rho(x,u))du}d\tau\right],$$

where $H(t) = e^{-\lambda t}t^{\alpha-1}E_{\alpha,\alpha}(\lambda^\alpha t^\alpha)$ and its Laplace transform $\hat{H}(s) = \frac{1}{(s+\lambda)^\alpha-\lambda^\alpha}$. Here $E_{\alpha,\alpha}(\cdot)$ is the two-parameter Mittag-Leffler function [Podlubny (1999)], which is defined as $E_{\alpha,\beta}(z) = \sum_{k=0}^\infty \frac{z^k}{\Gamma(\alpha k+\beta)}$ with $\alpha > 0$ and $\beta > 0$.

6.2.2 Forward Feynman-Kac Equations for Nonlinear Reaction Rate $r(t)$

Following the similar derivation procedure in the previous subsection, we obtain the forward Feynman-Kac equation with linear reaction rate $r(t)$, i.e.,

$$\frac{\partial \rho(x, p, t)}{\partial t} =$$

$$\int_{\mathbb{R}} \int_0^t K(t - \tau)\rho(x - z, p, \tau)e^{ip\int_\tau^t U(x-z,t')dt'}e^{\int_\tau^t r(u)du}w(z)d\tau dz -$$

$$\int_0^t K(t - \tau)\rho(x, p, \tau)e^{ip\int_\tau^t U(x,t')dt'}e^{\int_\tau^t r(u)du}d\tau + [ipU(x, t) + r(t)]\rho(x, p, t),$$

$$(6.58)$$

where $K(t)$ is defined by Eq. (6.50). By simple calculations, it can be easily checked that $\rho(x, p, t) = n(x, p, t)e^{\int_0^t r(u)du}$ satisfies Eq. (6.58) if $n(x, p, t)$ is the solution of Eq. (6.52), which means that for the reaction diffusion process with linear reaction rate $r(t)$, the effect of transport with memory and the linear reaction rate depending on time t can be decoupled. In fact, this also implies that the conclusion in [Fedotov (2010)] for the case of constant reaction rate still holds for the linear reaction rate, and even for describing the distribution of functionals.

6.2.3 Forward Feynman-Kac Equations for Nonlinear Reaction Rate $r(x)$

We further consider the linear reaction rate $r(x)$ and the traditional functional definition $A = \int_0^t U(x(\tau))d\tau$, in order to give another derivation of the corresponding forward Feynman-Kac equation. To start with, under these conditions, we have

$$\rho(x, A, t) = \rho_0(x)\delta(A - tU(x))e^{r(x)t}W(t)$$

$$+ \int_0^t j(x, A - (t - \tau)U(x), \tau)e^{(t-\tau)r(x)}W(t - \tau)d\tau,$$

$$(6.59)$$

and

$$j(x, A, t) = \int_{\mathbb{R}} \rho_0(x - z)\delta(A - tU(x - z))e^{r(x-z)t}w(z)\phi(t)dz$$

$$+ \int_0^t \int_{\mathbb{R}} j(x - z, A - (t - \tau)U(x - z), \tau)e^{(t-\tau)r(x-z)}w(z)\phi(t - \tau)dzd\tau.$$

$$(6.60)$$

Performing Fourier transform $A \to p$ on both sides of Eqs. (6.59) and (6.60) gives that

$$\rho(x,p,t) = \rho_0(x)e^{iptU(x)}e^{r(x)t}W(t)$$
$$+ \int_0^t j(x,p,\tau)e^{ip(t-\tau)U(x)}e^{(t-\tau)r(x)}W(t-\tau)d\tau, \qquad (6.61)$$

and

$$j(x,p,t) = \int_{\mathbb{R}} \rho_0(x-z)e^{iptU(x-z)}e^{r(x-z)t}w(z)\phi(t)dz$$
$$+ \int_0^t \int_{\mathbb{R}} j(x-z,p,\tau)e^{ip(t-\tau)U(x-z)}e^{(t-\tau)r(x-z)}w(z)\phi(t-\tau)dzd\tau. \qquad (6.62)$$

Conducting Laplace transform $t \to s$ on Eqs. (6.61) and (6.62), respectively, we obtain

$$\hat{\rho}(x,p,s) = \rho_0(x)\hat{W}(s-ipU(x)-r(x)) + \hat{j}(x,p,s)\hat{W}(s-ipU(x)-r(x)), \qquad (6.63)$$

and

$$\hat{j}(x,p,s) = \int_{\mathbb{R}} \rho_0(x-z)w(z)\hat{\phi}(s-ipU(x-z)-r(x-z))dz$$
$$+ \int_{\mathbb{R}} \hat{j}(x-z,p,s)w(z)\hat{\phi}(s-ipU(x-z)-r(x-z))dz. \qquad (6.64)$$

Performing Fourier transform $x \to k$ on Eqs. (6.63) and (6.64) gives

$$\hat{\tilde{\rho}}(k,p,s) = \hat{\tilde{W}}\left(s-ipU\left(-i\frac{\partial}{\partial k}\right) - r\left(-i\frac{\partial}{\partial k}\right)\right)[\tilde{\rho}_0(k)+\hat{\tilde{j}}(k,p,s)], \qquad (6.65)$$

and

$$\hat{\tilde{j}}(k,p,s) = \tilde{w}(k)\hat{\tilde{\phi}}\left(s-ipU\left(-i\frac{\partial}{\partial k}\right) - r\left(-i\frac{\partial}{\partial k}\right)\right)[\tilde{\rho}_0(k)+\hat{\tilde{j}}(k,p,s)], \qquad (6.66)$$

since we have the identity [Carmi *et al.* (2010)] $\mathcal{F}\{xf(x)\} = -i\frac{\partial}{\partial k}\tilde{f}(k)$. Substituting $\hat{\tilde{j}}(k,p,s) = \frac{\tilde{w}(k)\hat{\tilde{\phi}}\left(s-ipU\left(-i\frac{\partial}{\partial k}\right)-r\left(-i\frac{\partial}{\partial k}\right)\right)\tilde{\rho}_0(k)}{1-\tilde{w}(k)\hat{\tilde{\phi}}\left(s-ipU\left(-i\frac{\partial}{\partial k}\right)-r\left(-i\frac{\partial}{\partial k}\right)\right)}$ into Eq. (6.65), we get

$$\hat{\tilde{\rho}}(k,p,s) = \frac{\hat{\tilde{\Psi}}\left(s-ipU\left(-i\frac{\partial}{\partial k}\right)-r\left(-i\frac{\partial}{\partial k}\right)\right)\tilde{\rho}_0(k)}{1-\tilde{w}(k)\hat{\tilde{\phi}}\left(s-ipU\left(-i\frac{\partial}{\partial k}\right)-r\left(-i\frac{\partial}{\partial k}\right)\right)}. \qquad (6.67)$$

Furthermore, we specify three kinds of typical waiting time and jump length distributions and then calculate their asymptotic forms in the Laplace or Fourier domain as $s \to 0$ or $k \to 0$. Naturally, the inverse Fourier-Laplace transform technique is applied to write down the respective forward Feynman-Kac equations in the long time and large scale limits.

(1) Exponential waiting time and Gaussian jump length distributions. Suppose $\phi(t) = \frac{1}{\tau}\exp(-\frac{t}{\tau})$ and $w(x) = \frac{1}{\sqrt{2\pi\sigma^2}}e^{-\frac{x^2}{2\sigma^2}}$. Then $\hat{\phi}(s) = \frac{1}{1+\tau s} \sim 1 - \tau s$ and $\tilde{w}(k) = \int_{-\infty}^{+\infty}e^{ikx}w(x)dx = e^{-\frac{1}{2}\sigma^2 k^2} \sim 1 - \frac{\sigma^2 k^2}{2}$. Substituting the asymptotic forms of $\hat{\phi}(s)$ and $\tilde{w}(k)$ into Eq. (6.67) and conducting inverse Fourier-Laplace transform, we have

$$\frac{\partial\rho(x,p,t)}{\partial t} = \frac{\sigma^2}{2\tau}\frac{\partial^2\rho(x,p,t)}{\partial x^2} + [ipU(x) + r(x)]\rho(x,p,t). \qquad (6.68)$$

(2) Tempered power law waiting time and Gaussian jump length distributions. Plugging $\hat{\phi}(s) \sim 1 - \frac{\Gamma(1-\alpha)}{\alpha C_\tau}(s+\lambda)^\alpha + \frac{\Gamma(1-\alpha)}{\alpha C_\tau}\lambda^\alpha$ and $\tilde{w}(k) \sim 1 - \frac{\sigma^2 k^2}{2}$ into Eq. (6.67) and performing inverse Laplace-Fourier transform give that

$$\frac{\partial\rho(x,p,t)}{\partial t} = [r(x) + ipU(x)]\rho(x,p,t)$$
$$+ \frac{\alpha C_\tau\sigma^2}{2\Gamma(1-\alpha)}\frac{\partial^2}{\partial x^2}\left(\frac{\partial}{\partial t} - r(x) - ipU(x)\right)\int_0^t H(t-\tau;x)\rho(x,p,\tau)d\tau,$$

$$(6.69)$$

where $H(t;x)$ is defined by its Laplace transform

$$\hat{H}(s;x) = \frac{1}{(s+\lambda - r(x) - ipU(x))^\alpha - \lambda^\alpha}. \qquad (6.70)$$

The equation can also be written as

$$\frac{\partial\rho(x,p,t)}{\partial t} = \frac{\alpha C_\tau\sigma^2}{2\Gamma(1-\alpha)}\frac{\partial^2}{\partial x^2}D_t^{1-\alpha,\lambda,x}\rho(x,p,t) + [r(x) + ipU(x)]\rho(x,p,t)$$
$$+ [\lambda^\alpha D_t^{1-\alpha,\lambda,x} - \lambda][\rho(x,p,t) - e^{iptU(x)}e^{r(x)t}\rho_0(x)],$$

$$(6.71)$$

where the tempered fractional substantial derivative [Friedrich *et al.* (2006)] $D_t^{1-\alpha,\lambda,x}$ is defined in the Laplace domain as

$$\mathcal{L}\{D_t^{1-\alpha,\lambda,x}\rho(x,p,t)\} = (s+\lambda - r(x) - ipU(x))^{1-\alpha}\rho(x,p,s). \qquad (6.72)$$

And in the time domain,

$$D_t^{1-\alpha,\lambda,x}\rho(x,p,t)$$
$$= \frac{1}{\Gamma(\alpha)}\left[\frac{\partial}{\partial t} + \lambda - r(x) - ipU(x)\right]\int_0^t \frac{e^{(t-\tau)(r(x)+ipU(x)-\lambda)}}{(t-\tau)^{1-\alpha}}\rho(x,p,\tau)d\tau,$$

$$(6.73)$$

which is equivalent to

$$D_t^{1-\alpha,\lambda,x}\rho(x,p,t)$$
$$= \frac{e^{iptU(x)+r(x)t-\lambda t}}{\Gamma(\alpha)} \frac{\partial}{\partial t} \int_0^t \frac{e^{\lambda\tau-ip\tau U(x)-r(x)\tau}\rho(x,p,\tau)}{(t-\tau)^{1-\alpha}}d\tau. \tag{6.74}$$

It should be noted that Eqs. (6.69) and (6.71) are the generalizations of Eqs. (11) and (5) in [Wu *et al.* (2016)], respectively, to the reaction diffusion cases.

(3) Tempered power law waiting time and jump length distributions. In this case, we assume that the waiting time and jump length obey different tempered power law distributions. Let $w(x) = C_\varepsilon^{-1}e^{-\gamma|x|}|x|^{-1-\beta}\mathbf{1}_{[\varepsilon,+\infty)}(|x|)$ for $0 < \beta < 2$. The normalization factor C_ε is defined as $C_\varepsilon = \int_\varepsilon^{+\infty} e^{-\gamma x}x^{-1-\beta}dx + \int_{-\infty}^{-\varepsilon} e^{-\gamma|x|}|x|^{-1-\beta}dx$ to make sure $\int_{-\infty}^{+\infty} w(x)dx = 1$. Then it could be calculated that $\tilde{w}(k)$ has the asymptotic form $\tilde{w}(k) \sim 1 - \frac{2\Gamma(1-\beta)}{\beta C_\varepsilon}(k^2 + \gamma^2)^{\beta/2} + \frac{2\Gamma(1-\beta)}{\beta C_\varepsilon}\gamma^\beta$ for $k \to 0$. Substituting the asymptotic $\hat{\phi}(s)$ and $\tilde{w}(k)$ into Eq. (6.67) and conducting inverse Laplace-Fourier transform, we obtain

$$\frac{\partial\rho(x,p,t)}{\partial t} = [r(x) + ipU(x)]\rho(x,p,t)$$
$$+ \frac{2\alpha C_\tau \Gamma(1-\beta)}{\beta C_\varepsilon \Gamma(1-\alpha)}(\nabla_x^{\beta,\gamma} + \gamma^\beta)\left(\frac{\partial}{\partial t} - r(x) - ipU(x)\right)$$
$$\times \int_0^t H(t-\tau;x)\rho(x,p,\tau)d\tau, \tag{6.75}$$

where the tempered fractional Riesz derivative $\nabla_x^{\beta,\gamma}$ (see [Wu *et al.* (2016)] for more details) is defined in the Fourier domain as $\mathcal{F}\{\nabla_x^{\beta,\gamma}\rho(x,p,t)\} = -(k^2 + \gamma^2)^{\beta/2}\rho(k,p,t)$. Equivalently,

$$\frac{\partial\rho(x,p,t)}{\partial t} = \frac{2\alpha C_\tau \Gamma(1-\beta)}{\beta C_\varepsilon \Gamma(1-\alpha)}(\nabla_x^{\beta,\gamma} + \gamma^\beta)D_t^{1-\alpha,\lambda,x}\rho(x,p,t)$$
$$+ [\lambda^\alpha D_t^{1-\alpha,\lambda,x} - \lambda][\rho(x,p,t) - e^{iptU(x)}e^{r(x)t}\rho_0(x)] \tag{6.76}$$
$$+ [r(x) + ipU(x)]\rho(x,p,t),$$

where the operator $D_t^{1-\alpha,\lambda,x}$ in defined in Eq. (6.73) or Eq. (6.74). When $r(x) \equiv 0$, Eq. (6.76) reduces to Eq. (36) in [Wu *et al.* (2016)].

6.2.4 Derivation of Backward Feynman-Kac Equations

Let $\rho_{x_0}(A, t)$ be the PDF of the functional A at time t with the initial position x_0. The backward Feynman-Kac equation regarding $\rho_{x_0}(A, t)$ with the nonlinear reaction rate $r(\rho(x, t))$ can be written as follows

$$\rho_{x_0}(A, t) = \delta\left(A - \int_0^t U(x_0, t')dt'\right) e^{\int_0^t r(\rho(x_0, u))du} W(t)$$

$$+ \int_{\mathbb{R}} \int_0^t w(z)\phi(\tau)\rho_{x_0+z}\left(A - \int_0^\tau U(x_0, t')dt', t - \tau\right) e^{\int_0^\tau r(\rho(x_0, u))du} d\tau dz.$$

$$(6.77)$$

It should be noted that $\int_{-\infty}^{+\infty} \rho_{x_0}(A, t)dA = 1$. The first term on the right hand side of Eq. (6.77) indicates the motionless particles remaining at their initial position x_0 up to time t and reacting with the nonlinear rate $r(\rho(x_0, t))$ during the time interval $(0, t)$. Thus, the factors $e^{\int_0^t r(\rho(x_0, u))du}$ and $W(t)$ are multiplied respectively. The corresponding functional value $A = \int_0^t U(x_0, t')dt'$, since the particles stay at the initial position x_0 in the time interval $(0, t)$. The second term on the right hand side of Eq. (6.77) alternatively represents the particles that conduct their first jump to the location $x_0 + z$ at time τ ($\tau < t$) and meanwhile evolve according to the reaction rate $r(\rho(x_0, t))$ during the time interval $(0, \tau)$. Therefore, the jump length distribution $w(z)$, the waiting time distribution $\phi(\tau)$ and the factor $e^{\int_0^\tau r(\rho(x_0, u))du}$ are multiplied respectively. Next the particles proceed from the new initial position $x_0 + z$ and at time $t - \tau$ their functional value is $A - \int_0^\tau U(x_0, t')dt'$, since they stay at the position x_0 during the time interval $(0, \tau)$. Due to the arbitrary z and τ, the double integrals of z and τ are introduced. Similarly, conducting the Fourier transform $A \to p$, we obtain

$$\rho_{x_0}(p, t) = e^{ip \int_0^t U(x_0, t')dt'} e^{\int_0^t r(\rho(x_0, u))du} W(t)$$

$$+ \int_{\mathbb{R}} \int_0^t w(z)\phi(\tau)\rho_{x_0+z}(p, t - \tau)e^{ip \int_0^\tau U(x_0, t')dt'} e^{\int_0^\tau r(\rho(x_0, u))du} d\tau dz.$$

$$(6.78)$$

In what follows, we consider a special case of Eq. (6.78), which supposes the reaction rate is $r(x)$ and the functional is defined as $A = \int_0^t U[x(t')]dt'$. Thus, under these assumptions, Eq. (6.78) can be rewritten as

$$\rho_{x_0}(p, t) = e^{iptU(x_0)} e^{r(x_0)t} W(t)$$

$$+ \int_{\mathbb{R}} \int_0^t w(z)\phi(\tau)\rho_{x_0+z}(p, t - \tau)e^{ip\tau U(x_0)} e^{r(x_0)\tau} d\tau dz.$$

$$(6.79)$$

Consequently, the standard Laplace-Fourier transform technique is applicable to further simplify the concerned equation. Conducting the Laplace transform $t \to s$ and the Fourier transform $x_0 \to k$ of Eq. (6.79), we get

$$\hat{\rho}_k(p,s) = \frac{\hat{\tilde{W}}\left(s - r\left(-i\frac{\partial}{\partial k}\right) - ipU\left(-i\frac{\partial}{\partial k}\right)\right)\tilde{\delta}(k)}{1 - \hat{\tilde{\phi}}\left(s - r\left(-i\frac{\partial}{\partial k}\right) - ipU\left(-i\frac{\partial}{\partial k}\right)\right)\tilde{w}(k)}. \qquad (6.80)$$

Here we list three kinds of typical waiting time and jump length distributions and write down the respective backward equations in the continuum limit. We omit the power law waiting time distribution cases, since they could be recovered when the exponential tempering exponents are set to zeros.

(1) Exponential waiting time and Gaussian jump length distributions. Plugging $\hat{\phi}(s) \sim 1 - \tau s$ and $\tilde{w}(k) \sim 1 - \frac{\sigma^2 k^2}{2}$ into Eq. (6.80) and performing inverse Laplace-Fourier transform give that

$$\frac{\partial \rho_{x_0}(p,t)}{\partial t} = \frac{\sigma^2}{2\tau}\frac{\partial^2 \rho_{x_0}(p,t)}{\partial x_0^2} + [r(x_0) + ipU(x_0)]\rho_{x_0}(p,t). \qquad (6.81)$$

(2) Tempered power law waiting time and Gaussian jump length distributions. With the substitution of $\hat{\phi}(s) \sim 1 - \frac{\Gamma(1-\alpha)}{\alpha C_\tau}(s+\lambda)^\alpha + \frac{\Gamma(1-\alpha)}{\alpha C_\tau}\lambda^\alpha$ and $\tilde{w}(k) \sim 1 - \frac{\sigma^2 k^2}{2}$ into Eq. (6.80) and inverse Laplace-Fourier transform, we obtain

$$\frac{\partial \rho_{x_0}(p,t)}{\partial t} = [r(x_0) + ipU(x_0)]\rho_{x_0}(p,t)$$
$$+ \frac{\alpha C_\tau \sigma^2}{2\Gamma(1-\alpha)}\left(\frac{\partial}{\partial t} - r(x_0) - ipU(x_0)\right)\int_0^t H(t-\tau;x_0)\frac{\partial^2 \rho_{x_0}(p,\tau)}{\partial x_0^2}d\tau, \qquad (6.82)$$

where $H(t;x_0)$ is defined by its Laplace transform

$$\hat{H}(s;x_0) = \frac{1}{(s+\lambda - r(x_0) - ipU(x_0))^\alpha - \lambda^\alpha}. \qquad (6.83)$$

Equivalently,

$$\frac{\partial \rho_{x_0}(p,t)}{\partial t} = \frac{\alpha C_\tau \sigma^2}{2\Gamma(1-\alpha)}D_t^{1-\alpha,\lambda,x_0}\frac{\partial^2 \rho_{x_0}(p,t)}{\partial x_0^2} + [r(x_0) + ipU(x_0)]\rho_{x_0}(p,t)$$
$$+ [\lambda^\alpha D_t^{1-\alpha,\lambda,x_0} - \lambda][\rho_{x_0}(p,t) - e^{iptU(x_0)}e^{r(x_0)t}]. \qquad (6.84)$$

(3) Tempered power law waiting time and jump length distributions. Substituting $\hat{\phi}(s) \sim 1 - \frac{\Gamma(1-\alpha)}{\alpha C_\tau}(s+\lambda)^\alpha + \frac{\Gamma(1-\alpha)}{\alpha C_\tau}\lambda^\alpha$ and $\tilde{w}(k) \sim 1 - \frac{2\Gamma(1-\beta)}{\beta C_\varepsilon}(k^2+\gamma^2)^{\beta/2} + \frac{2\Gamma(1-\beta)}{\beta C_\varepsilon}\gamma^\beta$ into Eq. (6.80) and doing inverse Laplace-Fourier transform lead to

$$\frac{\partial \rho_{x_0}(p,t)}{\partial t} = [r(x_0) + ipU(x_0)]\rho_{x_0}(p,t)$$
$$+ \frac{2\alpha C_\tau \Gamma(1-\beta)}{\beta C_\varepsilon \Gamma(1-\alpha)}\left(\frac{\partial}{\partial t} - r(x_0) - ipU(x_0)\right) \qquad (6.85)$$
$$\int_0^t H(t-\tau;x_0)(\nabla_{x_0}^{\beta,\gamma} + \gamma^\beta)\rho_{x_0}(p,\tau)d\tau,$$

where the operator $\nabla_{x_0}^{\beta,\gamma}$ is defined in the Fourier domain as

$$\mathcal{F}\{\nabla_{x_0}^{\beta,\gamma}\rho_{x_0}(p,t)\} = -(k^2+\gamma^2)^{\beta/2}\rho_k(p,t). \qquad (6.86)$$

Equivalently,

$$\frac{\partial \rho_{x_0}(p,t)}{\partial t} = \frac{2\alpha C_\tau \Gamma(1-\beta)}{\beta C_\varepsilon \Gamma(1-\alpha)}D_t^{1-\alpha,\lambda,x_0}(\nabla_{x_0}^{\beta,\gamma} + \gamma^\beta)\rho_{x_0}(p,t)$$
$$+ [\lambda^\alpha D_t^{1-\alpha,\lambda,x_0} - \lambda][\rho_{x_0}(p,t) - e^{iptU(x_0)}e^{r(x_0)t}] \qquad (6.87)$$
$$+ [r(x_0) + ipU(x_0)]\rho_{x_0}(p,t).$$

6.2.5 *Distribution of Occupation Time in Half Space and its Fluctuations*

In this subsection and what follows, we present the distributions of specific functionals of the paths of particles performing temporal-tempered anomalous dynamics with piecewise constant reaction rate. We analytically solve the corresponding backward Feynman-Kac equations in order to obtain the distributions, the moments, and other properties of the functionals. As pointed out in [Wu et al. (2016)], our analysis is mainly based on the derived backward Feynman-Kac equations, since here we are more concerned with the functional distributions than the particles' positions. From this perspective, backward equations are more convenient, although the forward equations could lead to the same conclusions with the extra integration over x step.

In what follows, we assume the reaction rate function $r(x)$ satisfying

$$r(x) = \begin{cases} \kappa_1, & x > 0; \\ \kappa_2, & x < 0, \end{cases}$$

where κ_1 and κ_2 are negative constants. That is to say, the reaction rate varies with the particle's position. Possibly due to temperature or pressure difference, the reaction rate is κ_1 in the positive half space and changes to κ_2 in the negative half space. Here we confine our analysis to negative reaction rates in order to model the spontaneous evanescent process in which the particles are destroyed or removed at different constant rates depending on their positions. As for the reproduction process with positive reaction rates, one must specify the rules regarding the waiting time of the newborn particles [Fedotov (2010); Abad *et al.* (2010)], which is a problem to be explored in the future.

Define the occupation time of a particle in the positive half space as $T^+ = \int_0^t \theta(x(\tau))d\tau$, where $\theta(x) = 1$ for $x \geq 0$ and $\theta(x) = 0$ for $x < 0$. Since obviously $T^+ \geq 0$, we rely on the Laplace transform $T^+ \to p$, instead of the Fourier transform. In order to find the PDF of T^+, we consider the backward equation, conduct Laplace transform $t \to s$, and substitute the assumed $\theta(x_0)$ and $r(x_0)$. Using the notation $K_\alpha = \frac{\alpha C_\tau \sigma^2}{2\Gamma(1-\alpha)}$, we arrive at

$$\hat{\rho}_{x_0}(p,s) = \begin{cases} \dfrac{K_\alpha}{(s+\lambda+p-\kappa_1)^\alpha - \lambda^\alpha} \dfrac{\partial^2 \hat{\rho}_{x_0}(p,s)}{\partial x_0^2} + \dfrac{1}{s+p-\kappa_1}, & x_0 > 0; \\[3mm] \dfrac{K_\alpha}{(s+\lambda-\kappa_2)^\alpha - \lambda^\alpha} \dfrac{\partial^2 \hat{\rho}_{x_0}(p,s)}{\partial x_0^2} + \dfrac{1}{s-\kappa_2}, & x_0 < 0. \end{cases} \tag{6.88}$$

Suppose $\hat{\rho}_{x_0}(p,s) < \infty$ for $|x_0| \to \infty$, $\hat{\rho}_{x_0}(p,s)$ and its first derivative are continuous at $x_0 = 0$ and the particle starts from $x_0 = 0$. With the notations $C_1(p,s) = [(s+\lambda+p-\kappa_1)^\alpha - \lambda^\alpha]^{1/2}$ and $C_2(s) = [(s+\lambda-\kappa_2)^\alpha - \lambda^\alpha]^{1/2}$, we have

$$\hat{\rho}_0(p,s) = \frac{(s-\kappa_2)C_1(p,s) + (s+p-\kappa_1)C_2(s)}{(s+p-\kappa_1)(s-\kappa_2)[C_1(p,s) + C_2(s)]}, \tag{6.89}$$

which describes the PDF of T^+ in the Laplace domain ($T^+ \to p$ and $t \to s$). When both κ_1 and κ_2 are set to zeros in Eq. (6.89), Eq. (47) in [Wu *et al.* (2016)] is recovered, describing the distribution of T^+ for tempered anomalous motions without reactions. However, it seems difficult to invert Eq. (6.89) analytically, even for the $\kappa_1 = \kappa_2 = 0$ case as reported in [Wu *et al.* (2016)]. As shown in [Carmi *et al.* (2010)], the occupation fraction T^+/t obeys Lamperti distribution for anomalous motions without exponential tempering ($\lambda = 0$) and reactions ($\kappa_1 = \kappa_2 = 0$). Specially, when $\alpha = 1$, $\hat{\rho}_0(p,s) = (s-\kappa_2)^{-1/2}(s+p-\kappa_1)^{-1/2}$, apparently different from the arcsine law of the occupation fraction for Brownian motion [Majumdar (2005)].

Furthermore, in order to evaluate the expectation of the occupation time, we calculate the first moment of T^+ in the Laplace domain ($t \to s$)

from Eq. (6.89)

$$\langle T^+ \rangle_s = \frac{C_1(s)}{(s-\kappa_1)^2[C_1(s)+C_2(s)]}$$
$$+ \frac{\alpha(\kappa_2-\kappa_1)(s+\lambda-\kappa_1)^{\alpha-1}C_2(s)}{2(s-\kappa_1)(s-\kappa_2)C_1(s)[C_1(s)+C_2(s)]^2},$$

(6.90)

where we denote $C_1(p=0,s)$ as $C_1(s)$ for simplicity. When $\kappa_1 = \kappa_2$ in Eq. (6.90), $\langle T^+ \rangle_s = 1/2(s-\kappa_1)^2$, and correspondingly in the time domain $\langle T^+ \rangle = te^{\kappa_1 t}/2$, or $\langle T^+/t \rangle = e^{\kappa_1 t}/2$. Here $e^{\kappa_1 t}$ can be interpreted as the survival probability. This result is different from the previous studies, for example [Wu *et al.* (2016)], which presents $\langle T^+ \rangle = t/2$, or $\langle T^+/t \rangle = 1/2$ for (tempered) anomalous diffusion processes.

We calculate the second moment of T^+ in the Laplace domain $(t \to s)$ to measure the fluctuation of the occupation time as follows

$$\langle (T^+)^2 \rangle_s = \frac{2C_1(s)}{(s-\kappa_1)^3[C_1(s)+C_1(s)]} - \frac{\alpha(s+\lambda-\kappa_1)^{\alpha-1}C_2(s)}{(s-\kappa_1)^2C_1(s)[C_1(s)+C_2(s)]^2}$$
$$+ \frac{\alpha(\kappa_2-\kappa_1)(s+\lambda-\kappa_1)^{\alpha-2}C_2(s)\left[B_1(s)C_1(s)+B_2(s)C_2(s)\right]}{4(s-\kappa_1)(s-\kappa_2)\left[C_1(s)(C_1(s)+C_1(s))\right]^3}$$

(6.91)

with $B_1(s) = 2(\alpha-1)\lambda^\alpha + (2+\alpha)(s+\lambda-\kappa_1)^\alpha$ and $B_2(s) = 2(\alpha-1)\lambda^\alpha + (2-\alpha)(s+\lambda-\kappa_1)^\alpha$. Substitution of $\kappa_1 = \kappa_2 = 0$ into Eq. (6.91) gives the following special case of our derivation

$$\langle (T^+)^2 \rangle_s = \frac{1}{s^3} - \frac{\alpha(s+\lambda)^{\alpha-1}}{4s^2[(s+\lambda)^{\alpha-1}-\lambda^\alpha]},$$

which is exactly in agreement with Eq. (54) in [Wu *et al.* (2016)]. Particularly, assuming $\kappa_1 = \kappa_2$ and talking inverse Laplace transform of Eq. (6.91) give

$$\langle (T^+)^2 \rangle = \frac{t^2 e^{\kappa_1 t}}{2} - \frac{\alpha e^{\kappa_1 t}}{4} \int_0^t (t-\tau)e^{-\lambda\tau}E_{\alpha,1}(\lambda^\alpha\tau^\alpha)d\tau.$$

When $t \to 0$, $\langle (T^+)^2 \rangle \sim \frac{4-\alpha}{8}t^2$, which is the expected result [Wu *et al.* (2016)] since initially tempering and reaction terms have negligible influence on the process.

6.2.6 *Distribution of First Passage Time*

The time T_f when a particle starting at $x_0 = -b$ $(b > 0)$ passes $x = 0$ for the first time is called the first passage time [Redner (2001)]. According to [Kac (1951); Wu *et al.* (2016)], the relation between the distribution of T_f and the PDF of the occupation time in half space T^+ (in the Laplace domain $T^+ \to p$) satisfies $P\{T_f > t\} = \lim_{p \to \infty} \rho_{-b}(p, t)$. Denote the PDF of T_f as $f(t)$ and in the Laplace domain $t \to s$,

$$\hat{f}(s) = \frac{-\kappa_2}{s - \kappa_2} + \frac{s}{s - \kappa_2} e^{-b\sqrt{\frac{(s+\lambda-\kappa_2)^\alpha - \lambda^\alpha}{K_\alpha}}}. \tag{6.92}$$

When $s \to \infty$, $\hat{f}(s) \sim \exp\left(-\frac{b}{\sqrt{K_\alpha}} s^{\frac{\alpha}{2}}\right)$, which is equivalent to the one sided Lévy law in the time domain. Consequently, we assert that $f(t)$ decays very fast to zero when $t \to 0$, behaves as $t^{-1-\alpha/2}$ for short but not too short time scales.

6.2.7 *Distribution of Occupation Time in Half Interval*

In this subsection, we suppose that the particle motion is restricted to the interval $(-L, L)$ with the absorbing or reflecting boundaries. The occupation time T_a^+ or T_r^+ (corresponding to the absorbing or reflecting boundary conditions, respectively) in the positive half interval is defined as $T_a^+ = T_r^+ = \int_0^t U[x(\tau)]d\tau$, where $U(x) = 1$ for $0 \leq x < L$ and $U(x) = 0$ for $-L < x < 0$. Thus, we have

$$\hat{\rho}_{x_0}(p, s) = \begin{cases} \frac{K_\alpha}{(s+\lambda+p-\kappa_1)^\alpha - \lambda^\alpha} \frac{\partial^2 \hat{\rho}_{x_0}(p,s)}{\partial x_0^2} + \frac{1}{s+p-\kappa_1}, & 0 < x_0 < L; \\ \frac{K_\alpha}{(s+\lambda-\kappa_2)^\alpha - \lambda^\alpha} \frac{\partial^2 \hat{\rho}_{x_0}(p,s)}{\partial x_0^2} + \frac{1}{s-\kappa_2}, & -L < x_0 < 0. \end{cases} \tag{6.93}$$

(1) Absorbing boundary conditions

Solving Eq. (6.93) in each interval, respectively, we obtain

$$\hat{\rho}_{x_0}(p, s) =$$
$$\begin{cases} D_1 e^{x_0 C_1(p,s)/\sqrt{K_\alpha}} + D_2 e^{-x_0 C_1(p,s)/\sqrt{K_\alpha}} + (s+p-\kappa_1)^{-1}, & 0 < x_0 < L; \\ D_3 e^{x_0 C_2(s)/\sqrt{K_\alpha}} + D_4 e^{-x_0 C_2(s)/\sqrt{K_\alpha}} + (s-\kappa_2)^{-1}, & -L < x_0 < 0. \end{cases}$$

With the absorbing boundary condition $\hat{\rho}_{x_0}(p, s)|_{x_0 = \pm L} = 0$ and the continuities of $\hat{\rho}_{x_0}(p, s)$ and its first derivative at $x_0 = 0$, we can determine the constants. Suppose the particle departs at $x_0 = 0$ and then we have

$$\hat{\rho}_0(p, s) =$$
$$\frac{\frac{1}{s-\kappa_2}\left[-1 - \frac{C_2(s)\tanh[LC_1(p,s)]}{C_1(p,s)\sinh[LC_2(s)]}\right] - \frac{1}{s+p-\kappa_1}\left[\frac{1}{\cosh[LC_1(p,s)]} - 1\right]}{1 + \frac{C_2(s)\tanh[LC_1(p,s)]}{C_1(p,s)\tanh[LC_2(s)]}} + \frac{1}{s-\kappa_2}. \tag{6.94}$$

(2) Reflecting boundary conditions

Solving Eq. (6.93) with $\frac{\partial \hat{\rho}_{x_0}(p,s)}{\partial x_0}|_{x_0 = \pm L} = 0$ gives

$$\hat{\rho}_{x_0}(p,s) = \begin{cases} D_1 \cosh\left[(L - x_0)C_1(p,s)/\sqrt{K_\alpha}\right] + (s + p - \kappa_1)^{-1}, & x_0 > 0; \\ D_2 \cosh\left[(L + x_0)C_1(p,s)/\sqrt{K_\alpha}\right] + (s - \kappa_2)^{-1}, & x_0 < 0. \end{cases}$$

Similarly, in order to determine the constants D_1 and D_2, we assume the continuities of $\hat{\rho}_{x_0}(p,s)$ and its first derivative at $x_0 = 0$, which implies that

$$\begin{cases} D_1 = -F(p,s)D_2, \\ D_2 = \frac{1}{F(p,s)\cosh[LC_1(p,s)] + \cosh[LC_2(s)]} \cdot \frac{\kappa_1 - \kappa_2 - p}{(s + p - \kappa_1)(s - \kappa_2)}, \end{cases}$$

with $F(p,s) = \frac{C_2(s)\sinh[LC_2(s)]}{C_1(p,s)\sinh[LC_1(p,s)]}$ for simplicity. If the particle starts from $x_0 = 0$, we obtain

$$\begin{aligned} \hat{\rho}_0(p,s) = & \frac{C_1(p,s)\tanh[LC_1(p,s)]}{(s + p - \kappa_1)[\tanh[LC_1(p,s)] + \tanh[LC_2(s)]]} \\ & + \frac{C_2(s)\tanh[LC_2(s)]}{(s - \kappa_2)[\tanh[LC_1(p,s)] + \tanh[LC_2(s)]]}. \end{aligned} \tag{6.95}$$

Especially, when $\lambda = \kappa_1 = \kappa_2 = 0$, we recover

$$\hat{\rho}_0(p,s) = \frac{(s + p)^{\frac{\alpha}{2} - 1}\tanh\left[(s + p)^{\frac{\alpha}{2}}L/\sqrt{K_\alpha}\right] + s^{\frac{\alpha}{2} - 1}\tanh\left[s^{\frac{\alpha}{2}}L/\sqrt{K_\alpha}\right]}{(s + p)^{\frac{\alpha}{2}}\tanh\left[(s + p)^{\frac{\alpha}{2}}L/\sqrt{K_\alpha}\right] + s^{\frac{\alpha}{2}}\tanh\left[s^{\frac{\alpha}{2}}L/\sqrt{K_\alpha}\right]},$$

which was previously derived in [Carmi and Barkai (2011)] using a similar method.

Chapter 7

Renewal Theory for Fractional Poisson Process: Typical versus Rare

Renewal processes with heavy-tailed power law distributed sojourn times are commonly encountered in physical modeling and so typical fluctuations of observables of interest have been investigated in detail. To describe rare events, the rate function approach from large deviation theory does not hold and new tools must be considered. Here, we investigate the large deviations of the number of renewals, the forward and backward recurrence times, the occupation time, and the time interval straddling the observation time. We show how non-normalized densities describe these rare fluctuations and how moments of certain observables are obtained from these limiting laws. Numerical simulations agree with our results, showing the deviations from arcsine, Dynkin, Darling-Kac, Lévy, and Lamperti laws. The organization of the chapter is as follows. In Sec. 7.2, we outline the model and give the necessary definitions. The behavior of the probability of observing N renewals in the interval $(0, t)$, $p_N(t)$ is analyzed in Sec. 7.3. In Secs. 7.4, 7.5 and 7.6, the densities of the forward and backward recurrence time, and the time interval straddling t, denoted as F, B and Z respectively, are derived. In order to see the effects of the typical fluctuations and large deviations, the fractional moments, e.g., $\langle F^q \rangle$, are considered and bi-fractal behavior is found. In Sec. 7.7, the behavior of the occupation time T^+ is studied. In the final section, we conclude this chapter with some discussions.

7.1 Introduction

Renewal processes [Godrèche and Luck (2001); Mainardi *et al.* (2004, 2007); Niemann *et al.* (2016); Miyaguchi *et al.* (2016); Wang *et al.* (2018a)] are simple stochastic models for events that occur on the time axis when the time intervals between events are IID random variables. This idealized approach has many applications, ranging from the analysis of photon arrival times to queuing theory. In some models the sojourn time PDF has heavy-tails, and this leads to fractal time renewal processes. In the case when the variance of the sojourn time diverges, we have deviations from the normal central limit theorem and/or the law of large numbers. Such heavy-tailed processes are observed in many systems, ranging from blinking quantum dots [Bianco *et al.* (2005)] to diffusion of particles in polymer networks [Edery *et al.* (2018)], or diffusion of particles on the membrane of cell [Weron *et al.* (2017)] to name a few. In these systems the renewal process is triggering jumps in intensity or in space. The CTRW model [Metzler and Klafter (2000)], the annealed trap model, the zero crossing of Brownian motion, the velocity zero crossing of cold atoms diffusing in momentum space [Barkai *et al.* (2014)], are all well known models which use this popular renewal approach (see however [Boettcher *et al.* (2018); Nyberg *et al.* (2018)]). Heavy-tailed renewal theory is also used in the context of localization in random wave guides. The number of renewals, under certain conditions, is described by Lévy statistics, and the fluctuations in these processes are large. Hence it is important to explore the rare events or the far tails of the distributions of observables of interest. As mentioned in [Touchette (2009); Whitelam (2018)] the large deviation principle, with its characteristic exponential decay of large fluctuations, does not describe this case, and instead the big jump principle is used to evaluate the rare events in Lévy type of processes.

The main statistical tools describing observables of interest are non-normalized states, being limiting laws with which we may obtain statistical information on the system, including for example the variance, which in usual circumstances is the way we measure fluctuations. These non-normalized states are previously investigated, in the context of Lévy walks [Rebenshtok *et al.* (2014a)], spatial diffusion of cold atoms [Aghion *et al.* (2017)], and very recently for Boltzmann-Gibbs states when the underlying partition function of the system diverges [Aghion *et al.* (2019)]. These functions describing the statistical behavior of the system are sometimes called infinite densities or infinite covariant densities, and they appear constantly in infinite ergodic theory [Aaronson (1997)].

Our goal in this chapter is to investigate the statistics of rare events in renewal theory. Consider for example a non-biased ordinary random walk on the integers. The spatial jump process is Markovian hence the zero crossing, where the zero is the origin, is a renewal process. Here like Brownian motion, the waiting time PDF between the zero crossings is heavy-tailed, in such a way the mean return time diverges. The distribution of the occupation time $0 < T^+ < t$, namely the time the random walker spends in the positive domain is well investigated [Redner (2001); Godrèche and Luck (2001)]. Naively one would expect that when the measurement time t is long the particle will spend half of its time on the right side of the origin. Instead one finds that this is the least likely scenario, and the PDF of the properly scaled occupation time reads

$$\lim_{t\to\infty} f_{T^+/t}(x) = \frac{1}{\pi\sqrt{x(1-x)}}. \tag{7.1}$$

Here and all along this chapter the subscript denotes the observable of interest, e.g., we consider the PDF of T^+/t which attains values $0 < x < 1$. This arcsine law, which describes also other features of Brownian motion [Mörters and Peres (2010); Akimoto and Yamamoto (2016); Sadhu *et al.* (2018)], exhibits divergences on $x \to 0$ or $x \to 1$. Here a particular scaling of $T^+ \propto t$ is considered. However, in cases studied below we show that other limiting laws are found when a second time scale is considered and these may modify the statistical properties of the occupation time when T^+ is either very small or very large. This in turn influences the anticipated blow up of the arcsine law at its extremes. Notice that here the least likely event, at least according to this law is the case $x = 1/2$, so our theory is not dealing with corrections to the least likely event, but rather corrections to the most likely events. This is because of the heavy-tailed waiting times, which make the discussion of deviations from familiar limiting laws a case study in its own right. While the theory deals with most likely events, from the sampling point of view these are still rare, as the probability of finding the occupation time in a small interval close to the extremes of the arcsine law is still small.

7.2 Model

Renewal process, an idealized stochastic model for events that occur randomly in time, has a very rich and interesting mathematical structure and can be used as a foundation for building more realistic models [Metzler and Klafter (2004); Brokmann *et al.* (2003)]. As mentioned, the basic mathematical assumption is that the time between the events are IID random variables. Moreover, renewal processes are often found embedded in other stochastic processes, most notably Markov chains.

Now, we briefly outline the main ingredients of the renewal process [Godrèche and Luck (2001)]. It is defined as follows: events occur at the random epochs of time t_1, t_2, ..., t_N, ..., from some time origin $t = 0$. When the time intervals between events, $\tau_1 = t_1$, $\tau_2 = t_2 - t_1$, ..., $\tau_N = t_N - t_{N-1}$, ..., are IID random variables with a common PDF $\phi(\tau)$, the process thus formed is a renewal process (see the top panel of the Fig. 7.1). We further consider the alternating renewal process $I(t)$ in which the process alternates between '+' and '−' states. A classical example is Brownian motion $x(t)$ in one dimension, where we denote $x(t) > 0$ with state '+' and $x(t) < 0$ with state '−'. Generically, we imagine that a device, over time, alternates between 'on' and 'off' states, like a blinking dot [Bianco *et al.* (2005); Margolin *et al.* (2005); Wang *et al.* (2018a)]. Here we suppose the process starts in '+' state and stays in that state for a period of time τ_1, then goes to '−' state and remains for time τ_2; see bottom panel of Fig. 7.1. Clearly, it is natural to discuss the total time in state '+' or '−'. T^+ and T^- are called the occupation times in the '+' and '−' states, respectively and $T^+ + T^- = t$. For Brownian motion, $\phi(\tau) \sim \tau^{-3/2}$ and the distribution of time in state '+' is the well known arcsine law.

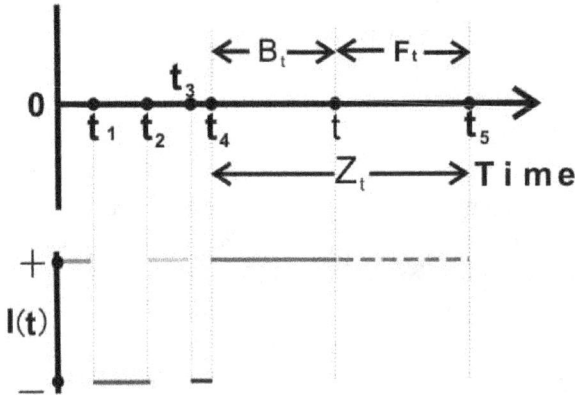

Fig. 7.1   Illustration of a renewal process. The t_i denotes the time when the i-th event occurs. B_t and F_t present the backward and the forward recurrence time, respectively. In addition, the time interval straddling time t is denoted with Z_t. The process $I(t)$, an alternating renewal process, is represented in the bottom of the figure. Here we suppose the initial state of the particle is '+'. We see that the occupation time in the '+' state is equal to $t_1 + t_3 - t_2 + B_t$.

In the following we will draw on the research literature [Godrèche and Luck (2001)], which is recommended for an introduction. The number of renewal events in the time interval between 0 and time t is

$$N(t) = \max[N, t_N \le t]. \tag{7.2}$$

Then we have the following relation $t_N = \tau_1 + \ldots + \tau_N$. Now we introduce the forward recurrence time F_t, the time between t and the next event

$$F_t = t_{N+1} - t;$$

see Fig. 7.1. While the corresponding backward recurrence time, the length between the last event before t and the observation time t, is defined by

$$B_t = t - t_N.$$

Utilizing the above two equations, we get the time interval straddling time t, i.e., Z_t, which is

$$Z_t = B_t + F_t.$$

For simplification, we drop the subscript, denoting the time dependence of the random quantities, from here on.

We notice that some equations can be further simplified for a specific $\phi(\tau)$ (see below), for example the Mittag-Leffler distribution [Podlubny (1999); Kozubowski (2001)]. In order to do so, we consider

$$\phi(\tau) = \tau^{\alpha-1} E_{\alpha,\alpha}(-\tau^\alpha) \tag{7.3}$$

with $0 < \alpha < 1$. In Laplace space, $\widehat{\phi}(s)$ has the specific form

$$\widehat{\phi}(s) = \frac{1}{1 + s^\alpha}. \tag{7.4}$$

This distribution can be considered as the positive counterpart of Pakes's generalized Linnik distribution [Jose *et al.* (2010)] with the PDF having the form $(1 + s^\alpha)^{-\beta}$, $0 < \alpha < 2$, $\beta > 0$.[1]

---

[1] When generating the random variables with the PDF $\ell_\alpha(\xi)$ or $\xi^{\alpha-1} E_{\alpha,\alpha}(-\xi^\alpha)$ needed to simulate the renewal process, the Monte Carlo statistical methods [Robert and Casella (2004)] are used. Chambers *et al.* [Chambers *et al.* (1976)] showed how to obtain a random variable drawn from the stable Lévy distribution with $0 < \alpha < 1$. Furthermore, Kozubowski constructed the following structural representation of a $\phi(\xi) = \xi^{\alpha-1} E_{\alpha,\alpha}(-\xi^\alpha)$ distributed random variable ξ as [Kozubowski (2001)]

$$\xi = \sigma \eta^{1/\alpha},$$

where σ is a random number from the exponential distribution with mean parameter 1, and η has the PDF

$$f(\eta) = \frac{\sin(\pi\alpha)}{\alpha\pi(\eta^2 + 2\eta\cos(\pi\alpha) + 1)}$$

with $0 < \alpha < 1$ and $\eta > 0$.

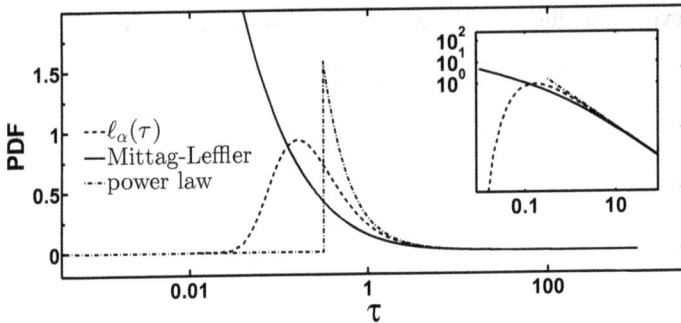

Fig. 7.2   The trends of three types of PDFs, namely one sided Lévy distribution, Mittag-Leffler function and power law distribution. Here we choose $\alpha = 1/2$ and $\tau_0 = 0.3183$. When τ is small, the behaviors of them are totally different. With the increase of τ, their asymptotic behaviors are the same, which is shown in log-log scale; see the inset.

7.3   Number of Renewals between 0 and t

Recall that in Sec. 1.3 we have introduced the probability of jumping N steps up to time t, i.e., $p_N(t)$. For simplification, we directly give the final result

$$\widehat{p}_N(s) = \widehat{\phi}^N(s)\frac{1 - \widehat{\phi}(s)}{s};$$

for more details see Eq. (1.38).

7.3.1   *Number of Renewals between* 0 *and* t *with* $0 < \alpha < 1$

Rewriting Eq. (1.38), using the convolution theorem of Laplace transform, and performing inverse Laplace transform w.r.t. s, we get a formal solution

$$p_N(t) = \int_0^t \mathcal{L}_\tau^{-1}[\widehat{\phi}^N(s)]d\tau - \int_0^t \mathcal{L}_\tau^{-1}[\widehat{\phi}^{N+1}(s)]d\tau, \qquad (7.5)$$

where N is a discrete random variable and $\mathcal{L}_\tau^{-1}[\widehat{\phi}^N(s)]$ means the inverse Laplace transform, from the Laplace space s to real space τ.

Summing the infinite series (summation over N), the normalization condition $\sum_{N=0}^{\infty} p_N(t) = 1$ is discovered as expected. We notice that Eq. (7.5) can be further simplified when $\phi(\tau)$ is one sided Lévy distribution Eq. (1.29). Then the inverse Laplace transform of Eq. (7.5) gives

$$p_N(t) = \int_{t/(N+1)^{1/\alpha}}^{t/N^{1/\alpha}} \ell_\alpha(y)dy. \qquad (7.6)$$

As usual the large time limit is investigated with the small s behavior of $\widehat{p}_N(s)$. Utilizing Eq. (1.29), the behavior of Eq. (1.38) in the large N limit and small s is

$$\widehat{\overline{p}}_N(s) \to b_\alpha s^{\alpha-1} \exp(-N b_\alpha s^\alpha)$$

$$= -\frac{1}{N\alpha}\frac{\partial}{\partial s}\exp(-N b_\alpha s^\alpha).$$

Here note that $\int_0^\infty b_\alpha s^{\alpha-1}\exp(-N b_\alpha s^\alpha)dN = 1/s$. This means that with this approximation N is treated as a continuous variable, which is fine since in fact we consider a long time limit, and the limiting PDF of N/t^α is approaching a smooth function. Hence, we have $\overline{p}_N(t)$ to denote the continuous approximation.

First, using the property of Laplace transform, i.e., $-\int_0^\infty \exp(-s\tau)$ $\tau f(\tau)d\tau = \frac{\partial}{\partial s}\widehat{f}(s)$, secondly, performing the inverse Laplace transform of the above equation, we find the well known result [Aaronson (1997); Godrèche and Luck (2001); Schulz *et al.* (2014)]

$$\overline{p}_N(t) \sim \frac{t}{\alpha N^{1+1/\alpha}b_\alpha^{1/\alpha}}\ell_\alpha\left(\frac{t}{(Nb_\alpha)^{1/\alpha}}\right). \tag{7.7}$$

Equation (7.7) is customarily called the inverse Lévy PDF. Furthermore, using $\widehat{\overline{p}}_N(s)$, we find that $\widehat{\overline{p}}_u(t)$ can be expressed as Mittag-Leffler probability density

$$\widehat{\overline{p}}_u(t) \sim t^{\alpha-1}E_{\alpha,\alpha}(-ut^\alpha/b_\alpha),$$

where $\widehat{\overline{p}}_u(t)$ is the Laplace transform of $\overline{p}_N(t)$ w.r.t. N and a two-parameter function of the Mittag-Leffler type is defined by the series expansion [Podlubny (1999)]

$$E_{\gamma,\nu}(z) = \sum_{n=0}^{\infty}\frac{z^n}{\Gamma(\gamma n + \nu)}$$

with $\gamma > 0$ and $\nu > 0$. Equation (7.7) describes statistics of functionals of certain Markovian processes, according to the Darling-Kac theorem. It is also investigated in the context of infinite ergodic theory [Aaronson (1997)] and CTRW.

The well known limit theorem Eq. (7.7) is valid when N and t are large and the ratio N/t^α is kept fixed. Now we consider rare events when N is kept fixed and finite, say $N \sim 0, 1, 2, 3$ and t is large. Using Eq. (7.5) we find

$$\lim_{t\to\infty}t^\alpha p_N(t) = \frac{b_\alpha}{\Gamma(1-\alpha)}. \tag{7.8}$$

Note that $0 < p_N(t) < 1$ is a probability, while $\bar{p}_N(t)$ is a PDF. To make a comparison between Eq. (7.7) and Eq. (7.8) we plot in Fig. 7.3, the probability that N is in the interval $(0, N_1)$ versus N_1 and compare these theoretical predictions with numerical simulations. Integrating Eq. (7.7) between 0 and N_1 gives what we call the typical fluctuations. While the result shown in Eq. (7.8) exhibits a staircase since according to this approximation

$$P(0 \leq N < N_1) \sim \sum_{N=0}^{\lfloor N_1 \rfloor} \frac{b_\alpha}{\Gamma(1-\alpha)t^\alpha}$$

$$\sim (\lfloor N_1 \rfloor + 1)\frac{b_\alpha}{\Gamma(1-\alpha)t^\alpha}, \quad (7.9)$$

where $\lfloor z \rfloor$ gives the greatest integer less than or equal to z. From Fig. 7.3 we see that, besides the obvious discreteness of the probability, deviations between the two results can be considered marginal and non-interesting. Luckily this will change in all the examples considered below, as the statistical description of rare events deviates considerably from the known limit theorems of the field.

Let us consider another interesting observable, i.e., the q-th moment of N. Using Eqs. (7.7) and (7.39)

$$\langle N^q \rangle \sim \int_0^\infty \frac{tN^q}{\alpha N^{1+1/\alpha}b_\alpha^{1/\alpha}} \ell_\alpha\left(\frac{t}{(Nb_\alpha)^{1/\alpha}}\right)dN$$

$$= \frac{\int_0^\infty \xi^{-\alpha q}\ell_\alpha(\xi)d\xi}{(b_\alpha)^q}t^{\alpha q}. \quad (7.10)$$

In the particular case $q \to 0$, the normalized condition is found, namely, $\langle N^0 \rangle = 1$.

7.3.2 *Number of Renewals between* 0 *and* t *with* $1 < \alpha < 2$

Based on Eq. (1.38), we obtain a useful expression

$$\widehat{\bar{p}}_u(s) = \frac{1 - \widehat{\phi}(s)}{s}\int_0^\infty \exp(-uN + N\log(\widehat{\phi}(s)))dN. \quad (7.11)$$

Here we consider the random variable, $\varepsilon = N - t/\langle \tau \rangle$, and explore its PDF denoted as $\bar{p}_\varepsilon(t)$. Applying Fourier-Laplace transform, $\varepsilon \to k$ and $t \to s$, the PDF of ε in Fourier-Laplace space is

$$\widehat{\widetilde{p}}_k(s) = \frac{1 - \widehat{\phi}\left(s + \frac{ik}{\langle \tau \rangle}\right)}{s + \frac{ik}{\langle \tau \rangle}}\frac{1}{-ik - \log\left(\widehat{\phi}\left(s + \frac{ik}{\langle \tau \rangle}\right)\right)}. \quad (7.12)$$

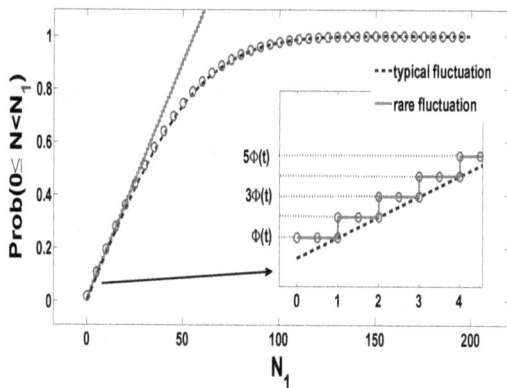

Fig. 7.3 Comparison of analytical prediction Eq. (7.9) solid line for $P(0 \leq N < N_1)$ with typical fluctuations with $\alpha = 1/2$. We choose $t = 1000$, waiting time PDF Eq. (1.30), and 10^7 trajectories. The typical fluctuations, plotted by dashed lines, are obtained from Eq. (7.7). The rare fluctuations are given by Eq. (7.9) and they describe the probability very well for small N_1 (see inset).

First, we consider the limit of small s and small k, and the ratio $s/|k|^\alpha$ is fixed. As we discuss below this leads to the description of what we call bulk or typical fluctuations, described by standard central limit theorem. Substituting $\widehat{\phi}(s)$ into the above equation and taking inverse Laplace transform

$$\widetilde{\overline{p}}_k(t) \sim \exp\left(\frac{b_\alpha}{\langle\tau\rangle}\left(\frac{ik}{\langle\tau\rangle}\right)^\alpha t\right). \tag{7.13}$$

Fourier inversion of the above equation yields the PDF $\overline{p}_\varepsilon(t)$, written in a scaling form [Godrèche and Luck (2001)]

$$\overline{p}_\varepsilon(t) \sim \frac{1}{C_{ev}t^{1/\alpha}}l_{\alpha,1}(\xi) \tag{7.14}$$

with $C_{ev} = (b_\alpha/\langle\tau\rangle^{1+\alpha})^{1/\alpha}$ and $\xi = \varepsilon/(C_{ev}t^{1/\alpha})$; see Fig. 7.4. We see that for fixed observation time t the parameter C_{ev} measures the PDF's width. Furthermore, the function $l_{\alpha,1}(x)$ is defined by

$$l_{\alpha,1}(x) = \frac{1}{2\pi}\int_{-\infty}^{\infty} \exp(-ikx)\exp[(ik)^\alpha]dk,$$

where $l_{\alpha,1}(x)$ is the asymmetric Lévy PDF; see Sec. 7.8. Compared with the one sided Lévy distribution, $l_{\alpha,1}(x)$ holds two sides with the right hand side decaying rapidly. Moreover, the second moment of $l_{\alpha,1}(x)$ diverges for $1 < \alpha < 2$.

As well known the central limit theorem (here of the Lévy form) describes the central part of the distribution, but for finite though large t it does not describe the rare events, i.e., the far tail of the distribution. So far we investigated the typical or bulk statistics and as we showed they are found for $N - t/\langle\tau\rangle \sim t^{1/\alpha}$. Technically this was obtained using the exact Laplace-Fourier transform, and then searching for a limit where s and $|k|^\alpha$ are small their ratio finite, as mentioned. However, it turns out that this limit is not unique. As we now show we can use the exact solution, assume both k and s are small, but their ratio $s/|k|$ is finite and we obtain a second meaningful solution. This in turn, leads to the description of rare events, i.e., the far tail of the distribution of the random variable N. Roughly speaking, in this problem (and similarly all along the chapter) we have two scales, one was just obtained and it grows like $t^{1/\alpha}$, the second (with this example) is $t/\langle\tau\rangle$, as we now show. This means that we have two ways to scale data, one emphasizing the bulk fluctuations (explained already) and the second the rare events.

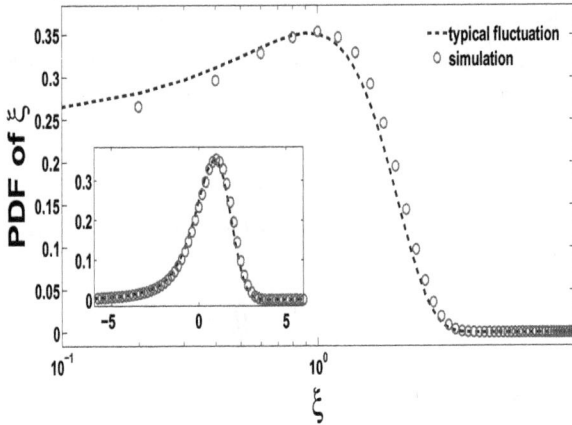

Fig. 7.4   Simulation of the number of renewals using the rescaled variable $\xi = (N - t/\langle\tau\rangle)/(C_{ev}t^{1/\alpha})$. We show that the distribution of ξ obtained from numerical simulations, converges to the Lévy stable density $l_{\alpha,1}(\xi)$ Eq. (7.14). This law describes the typical fluctuations, when $N - t/\langle\tau\rangle$ is of the order $t^{1/\alpha}$. The simulations use 2×10^6 realizations, $\alpha = 1.5$, and $\phi(\tau)$ given in Eq. (1.26). As the figure shows, the right hand side of $l_{\alpha,1}(\xi)$ tends toward to zero rapidly.

We now consider a second limiting law validly capturing the rare events

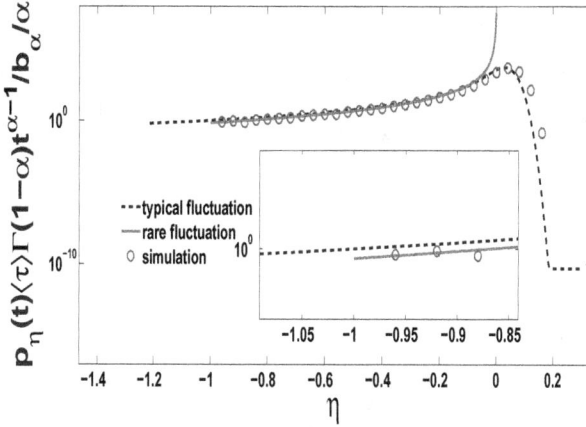

Fig. 7.5 Simulation of renewal process with Eq. (1.26) to yield N. We show how the scaled PDF of $\eta = (N - t/\langle\tau\rangle)/(t/\langle\tau\rangle)$, obtained from numerical simulations of the renewal process, converges to the non-normalized state Eq. (7.16). The bulk description Eq. (7.14) shown by dashed line extends to $\eta < -1$, which is certainly not a possibility, and further it is not a valid approximation for $\eta \to -1$. Here we choose $\alpha = 1.5$ and $\tau_0 = 0.1$. Deviations from typical fluctuations are clearly illustrated in the inset.

when ε is of the order $t/\langle\tau\rangle$. From Eq. (7.12), we have

$$\widetilde{\widetilde{p}}_k(s) \sim \frac{1}{s} - \frac{b_\alpha}{\langle\tau\rangle s}\left(s + \frac{ik}{\langle\tau\rangle}\right)^{\alpha-1} + \frac{b_\alpha}{\langle\tau\rangle s^2}\left(s + \frac{ik}{\langle\tau\rangle}\right)^{\alpha}.$$

Keep in mind that s and k are small and they are of the same order. After performing inverse Fourier-Laplace transform, the asymptotic behavior of $\bar{p}_\varepsilon(t)$ is

$$\bar{p}_\varepsilon(t) \sim -\frac{b_\alpha\alpha}{\Gamma(1-\alpha)}\left(t(-\varepsilon\langle\tau\rangle)^{-\alpha-1} + \frac{1-\alpha}{\alpha}(-\varepsilon\langle\tau\rangle)^{-\alpha}\right). \qquad (7.15)$$

Here the above equation is only valid for negative ε. We see that $\bar{p}_\varepsilon(t)$ decays like $t(-\varepsilon)^{-\alpha-1}$ for small negative ε. Moreover, the scaling behavior of $\eta = \varepsilon\langle\tau\rangle/t$ yields

$$\bar{p}_\eta(t) \sim \frac{b_\alpha(-\eta)^{-\alpha-1}}{\langle\tau\rangle\Gamma(-\alpha)t^{\alpha-1}}\left(1 - \frac{1-\alpha}{\alpha}\eta\right) \qquad (7.16)$$

with $-1 < \eta < 0$; see Fig. 7.5. It means that $\bar{p}_\eta(t)$ decays like $(-\eta)^{-\alpha-1}$ for $\eta \to 0$, thus $\bar{p}_\eta(t)$ is not normalized. Furthermore, for $\eta \to 0$, the dominating term $(-\eta)^{-1-\alpha}$ matches the left tail of Eq. (7.14); see Eq. (7.100) in Sec. 7.8.

For fixed observation time t, the central part of the PDF $\bar{p}_\varepsilon(t)$ is well illustrated by the typical fluctuations Eq. (7.14). While, its tail is described by Eq. (7.16), exhibiting the rare fluctuations. In order to discuss the effect of typical fluctuations and large deviations, we further consider the absolute moment of ε [Schulz and Barkai (2015)], defined by

$$\langle |\varepsilon|^q \rangle = \int_{-\infty}^{\infty} |\varepsilon|^q \bar{p}_\varepsilon(t) d\varepsilon. \tag{7.17}$$

Utilizing Eqs. (7.14), (7.16), and (7.17), we have

$$\langle |\varepsilon|^q \rangle \sim \begin{cases} (C_{ev})^q t^{q/\alpha} \int_{-\infty}^{\infty} |z|^q l_{\alpha,1}(z) dz, & q < \alpha; \\[2mm] \dfrac{b_\alpha q t^{q+1-\alpha}}{|\Gamma(1-\alpha)| \langle \tau \rangle^{q+1} (1+q-\alpha)(q-\alpha)}, & q > \alpha. \end{cases} \tag{7.18}$$

Here we use the fact that $\int_{-\infty}^{\infty} |z|^q l_{\alpha,1}(z) dz$ is a finite constant for $q < \alpha$. Note that to derive Eq. (7.18) we use the non-normalized solution Eq. (7.15) for $q > \alpha$, indicating that Eq. (7.15) while not being a probability density, does describe the high order moments. In the particular case $q = 2$ (high order moment), we have

$$\langle |\varepsilon|^2 \rangle = \left\langle \left(n - \frac{t}{\langle \tau \rangle} \right)^2 \right\rangle$$

$$\sim \langle n^2 \rangle - \langle n \rangle^2$$

$$\sim \frac{2\tau_0^\alpha}{\langle \tau \rangle^3 (\alpha - 2)(\alpha - 3)} t^{3-\alpha}.$$

While this result is known [Godrèche and Luck (2001)], our work shows that the second moment, in fact any moment of order $q > \alpha$, stems from the non-normalized density describing the rare fluctuations Eq. (7.15). Other examples of such infinite densities will follow.

7.4   Forward Recurrence Time

Several authors investigate the distributions of F both for $F \propto t$, meaning F is of the order t, for $0 < \alpha < 1$ and also $F \propto t^0$ for $\alpha > 1$; see [Dynkin (1961); Feller (1971); Schulz et al. (2014)]. These works consider the typical fluctuations of F, while we focus on the events of large deviations. This means that we consider $F \propto t^0$ for $\alpha < 1$ and $F \propto t$ for $1 < \alpha$. The forward recurrence time is an important topic of many stochastic processes, such as ACTRW [Schulz et al. (2014); Kutner and Masoliver (2017)], sign renewals of Kardar-Parisi-Zhang fluctuations [Takeuchi and Akimoto (2016)] and so on. The forward recurrence time, also called the excess time (see schematic Fig. 7.1), is the time interval between next renewal event and t. In ACTRW, we are interested in the time interval that the particle has to wait before next jump if the observation is made at time t. Note that an important equality is the conditional probability density of the forward recurrence time F given that exactly N events occurred before time t, defined by

$$f_N(t, F) = \int_0^t Q_N(\tau)\phi(t - \tau + F)d\tau. \tag{7.19}$$

Then the PDF of the forward recurrence time is given by

$$f_F(t, F) = \sum_{N=0}^{\infty} \int_0^t Q_N(\tau)\phi(t - \tau + F)d\tau. \tag{7.20}$$

The Laplace transform of $f_F(t, F)$ w.r.t. t follows from the shift theorem of Laplace transform and reads

$$\widehat{f}_F(s, F) = \frac{1}{1 - \widehat{\phi}(s)} \exp(sF) \int_F^{\infty} \phi(z) \exp(-sz)dz.$$

Then, taking Laplace transform and using partial integration, lead to the final result Eq. (7.21). In double Laplace space, the PDF of F [Godrèche and Luck (2001)] is

$$\widehat{f}_F(s, u) = \frac{\widehat{\phi}(u) - \widehat{\phi}(s)}{s - u} \frac{1}{1 - \widehat{\phi}(s)}. \tag{7.21}$$

Based on the above equation, we will consider its analytic forms and asymptotic ones. In general, the inversion of Eq. (7.21) is a function that depends on F and t. While, for $\phi(\tau) = \exp(-\tau)$, the above equation can be simplified as $f_F(t, F) = \exp(-F)$, which is independent of the observation time t. As expected, for this example we do not have an infinite density, neither multi-scaling of moments, since $\exp(-F)$ and more generally thin tailed PDFs, do not have large fluctuations like Lévy statistics.

7.4.1 *Forward Recurrence Time with $0 < \alpha < 1$*

First, we are interested in the case of $F \ll t$. In Laplace space, this corresponds to $s \ll u$. From Eq. (7.21)

$$\widehat{f}_F(s, u) \sim \frac{1 - \widehat{\phi}(u)}{u} \frac{1}{1 - \widehat{\phi}(s)}. \tag{7.22}$$

Plugging Eq. (7.4) into Eq. (7.22) leads to

$$\widehat{f}_F(s, u) \sim \frac{u^{\alpha-1}}{1 + u^\alpha} \frac{1}{s^\alpha}.$$

Taking the double inverse Laplace transforms yields

$$f_F(t, F) \sim \frac{1}{\Gamma(\alpha)} E_{\alpha,1}(-F^\alpha) t^{\alpha-1}; \qquad (7.23)$$

see Fig. 7.6. Notice that $E_{\alpha,1}(0) = 1$, so for $t > 0$, the PDF of F for $F \to 0$ gives $f_F(t, 0) \sim t^{\alpha-1}/\Gamma(\alpha)$.

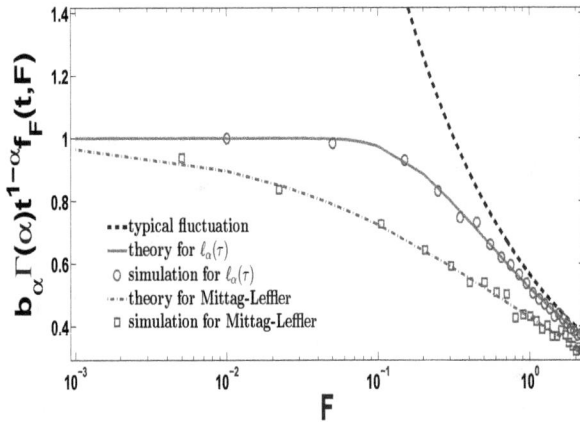

Fig. 7.6   The behavior of $f_F(t, F)$ for small F with $\alpha = 0.5$. The full and the dash-dot lines describing the large deviations are the analytical results Eqs. (7.24) and (7.23), respectively. The dashed line given by Eq. (7.28) is Dynkin's limit theorem which gives the PDF when F is of the order t, and t is large. Simulations are obtained by averaging 2×10^7 trajectories with $t = 2000$. Note that $b_\alpha \Gamma(\alpha) t^{1-\alpha} f_F(t, F)$ approaches to one for $F \to 0$.

More generally, using Eq. (7.22), we have

$$\widehat{f}_F(s, u) \sim \frac{1 - \widehat{\phi}(u)}{u} \frac{1}{b_\alpha s^\alpha}.$$

Performing inverse double Laplace transforms leads to the main result of this section, and the density describing the large deviations is

$$f_F(t, F) \sim \frac{\int_F^\infty \phi(y) dy}{\langle \tau^* \rangle}, \qquad (7.24)$$

which exhibits interesting aging effects [Schulz *et al.* (2014)]. Here $\langle \tau^* \rangle$ for large t is equal to $(\Gamma(2-\alpha)\Gamma(\alpha)/\alpha) \int_0^t \tau' \phi(\tau') d\tau' \sim b_\alpha \Gamma(\alpha) t^{1-\alpha}$, namely

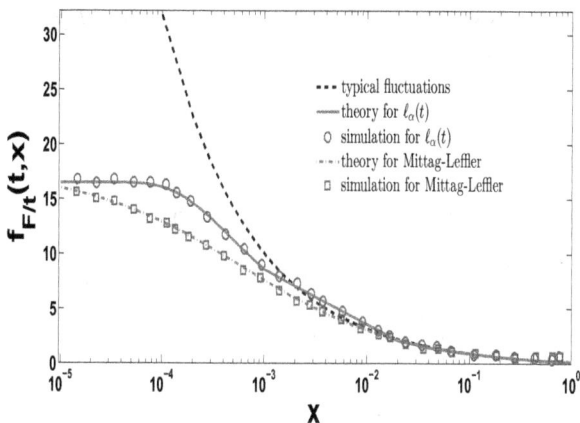

Fig. 7.7 The behaviors of $f_F(t, F)$ with the scaling variable $x = F/t$ for $\alpha = 0.5$ generated by 10^8 trajectories with $t = 2000$. The symbols are the simulation results. For $\alpha = 0.5$, based on Eq. (7.28), we get $f_{F/t}(t, x) \sim (\pi(1 + x)\sqrt{x})^{-1}$, which is shown by the dashed line. Here we use Eqs. (7.26) and (7.27) to predict theoretical results. When $x \to 0$, the PDF of F depends on particular properties of $\phi(\tau)$, while for large x, the details of the PDF become unimportant, besides the value of α.

$\langle \tau^* \rangle$ is increasing with measurement time t, and for reasons that become clear later we may call it the effective average waiting time (recall the $\langle \tau \rangle$ is a constant only if $\alpha > 1$). The large deviations show that for large F the forward recurrence time $f_F(t, F)$ decays as $F^{-\alpha}$. Furthermore, the integration of Eq. (7.24) over F diverges since $F^{-\alpha}$ is not integrable for large F. Hence Eq. (7.24) is not a normalized density. For that reason, we may call $f_F(t, F)$ in Eq. (7.24) an 'infinite' density [Rebenshtok et al. (2014a)], the term infinite means non-normalizable, hence this is certainly not a probability density. Even though $f_F(t, F)$ shown in Eq. (7.24) is not normalized, it is used to obtain certain observables, such as averages of observables integrable w.r.t. this non-normalized state. Besides, infinite densities play an important role in infinite ergodic theory [Thaler and Zweimüller (2006); Akimoto (2012)] and intermittent maps [Korabel and Barkai (2009)].

Using Eq. (7.21), we find a formal solution to the problem

$$f_F(t, F) = \phi(t + F) *_t \mathcal{L}_t^{-1} \left[\frac{1}{1 - \widehat{\phi}(s)} \right]. \tag{7.25}$$

The double Laplace transforms of the function $f(t + F)$ is

$$\int_0^\infty \int_0^\infty \exp(-st - uF) f(t + F) dt dF = \frac{\widehat{f}(u) - \widehat{f}(s)}{s - u}.$$

We further discuss a special choice of $\phi(\tau)$, i.e., $\phi(\tau) = \ell_\alpha(\tau)$. After some simple calculations, Eq. (7.25) gives

$$f_F(t, F) = \sum_{n=1}^{\infty} \frac{1}{n^{1/\alpha}} \int_0^t \ell_\alpha(t - \tau + F) \ell_\alpha \left(\frac{\tau}{n^{1/\alpha}} \right) d\tau + \ell_\alpha(t + F). \quad (7.26)$$

For Mittag-Leffler waiting time Eq. (7.3), we obtain

$$\begin{aligned}
f_F(t, F) &= (t + F)^{\alpha-1} E_{\alpha,\alpha}(-(t + F)^\alpha) \\
&+ \frac{1}{\Gamma(\alpha)} \int_0^t (\tau + F)^{\alpha-1}(t - \tau)^{\alpha-1} E_{\alpha,\alpha}(-(\tau + F)^\alpha) d\tau,
\end{aligned} \quad (7.27)$$

from which we get the PDF of $x = F/t$ plotted in Fig. 7.7.

We now focus on the typical fluctuations, namely the case $F \propto t$ and both F and t are large. This means that s and u are small but of the same order. Plugging Eq. (1.27) into Eq. (7.21) and taking double inverse Laplace transforms, lead to the normalized solution [Godrèche and Luck (2001); Dynkin (1961)]

$$f_F(t, F) \sim \frac{\sin(\pi\alpha)}{\pi} \frac{1}{(\frac{F}{t})^\alpha (t + F)}, \quad (7.28)$$

which is plotted by the dashed lines in Figs. 7.6 and 7.7. The well known solution Eq. (7.28) describes the typical fluctuations when $F \sim t$.

To summarize, the forward recurrence time shows three distinct behaviors: for $0 < F \propto t^0$, the infinite density Eq. (7.24) rules, and only in this range, the PDF of F depends on the behavior of $\phi(\tau)$; for $t^0 \ll F \ll t$, both Eqs. (7.24) and (7.28) are valid and predict $f_F(t, F) \sim F^{-\alpha}$; for $F \gg t$, we use Eq. (7.28) and then $f_F(t, F) \sim F^{-\alpha-1}$. Note that for certain observables, for example B and T^+, when $B, T^+ \to t$, their PDFs are also governed by the shape of $\phi(\tau)$; see below.

We further consider another observable $P(F_1, F_2)$, the probability that F is between F_1 and F_2, defined by

$$P(F_1, F_2) = \int_{F_1}^{F_2} f_F(t, F) dF \quad (7.29)$$

with $F_1, F_2 \gg 0$. For $F_1, F_2 \gg 1$, using Eq. (7.28), there exists

$$\begin{aligned}
P(F_1, F_2) &\sim \frac{\sin(\pi\alpha)t^\alpha}{\pi} \int_{F_1}^{F_2} \frac{1}{F^\alpha(t+F)} dF \\
&\sim \frac{\sin(\pi\alpha)}{\pi} \left(\int_0^{t/F_1} y^{\alpha-1}(1 + y)^{-1} dy - \int_0^{t/F_2} y^{\alpha-1}(1 + y)^{-1} dy \right).
\end{aligned}$$

After calculating integrals, we obtain

$$P(F_1, F_2) \sim \frac{\sin(\pi\alpha)}{\pi} \left(\left(\frac{t}{F_1}\right)^\alpha F\left(1, \alpha, 1+\alpha; -\frac{t}{F_1}\right) \right.$$
$$\left. - \left(\frac{t}{F_2}\right)^\alpha F\left(1, \alpha, 1+\alpha; -\frac{t}{F_2}\right) \right), \tag{7.30}$$

which gives $P(0, \infty) = 1$ as expected. Here $F(\alpha, \beta, \gamma, x)$ is the hypergeometric function[2]. Using the relation Eq. (7.32), for large F_1 and F_2, the asymptotic behavior of Eq. (7.30) is

$$P(F_1, F_2) \sim \frac{\alpha \sin(\pi\alpha)}{(1-\alpha)\pi t^{1-\alpha}} \left(F_1^{1-\alpha} - F_2^{1-\alpha}\right). \tag{7.33}$$

Note that the above equation is effective for the distributions with power law tails under the conditions that $t/F_1, t/F_2 \gg 1$. From Eq. (7.33), we find that $P(F_1, F_2)$ has a power law attenuation for large F_1 and F_2. However, for $F_1, F_2 \ll 1$, $P(F_1, F_2)$ can be represented as

$$P(F_1, F_2) = \frac{1}{b_\alpha \Gamma(\alpha) t^{1-\alpha}} \int_{F_1}^{F_2} \int_F^\infty \phi(y) dy dF. \tag{7.34}$$

Using a change of variable, we get

$$f_{F/t}(t, x) = t f_F(t, xt). \tag{7.35}$$

Still as for other examples in this chapter, this trick is used in order to compare with the scaled form of the bulk fluctuations.

---

[2]The hypergeometric function $F(\alpha, \beta, \gamma; x)$ [Abramowitz and Stegun (1984)], called the first hypergeometric function, is a solution of the Gaussian hypergeometric equation [Seaborn (1991)]. In general, hypergeometric function arises most frequently in physical problems. For $\gamma \neq 0, -1, -2, \ldots$, $F(\alpha, \beta, \gamma; x)$ can be can be expressed in terms of the hypergeometric series. The right hand side of Eq. (7.42) certainly converges for $|x| < 1$. When $x \to 0$, from Eq. (7.42), there exists

$$F(\alpha, \beta, \gamma; x) \sim 1 + \frac{\alpha\beta}{\gamma} x. \tag{7.31}$$

While for $x \to \infty$, we have

$$F(\alpha, \beta, \gamma; -x) \sim \frac{\Gamma(-\alpha+\beta)\Gamma(\gamma)}{\Gamma(\beta)\Gamma(-\alpha+c)} x^{-\alpha} + \frac{\Gamma(\alpha-\beta)\Gamma(\gamma)}{\Gamma(\alpha)\Gamma(\gamma-\beta)} x^{-\beta}. \tag{7.32}$$

Furthermore, many of the special functions of mathematical physics can be expressed in terms of specific forms of the hypergeometric function, for example

$$F(1, 1, 1, z) = \frac{1}{1-z}.$$

7.4.2 *Forward Recurrence Time with $1 < \alpha < 2$*

For $F \ll t$, according to Eq. (7.21)

$$\widehat{f}_F(s, u) \sim \frac{1 - \widehat{\phi}(u)}{u \langle \tau \rangle s},$$

where as mentioned $\langle \tau \rangle$ is finite. This can be finally inverted, yielding the typical fluctuations [Feller (1971); Tunaley (1974); Godrèche and Luck (2001)]

$$f_F(t, F) \sim \frac{1}{\langle \tau \rangle} \int_F^\infty \phi(y) dy. \tag{7.36}$$

Since $1 < \alpha < 2$, Eq. (7.36) is a normalized PDF and independent of the observation time t, which is different from Eq. (7.24), but they have similar forms. This is the reason why in the previous subsection we called $\langle \tau^* \rangle$ the effective average waiting time.

Next we discuss the uniform approximation, which is valid for varieties of F and large t, namely within uniform approximation, we only need the condition that t is large but the ratio of F and t is arbitrary. It can be noticed that Eq. (7.21) can be arranged into the following formula

$$\widehat{f}_F(s, u) = \frac{\widehat{\phi}(u) - 1}{(s - u)\left(1 - \widehat{\phi}(s)\right)} + \frac{1}{s - u}.$$

For $F \neq t$, we may neglect the second term, then using $1 - \widehat{\phi}(s) \sim \langle \tau \rangle s$ and inverting we get

$$f_F(t, F) \simeq \frac{1}{\langle \tau \rangle} \int_F^{t+F} \phi(y) dy, \tag{7.37}$$

which captures both the infinite density and the bulk fluctuations; see Fig. 7.8. Here, Eq. (7.37) is true for large t without considering the relation between t and F. If $F \ll t$, Eq. (7.37) can be approximated by Eq. (7.36).

For the rare fluctuations, i.e., both s and u are small and comparable, inserting Eq. (1.27) into Eq. (7.21), yields

$$\widehat{f}_F(s, u) \sim \frac{1}{s} + \frac{b_\alpha}{\langle \tau \rangle} \frac{u^\alpha - s^\alpha}{(s - u)s}.$$

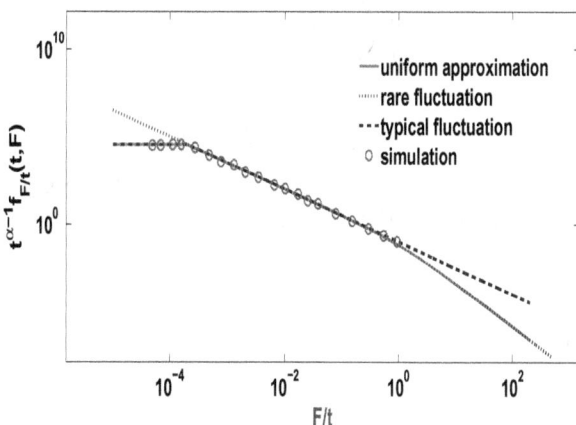

Fig. 7.8 The PDF of $x = F/t$ multiplied by $t^{\alpha-1}$ versus F/t generated by 3×10^6 trajectories with Eq. (1.26). We choose $\alpha = 1.5$ and $\tau_0 = 0.1$. To obtain our theoretical results we use Eqs. (7.37), (7.38) and (7.36). As the figure shows, Eq. (7.38), describing the large deviations is valid here for large F/t, though we experience a sampling problem in simulation.

For $F > 0$, taking double inverse Laplace transforms, we find [3]

$$f_F(t, F) \sim \frac{b_\alpha}{|\Gamma(1-\alpha)|\langle\tau\rangle}(F^{-\alpha} - (F+t)^{-\alpha}), \qquad (7.38)$$

which is consistent with Eq. (7.36) for large F and $t \to \infty$. Besides, for $t \to \infty$, $f_F(t, F)$ decays as $F^{-\alpha}$ independent of the observation time t. On the other hand, if $F \gg t$, $f_F(t, F)$ grows linearly with t, namely $f_F(t, F) \sim t/F^{-1-\alpha}$. In addition, using the asymptotic behavior of $\phi(F)$, for large F the uniform approximation Eq. (7.37) reduces to Eq. (7.38). As for other examples in this chapter, we may still use Eq. (7.38) to calculate a class of high order moments (for example $\alpha > q > \alpha - 1$), i.e., those moments which are integrable w.r.t. this infinite density.

With the help of the above equations, now we turn our attention to the fractional moment, defined by

$$\langle F^q \rangle = \int_0^\infty F^q f_F(t, F) dF. \qquad (7.39)$$

---

[3] For simulations presented in Fig. 7.8, we use 3×10^6 particles on a standard workstation, taking about 1 day. We see that in this case we do not sample the rare events. In [Rebenshtok *et al.* (2014a)], simulations of the Lévy walk process with 10^{10} particles are performed, in order to explore graphically the far tails of the propagator of the Lévy walk. When increasing the number of particles, we will observe rare events, however clearly in our case 3×10^6 realizations are simply not sufficient for meaningful sampling.

Using the calculated results of $f_F(t, F)$, we study fractional moment $\langle F^q \rangle$. First, we obtain the low order moment with $\alpha > 1$, i.e., $q < \alpha - 1$. Using Eqs. (7.36) and (7.39), and utilizing integration by parts

$$\langle F^q \rangle \sim \frac{1}{(q+1)\langle \tau \rangle} \int_0^\infty F^{q+1} \phi(F) dF. \tag{7.40}$$

We notice that the right hand side of Eq. (7.40) is a finite number due to $q - \alpha < -1$. Then, we discuss the case of $q > \alpha - 1$. According to Eq. (7.38)

$$\langle F^q \rangle \sim \lim_{z \to 0} \int_z^\infty (F^{-\alpha} - (F+t)^{-\alpha}) \frac{b_\alpha F^q}{|\Gamma(1-\alpha)|\langle \tau \rangle} dF$$

$$\sim \lim_{z \to 0} \frac{b_\alpha z^{1-\alpha+q}}{|\Gamma(1-\alpha)|\langle \tau \rangle (1+q-\alpha)} F\left(\alpha, -1+\alpha-q, \alpha-q; -\frac{t}{z}\right), \tag{7.41}$$

where $F(\alpha, \beta, \gamma; x)$ is the hypergeometric function [Abramowitz and Stegun (1984); Seaborn (1991)], defined by

$$F(\alpha, \beta, \gamma; x) = 1 + \sum_{n=1}^\infty \frac{(\alpha)_n (\beta)_n}{(\gamma)_n} \frac{x^n}{n!} \tag{7.42}$$

with $(\alpha)_n = \alpha(\alpha+1)\dots(\alpha+n-1)$. Note that the asymptotic behavior of $F(\alpha, \beta, \gamma; -x)$ is

$$F(\alpha, \beta, \gamma; -x) \sim x^{-\alpha} \frac{\Gamma(\beta-\alpha)\Gamma(\gamma)}{\Gamma(\beta)\Gamma(\gamma-\alpha)} + x^{-\beta} \frac{\Gamma(\alpha-\beta)\Gamma(\gamma)}{\Gamma(\alpha)\Gamma(\gamma-\beta)} \tag{7.43}$$

with $x > 0$. Using Eq. (7.43), the dominant term of Eq. (7.41) gives

$$\langle F^q \rangle \sim \frac{b_\alpha \Gamma(1+q)\Gamma(\alpha-q)}{|\Gamma(1-\alpha)|\langle \tau \rangle (1+q-\alpha)\Gamma(\alpha)} t^{1-\alpha+q}. \tag{7.44}$$

Keep in mind that substituting uniform approximation Eq. (7.37) into Eq. (7.39) yields the same result as Eq. (7.44). Thus, we have

$$\langle F^q \rangle \sim \begin{cases} \dfrac{\int_0^\infty F^{q+1} \phi(F) dF}{(q+1)\langle \tau \rangle}, & q < \alpha - 1; \\[3mm] \dfrac{b_\alpha \Gamma(\alpha-q)\Gamma(1+q) t^{q+1-\alpha}}{\langle \tau \rangle |\Gamma(1-\alpha)|\Gamma(\alpha)(1+q-\alpha)}, & \alpha - 1 < q < \alpha; \\[3mm] \infty, & q > \alpha. \end{cases} \tag{7.45}$$

This is to say, for $q < \alpha - 1$, $\langle F^q \rangle$ is a constant, namely, it does not depend on the observation time t. Moments of order $q < \alpha$ are determined by the known result Eq. (7.36), which describes typical fluctuations when F is of the order t^0. The rare fluctuations, described by Eq. (7.38), give information of events with $F \propto t$, and this non-normalized density Eq. (7.38) yields the moment of $\alpha - 1 < q < \alpha$. Especially, if $q = 1$, $\langle F \rangle \sim b_\alpha (\langle \tau \rangle \Gamma(3-\alpha))^{-1} t^{2-\alpha}$, so in this case the mean is determined by the infinite density. When $q > \alpha$, $\langle F^q \rangle$ is divergent. This is expected since the moment of order $q > \alpha$ of $\phi(\tau)$ diverges.

7.5   Backward Recurrence Time

Compared with the forward recurrence time, one of the important difference is that B can not be larger than t. In some cases, B is called the age at time t. Because in the lightbulb lifetime example, it represents the age of the light bulb you find burning at time t. Similar to the derivation of the forward recurrence time

$$f_B(t,B) = \sum_{N=0}^{\infty} \int_0^t Q_N(\tau)\delta(t-\tau-B) \int_B^{\infty} \phi(y)dyd\tau.$$

In Laplace space, let $t \to s$ and $B \to u$. Using the convolution theorem of Laplace transform and Eq. (1.37), this gives

$$\widehat{f}_B(s,u) = \frac{1-\widehat{\phi}(s+u)}{s+u}\frac{1}{1-\widehat{\phi}(s)}, \qquad (7.46)$$

which is derived in [Godrèche and Luck (2001)] using a different method.

7.5.1   *Backward Recurrence Time with $0 < \alpha < 1$*

First of all, we study the behaviors of large deviations. For $B \ll t$, i.e., $s \ll u$

$$\widehat{f}_B(s,u) \sim \frac{1-\widehat{\phi}(u)}{u}\frac{1}{1-\widehat{\phi}(s)}. \qquad (7.47)$$

In the long time limit, i.e., $s \to 0$, $\widehat{f}_B(s,u) \sim (1-\widehat{\phi}(u))/(b_\alpha\,us^\alpha)$. Performing the double inverse Laplace transforms w.r.t. s and u, respectively, yields

$$f_B(t,B) \sim \frac{\int_B^{\infty}\phi(y)dy}{\langle\tau^*\rangle}. \qquad (7.48)$$

Note that $\int_0^{\infty} f_B(t,B)dB = \infty$, which means that Eq. (7.48) is non-normalized. Here $\langle\tau^*\rangle$ is defined below Eq. (7.24). We can see that $\lim_{B\to 0} f_B(t,B)\langle\tau^*\rangle \sim 1$, which is confirmed in Fig. 7.9.

Now we construct a uniform approximation, which is valid for a wider range of B, though t is large. We rewrite Eq. (7.46) as

$$\widehat{f}_B(s,u) = \left(\frac{1}{s+u} - \frac{\widehat{\phi}(s+u)}{s+u}\right)\frac{1}{1-\widehat{\phi}(s)}.$$

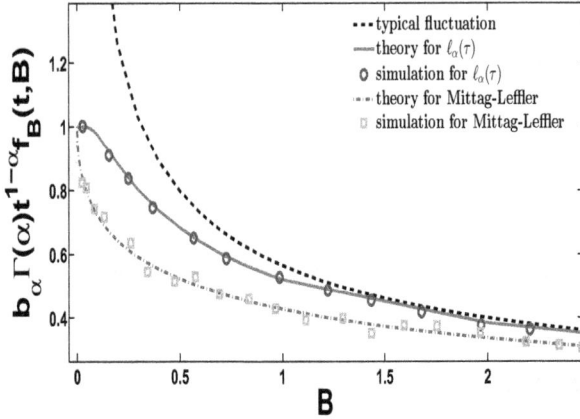

Fig. 7.9   The scaled PDF of the backward recurrence time B, when B is of the order unity. The parameters are $t = 1000$ and $\alpha = 0.5$. The full and dash-dot lines are the analytical result Eq. (7.48) depicting the large deviations. For the typical result, we use Eq. (7.55), i.e., $f_B(t, B) \sim 1/(\pi\sqrt{B(t-B)})$. Besides, the symbols are obtained by averaging 10^7 trajectories with one sided Lévy stable distribution Eq. (1.30) and Mittag-Leffler Eq. (7.3) time statistics, respectively.

For simplification, let $\phi(\tau)$ be the one sided Lévy stable distribution Eq. (1.29). Expanding the above equation, i.e., $1/(1 - \widehat{\phi}(s)) = \sum_{n=0}^{\infty} \widehat{\phi}^n(s)$, and then using the convolution theorem of the Laplace transform

$$f_B(t, B) = \delta(t - B) \int_t^{\infty} \ell_\alpha(y) dy + \theta(t - B)$$

$$\times \int_B^{\infty} \ell_\alpha(y) dy \sum_{n=1}^{\infty} \frac{1}{n^{1/\alpha}} \ell_\alpha\left(\frac{t - B}{n^{1/\alpha}}\right). \tag{7.49}$$

The $\theta(t - B)$ in Eq. (7.49) yields $B \leq t$ as expected. In addition, for $B \ll t$, Eq. (7.49) reduces to Eq. (7.48). Note that, for $\alpha = 1/2$, and comparable t and B, Eq. (7.49) is consistent with the arcsine law, while, let B go to either 0 or t (the extreme cases), the arcsine law does not work anymore; see Fig. 7.10.

Now we turn our attention to the case of $B \to t$, using the random variable $\varepsilon = t - B \to 0$. In Laplace space, the PDF of ε is

$$\widehat{f}_\varepsilon(s, u_\varepsilon) = \int_0^{\infty} \int_0^{\infty} \exp(-st - u_\varepsilon \varepsilon) f_\varepsilon(t, \varepsilon) dt d\varepsilon$$

$$= \widehat{f}_B(s + u_\varepsilon, -u_\varepsilon). \tag{7.50}$$

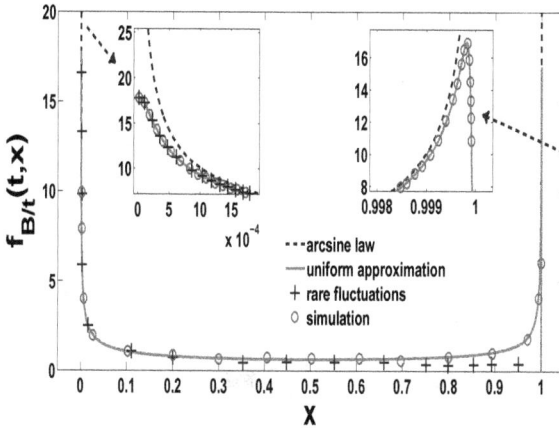

Fig. 7.10   The PDF $f_{B/t}(t, x)$ versus $x = B/t$ for a renewal process with $\phi(\tau)$ being a one sided Lévy density Eq. (1.30). Here we choose $t = 1000$, and $\alpha = 0.5$. The dashed, the full lines and the symbols (+) represent the arcsine law Eq. (7.56), the analytical result Eq. (7.49), and rare events Eq. (7.48), respectively. Notice what the arcsine law predicts here is a symmetric distribution, while our results describing the large deviations exhibit non-symmetry. Furthermore our theory does not blow up at $x \to 0$ and $x \to 1$, unlike the arcsine law.

According to Eq. (7.46)

$$\widehat{f}_\varepsilon(s, u_\varepsilon) = \frac{1 - \widehat{\phi}(s)}{s} \frac{1}{1 - \widehat{\phi}(s + u_\varepsilon)}. \tag{7.51}$$

For $s \ll u_\varepsilon$, performing the double inverse Laplace transforms and using $B = t - \varepsilon$

$$f_B(t, B) = \int_t^\infty \phi(y) dy \mathcal{L}_{t-B}^{-1} \left[\frac{1}{1 - \widehat{\phi}(u_\varepsilon)} \right]. \tag{7.52}$$

Let us consider a situation in which $\phi(\tau)$ is the Mittag-Leffler distribution Eq. (7.3) with $0 < \alpha < 1$. Next, plugging Eq. (7.4) into Eq. (7.52) yields

$$f_B(t, B) = \Phi(t)\delta(t - B) + \Phi(t)\frac{(t - B)^{\alpha-1}}{\Gamma(\alpha)}. \tag{7.53}$$

It demonstrates that $t^\alpha f_B(t, B)$ decays like $(t - B)^{\alpha-1}$. Thus, if B tends to the observation time t, we discover an interesting phenomenon that $t^\alpha f_B(t, B) \to \infty$, verified in Fig. 7.11. In general, Eq. (7.52) is not easy to calculate in real time exactly, though we use the numerical inversion of Laplace transform by MATLAB. Expanding the above equation, we find

$$f_\varepsilon(t, \varepsilon) = \int_t^\infty \phi(y) dy \left(\delta(\varepsilon) + \sum_{n=1}^\infty \mathcal{L}_\varepsilon^{-1} \left[\widehat{\phi}^n(u_\varepsilon) \right] \right).$$

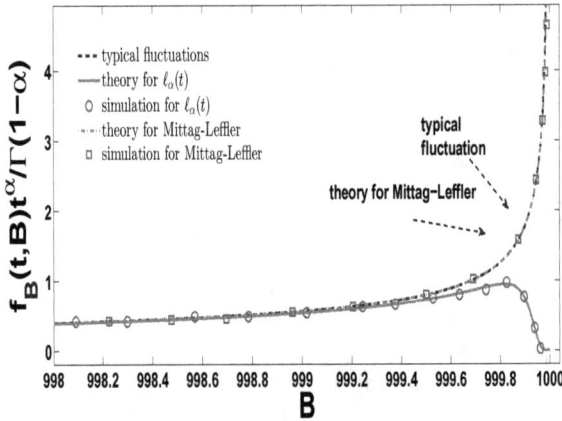

Fig. 7.11   The relation between $f_B(t, B)$ and large B. The parameters are $t = 1000$ and $\alpha = 0.5$. For the typical result we use Eq. (7.56) and for the large deviations we use Eqs. (7.53) and (7.54). The simulations, presented by symbols, are obtained by averaging 3×10^7 trajectories. It is difficult to distinguish between the typical result and the theoretical result with Mittag-Leffler time statistics, while for the choice of $\phi(\tau) = \ell_{1/2}(\tau)$ Eq. (1.30), the deviations are pronounced.

Consider a specific $\phi(\tau)$, namely one sided Lévy stable distribution

$$f_B(t, B) = \int_t^\infty \ell_\alpha(y)dy \left(\delta(t - B) + \sum_{n=1}^\infty \frac{1}{n^{1/\alpha}} \ell_\alpha \left(\frac{t - B}{n^{1/\alpha}} \right) \right), \quad (7.54)$$

which can be used for plotting. To summarize, large deviations are observed for $B \propto t^0$ and $B \to t$, Eq. (7.48) and Eqs. (7.53), (7.54) respectively (see Fig. 7.10), and these are non-symmetric for one sided Lévy distribution. Only when $B \sim t^0$, we find a non-normalized density, Eq. (7.48).

Next we discuss the typical fluctuations when $B \propto t$. Combining Eqs. (1.27) and (7.46), yields [Godrèche and Luck (2001)]

$$f_B(t, B) \sim \frac{\sin(\pi\alpha)}{\pi} \frac{1}{B^\alpha(t - B)^{1-\alpha}} \theta(t - B). \quad (7.55)$$

In a particular case $\alpha = 1/2$, Eq. (7.55) reduces to the arcsine law $f_B(t, B) \sim (\pi\sqrt{B(t - B)})^{-1}$, which is plotted by the dashed line in Figs. 7.9, 7.10 and 7.11. Let $x = B/t$. We get a well known formula [Godrèche and Luck (2001)]

$$f_{B/t}(x) \sim \frac{\sin(\pi\alpha)}{\pi} \frac{1}{x^\alpha(1 - x)^{1-\alpha}} \theta(1 - x). \quad (7.56)$$

In particular, for $\alpha = 1/2$, Eq. (7.56) reduces to the arcsine law [Godrèche and Luck (2001)] on $(0, 1)$, see Fig. 7.10.

We further study the probability of B when it is smaller than B_2 and larger than B_1. Combining Eq. (7.21), for large B, $P(B_1, B_2)$ can be shown by

$$P(B_1, B_2) \sim \frac{\sin(\pi\alpha)}{\pi} \left(B\left(\frac{B_2}{t}, 1 - \alpha, \alpha\right) - B\left(\frac{B_1}{t}, 1 - \alpha, \alpha\right) \right),$$

where $B(z, \alpha, \beta)$ is incomplete Beta function [Polyanin and Manzhirov (2007)], defined by $B(z, \alpha, \beta) = \int_0^z \tau^{\alpha-1}(1 - t)^{\beta-1}d\tau$. As expected $P(B_1, B_2) = 1$ when $B_1 \to 0$ and $B_2 \to t$. If B_1/t, $B_2/t \ll 1$, we obtain

$$P(B_1, B_2) \sim \frac{\sin(\pi\alpha)}{(1 - \alpha)\pi} \frac{B_2^{1-\alpha} - B_1^{1-\alpha}}{t^{1-\alpha}}. \tag{7.57}$$

Compared with the forward case, Eqs. (7.30) and (7.57) have the same power law behaviors but with different coefficients, which is similar to the relation between first moment of $f_B(t, B)$ and $f_F(t, F)$. Besides, for $B_1, B_2 \ll 1$, $P(B_1, B_2)$ is similar to Eq. (7.34).

We now study the fractional moment of B. Note that B^q with $q > \alpha$, are non-integrable w.r.t. the non-normalized density Eq. (7.48). We find that the fractional moment of B is governed by the typical fluctuations Eq. (7.55), namely

$$\begin{aligned} \langle B^q \rangle &= \int_0^\infty B^q f_B(B, t)dB \\ &\sim \frac{\sin(\pi\alpha)\Gamma(\alpha)\Gamma(1 - \alpha + q)}{\pi\Gamma(1 + q)}t^q. \end{aligned} \tag{7.58}$$

We check this result in the following: for a natural number q, expanding Eq. (7.46) as a Taylor series in u, and performing the inverse Laplace transform term by term, we obtain the corresponding moment, which is the same as Eq. (7.58).

To summarize, if $\alpha < 1$, then for all observables in this chapter, i.e., N, F, B, Z, and T^+, the moments (if they exist) are obtained by the PDF describing the typical fluctuations. Note that for F and Z, high order $(q > \alpha)$ moment diverges. One may wonder in what sense is Eq. (7.48) an infinite density? For that we consider the observable $\theta(B_1 < B < B_2)$ with $B_1, B_2 \ll t$, where $\theta(B_1 < B < B_2)$ is one if the condition holds. Then

$$\begin{aligned} \langle \theta(B_1 < B < B_2) \rangle &= \int_0^\infty \theta(B_1 < B < B_2)f_B(t, B)dB \\ &\sim \frac{1}{\langle \tau^* \rangle} \int_{B_1}^{B_2} \int_B^\infty \phi(y)dydB, \end{aligned}$$

where $\langle \tau^* \rangle$, defined below Eq. (7.24), is the effective average waiting time. In other words, the observable $\theta(B_1 < B < \theta_2)$ is integrable w.r.t. the non-normalized density, and hence the latter one is used for the calculation of the average $\theta(B_1 < B < \theta_2)$.

7.5.2 *Backward Recurrence Time with $1 < \alpha < 2$*

We again consider the limit $s \ll u$. Combining Eqs. (1.28) and (7.46)

$$\widehat{f}_B(s, u) \sim \frac{1 - \widehat{\phi}(u)}{u} \frac{1}{\langle \tau \rangle s},$$

which, by the double inverse Laplace transforms, yields the limiting result [Feller (1971)]

$$f_B(t, B) \sim \frac{\int_B^\infty \phi(y) dy}{\langle \tau \rangle}.$$

If B goes to 0, $f_B(t, B)$ reduces to $1/\langle \tau \rangle$.

Now we turn our attention to the case when $B \to t$. According to Eq. (7.51)

$$f_B(t, B) \sim \int_t^\infty \phi(y) dy \mathcal{L}_{t-B}^{-1} \left[\frac{1}{1 - \widehat{\phi}(u_\epsilon)} \right]. \tag{7.59}$$

For power law waiting time statistics, Eq. (7.59) reduces to

$$f_B(t, B) \sim \frac{b_\alpha}{|\Gamma(1 - \alpha)| t^\alpha} \mathcal{L}_{t-B}^{-1} \left[\frac{1}{1 - \widehat{\phi}(u_\epsilon)} \right].$$

The inverse Laplace transform gives the limiting law when $B \to t$.

Let us proceed with the discussion of rare fluctuations. Substituting $\widehat{\phi}(u)$ and $\widehat{\phi}(s)$ into Eq. (7.46), leads to

$$\widehat{f}_B(s, u) \sim \frac{1}{s} - \frac{b_\alpha}{\langle \tau \rangle s} (s + u)^{\alpha - 1} + \frac{b_\alpha}{\langle \tau \rangle} \frac{1}{s^{2-\alpha}},$$

when s and u are of the same order. By inversion of the above equation

$$f_B(t, B) \sim \frac{b_\alpha}{\langle \tau \rangle |\Gamma(1 - \alpha)|} B^{-\alpha} \theta(t - B) \tag{7.60}$$

with $B > 0$. We see that $f_B(t, B)$ blows up at $B \to 0$ and since $1 < \alpha < 2$ the solution Eq. (7.60) is non-integrable.

Next, utilizing Eq. (7.46), the uniform approximation is

$$f_B(t, B) \sim \frac{1}{\langle \tau \rangle} \theta(t - B) \int_B^\infty \phi(y) dy. \tag{7.61}$$

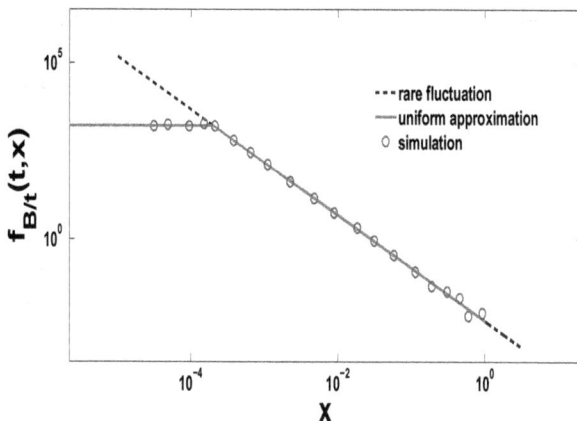

Fig. 7.12   The PDF $f_{B/t}(t, B)$ versus the scaling variable $x = B/t$ for the waiting time PDF Eq. (1.26). The parameters are $t = 800$, $\tau_0 = 0.1$ and $\alpha = 1.5$. The solid line is the theory Eq. (7.61), and the dashed line is Eq. (7.60), which gives the PDF when B is of the order t, and t is large. The simulations, presented by the symbols, are obtained by averaging 10^6 trajectories.

Note that B is limited by the observation time t. For large B, Eq. (7.61) reduces to Eq. (7.60).

The corresponding fractional moments are

$$
\langle B^q \rangle \sim
\begin{cases}
\dfrac{\int_0^\infty \phi(B)B^{q+1}dB}{\langle \tau \rangle (q+1)}, & q < \alpha - 1; \\[3ex]
\dfrac{b_\alpha t^{q-\alpha+1}}{\langle \tau \rangle |\Gamma(1-\alpha)| (q-\alpha+1)}, & q > \alpha - 1.
\end{cases}
\tag{7.62}
$$

Since $B < t$ all moments are finite, unlike the case of the forward recurrence time. The results show that the behaviors of fractional moments are divided into two parts. When $q < \alpha - 1$, $\langle B^q \rangle$ is determined by the typical fluctuations and it is a constant. The rare fluctuations, described by Eq. (7.60), give the information on events when $B \propto t$, and this non-normalized limiting law gives the moments of $q > \alpha - 1$.

7.6   Time Interval Straddling t

The time interval straddling the observation time t has been studied in [Barkai *et al.* (2014); Bertoin *et al.* (2006); Chung (1976); Getoor and Sharpe (1979)], where some results about the typical fluctuations are announced and discussed. To consider general initial ensemble in an annealed transit time model [Akimoto and Yamamoto (2016)], one has to consider the time interval straddling time t since the diffusion coefficient is governed by Z. Based on the previous result [Barkai *et al.* (2014)], the PDF of Z is given by the double Laplace inversions of

$$\widehat{f}_Z(s, u) = \frac{1}{1 - \widehat{\phi}(s)} \frac{\widehat{\phi}(u) - \widehat{\phi}(s + u)}{s}, \tag{7.63}$$

where u is the Laplace pair of Z, and s of t. One important feature of $f_Z(t, Z)$ is the discontinuity of its derivative at $Z = t$; see below.

7.6.1   *Time Interval Straddling t with $0 < \alpha < 1$*

Similar to previous sections, we first consider the events of large deviations, namely, Z is of the order t^0. Utilizing Eq. (7.63)

$$f_Z(t, Z) \sim \mathcal{L}_t^{-1} \left[\frac{1}{1 - \widehat{\phi}(s)} \right] Z\phi(Z), \tag{7.64}$$

which gives us an efficient way of calculation for $Z \ll t$. In particular, combining Eqs. (1.27) and (7.64), and taking the inverse Laplace transform lead to

$$f_Z(t, Z) \sim \frac{Z\phi(Z)}{\langle \tau^* \rangle}, \tag{7.65}$$

which is confirmed in Fig. 7.13. Note that $\langle \tau^* \rangle$ is the same as that defined in Eq. (7.24). Keep in mind that there is a difference among small F, B and Z. For small Z, $f_Z(t, Z)$ goes to 0, while for Eqs. (7.24) and (7.48) with $F, B \to 0$, $f_F(t, F)$ and $f_F(t, B)$ are equal to $1/\langle \tau^* \rangle$. In spite of these differences, the asymptotic behavior of $f_Z(t, Z)$ is consistent with the PDF of the forward recurrence time and the backward one with the increase of Z.

We further consider the PDF of Z more exactly. Taking inverse Laplace transforms of Eq. (7.63) w.r.t. u and s, respectively

$$f_Z(t, Z) = \phi(Z) \left(\mathcal{L}_t^{-1} \left[\frac{1}{s(1 - \widehat{\phi}(s))} \right] - \theta(t - Z) \mathcal{L}_{t-Z}^{-1} \left[\frac{1}{s(1 - \widehat{\phi}(s))} \right] \right). \tag{7.66}$$

In particular, for a Mittag-Leffler density Eq. (7.3), the inversion of Eq. (7.66) can be further simplified as

$$f_Z(t, Z) = \frac{\phi(Z)}{\Gamma(1+\alpha)} \left(t^\alpha - (t-Z)^\alpha \theta(t-Z)\right)$$
$$+ \phi(Z)(1 - \theta(t-Z)). \tag{7.67}$$

It is interesting to note that Eq. (7.67) is a uniform approximation for Mittag-Leffler sojourn time. In addition, we find that for $Z \ll t$ Eq. (7.67) reduces to Eq. (7.65). On the other hand, when $Z > t$, the above equation yields $f_Z(t, Z) \sim t^\alpha \phi(Z)/\Gamma(1+\alpha)$. For $t \to 0$, we see from Eq. (7.67) that $\lim_{t \to 0} f_Z(t, Z) = \phi(Z)$ as expected.

Let us proceed with the discussion of a general waiting time PDF $\phi(\tau)$. Expanding the term $(1 - \widehat{\phi}(s))^{-1}$ of Eq. (7.66) in powers of $\widehat{\phi}(s)$, and then taking inverse transform result in

$$f_Z(t, Z) = \phi(Z) \sum_{n=1}^{\infty} \left(\int_0^t \mathcal{L}_\tau^{-1}[\widehat{\phi}^n(s)] d\tau - \theta(t-Z) \right.$$
$$\left. \times \int_0^{t-Z} \mathcal{L}_\tau^{-1}[\widehat{\phi}^n(s)] d\tau \right) + \phi(Z)(1 - \theta(t-Z)). \tag{7.68}$$

For the one sided Lévy stable distribution

$$f_Z(t, Z) = \ell_\alpha(Z) \sum_{n=1}^{\infty} \frac{1}{n^{1/\alpha}} \left(\int_0^t \ell_\alpha\left(\frac{\tau}{n^{1/\alpha}}\right) d\tau - \theta(t-Z) \right.$$
$$\left. \times \int_0^{t-Z} \ell_\alpha\left(\frac{\tau}{n^{1/\alpha}}\right) d\tau \right) + \ell_\alpha(Z)(1 - \theta(t-Z)). \tag{7.69}$$

Note that Eq. (7.69) is valid for all kinds of t and Z. In Fig. 7.14, the scaling behaviors of $x = Z/t$ are displayed. If $Z > t$, Eq. (7.66) reduces to

$$f_Z(t, Z) \sim \frac{1}{\Gamma(1+\alpha)b_\alpha} t^\alpha \phi(Z). \tag{7.70}$$

For small and comparable s, u, substituting $\widehat{\phi}(s)$ and $\widehat{\phi}(u)$ into Eq. (7.63) yields

$$\widehat{f}_Z(s, u) \sim \frac{(s+u)^\alpha - u^\alpha}{s^{1+\alpha}},$$

and then taking double inverse Laplace transforms w.r.t. u and s, gives the typical fluctuations [Barkai *et al.* (2014); Dynkin (1961)]

$$f_Z(t, Z) \sim \frac{\sin(\pi\alpha)}{\pi} \frac{t^\alpha - (t-Z)^\alpha \theta(t-Z)}{Z^{1+\alpha}}. \tag{7.71}$$

7.6.2   *Time Interval Straddling Time t with* $1 < \alpha < 2$

For the typical fluctuations, i.e., $Z \sim t^0$. Based on Eq. (7.63),

$$f_Z(t, Z) \sim \frac{Z\phi(Z)}{\langle \tau \rangle}; \qquad (7.72)$$

see Fig. 7.15. Note that $f_Z(t, Z)$ tends to zero when $Z \to 0$. We now discuss the rare fluctuations, i.e., Z is of the order t. Plugging Eq. (1.27) into Eq. (7.63), then performing the inverse Laplace transform, lead to

$$f_Z(t, Z) \sim \frac{b_\alpha Z^{-1-\alpha}}{\langle \tau \rangle \Gamma(-\alpha)}(t - (t - Z)\theta(t - Z)) \qquad (7.73)$$

with $Z > 0$. According to Eq. (7.73), it gives us another representation of $f_Z(t, Z)$, namely

$$f_Z(t, Z) \sim \begin{cases} \dfrac{b_\alpha}{\langle \tau \rangle \Gamma(-\alpha)} Z^{-\alpha}, & Z < t; \\[3mm] \dfrac{b_\alpha}{\langle \tau \rangle \Gamma(-\alpha)} t Z^{-\alpha-1}, & Z > t. \end{cases}$$

We now construct a uniform approximation that interpolates between Eqs. (7.72) and (7.73). We restart from Eq. (7.63), but use Eq. (1.28) only to approximate $1/(1 - \widehat{\phi}(s)) \sim 1/(\langle \tau \rangle s)$. After performing double inverse Laplace transforms, we arrive at

$$f_Z(t, Z) = \frac{C(t)}{\langle \tau \rangle}(t\phi(Z) - \theta(t - Z)(t - Z)\phi(Z)), \qquad (7.74)$$

where we have added $C(t) = \langle \tau \rangle / (\int_0^t Z\phi(Z)dZ + \int_t^\infty t\phi(Z)dZ)$ as a normalizing factor, satisfying $\lim_{t \to \infty} C(t) = 1$. In the long time limit, Eq. (7.74) gives

$$f_Z(t, Z) \sim \frac{t}{\langle \tau \rangle}\phi(Z) \qquad (7.75)$$

with $Z > t$. It can be seen that Eq. (7.75) grows linearly with time t.

Similar to the calculations of $\langle F^q \rangle$ and $\langle B^q \rangle$

$$\langle Z^q \rangle \sim \begin{cases} \dfrac{1}{\langle \tau \rangle} \displaystyle\int_0^\infty Z^{q+1}\phi(Z)dZ, & q < \alpha - 1; \\[3mm] \dfrac{b_\alpha t^{q-\alpha+1}}{\langle \tau \rangle \Gamma(-\alpha)(q - \alpha + 1)(\alpha - q)}, & \alpha - 1 < q < \alpha; \\[3mm] \infty, & \alpha < q. \end{cases} \qquad (7.76)$$

As expected $\langle Z^0 \rangle = 1$. Similar to the previous examples, when $\alpha - 1 < q < \alpha$, the moments $\langle Z^q \rangle$ are obtained from Eq. (7.73) which is not a normalized PDF. In particular, expanding the right hand side of Eq. (7.63) to first order in u, and taking the inverse Laplace transform, lead to $\langle Z \rangle \sim (\langle \tau \rangle \Gamma(3 - \alpha))^{-1} b_\alpha \alpha t^{2-\alpha}$, which agrees with Eq. (7.76).

7.7 Occupation Time

The occupation time, the time spent by a process in a given subset of the state space during the interval of the observation, is widely investigated in mathematics and physics. It is a useful quantity with a large number of applications, for example the time spent by one dimensional Brownian motion in half space, the time spent in the bright state for blinking quantum dot models [Bianco *et al.* (2005); Majumdar and Comtet (2002)], and the time that a spin occupies in a state up [Majumdar (1999)]. Based on the alternating renewal process, here we focus on the study of the occupation time in the '+' state. Now, our aim is to obtain the PDF of the occupation time. Let $Q_N(t, T^+)$ be the PDF of the occupation time just arriving at T^+ at time t after finishing N steps. We have

$$Q_{N+2}^+(t, T^+) = \int_0^{T^+} \int_0^t Q_N(t - \tau - z, T^+ - z) \tag{7.77}$$
$$\times \phi(z)\phi(\tau)d\tau dz + q\delta(t)\delta(t - T^+)$$

and

$$Q_{N+2}^-(t, T^+) = \int_0^{T^+} \int_0^t Q_N(t - \tau - z, T^+ - z) \tag{7.78}$$
$$\times \phi(z)\phi(\tau)d\tau dz + (1 - q)\delta(t)\delta(T^+),$$

where $\pm$ in the superscript of $Q_{N+2}^\pm(t, T^+)$ means that the initial state of the particle is $\pm$. q is the probability that the initial state is '+', with $0 \le q \le 1$. In double Laplace space, representation of the above Eq. (7.77) takes an especially simple form

$$\widehat{Q}_{N+2}^+(s, u) = \frac{q}{1 - \widehat{\phi}(s + u)\widehat{\phi}(s)}. \tag{7.79}$$

Then the PDF $f_{T^+}(t, T^+)$ is

$$f_{T^+}(t, T^+) = \sum_{N=0}^{\infty} \left(\int_0^t \int_0^{T^+} Q_N^+(t - \tau - z, T^+ - z) \int_z^{\infty} \phi(y)dydzdt \right.$$
$$+ \int_0^t \int_0^{T^+} Q_N^+(t - \tau - z, T^+ - z)\phi(z) \int_\tau^{\infty} \phi(y)dydzdt$$
$$+ \int_0^t \int_0^{T^+} Q_N^-(t - \tau - z, T^+ - z)\phi(\tau) \int_z^{\infty} \phi(y)dydzdt$$
$$+ \left. \int_0^t Q_N^-(t - \tau, T^+) \int_\tau^{\infty} \phi(y)dydt \right).$$

Taking double Laplace transforms, summing the infinite terms, and then from Eq. (7.79) it follows that

$$\widehat{f}_{T^+}(s, u) = \frac{u(1 - q - q\widehat{\phi}(s + u))(1 - \widehat{\phi}(s))}{s(s + u)(1 - \widehat{\phi}(s)\widehat{\phi}(s + u))} + \frac{s}{s(s + u)}. \tag{7.80}$$

For $q = 1/2$, Eq. (7.80) reduces to

$$\widehat{f}_{T^+}(s, u) = \frac{2s + u}{2s(s + u)} + \frac{u(\widehat{\phi}(s + u) - \widehat{\phi}(s))}{2s(s + u)(1 - \widehat{\phi}(s + u)\widehat{\phi}(s))}; \tag{7.81}$$

Eq. (7.81) and the corresponding typical fluctuations are studied in [Godrèche and Luck (2001); Margolin *et al.* (2005)]. Note that in this model we start the process in the state 'up' and 'down' with equal probability. Utilizing Eq. (7.81) and taking $\epsilon = T^+ - t/2$, we detect that the PDF $f_\epsilon(t, \epsilon)$ is symmetric w.r.t. ϵ for a variety of $\phi(\tau)$. As usual the difficulty is to find the solution in real time, namely find the PDF $f_{T^+}(t, T^+)$.

7.7.1 *Occupation Time with* $0 < \alpha < 1$

We first consider the typical fluctuations, i.e., T^+ is of the order t. Substituting Eq. (1.27) into Eq. (7.81), and then taking the inverse double Laplace transforms, yield the PDF of T^+/t [Godrèche and Luck (2001); Lamperti (1958)]

$$\lim_{t \to \infty} f_{T^+/t}(x) \sim \frac{\sin(\pi\alpha)}{\pi} \frac{x^{\alpha-1}(1 - x)^{\alpha-1}}{x^{2\alpha} + (1 - x)^{2\alpha} + 2\cos(\pi\alpha)x^\alpha(1 - x)^\alpha} \tag{7.82}$$

with $0 < x < 1$. It implies that the probability distribution of the random variable $x = T^+/t$ will converge in the limit of long t, to a limiting distribution which is independent of t. In a particular case $\alpha = 1/2$, Eq. (7.82) reduces to the arcsine law on $(0, 1)$

$$\lim_{t \to \infty} f_{T^+/t}(x) \sim \frac{1}{\pi\sqrt{x(1 - x)}} \tag{7.83}$$

or

$$f_{T^+}(t, T^+) \sim \frac{1}{\pi\sqrt{T^+(t - T^+)}}. \tag{7.84}$$

Equation (7.82) is originally derived by Lamperti [Lamperti (1958)]; see also Darling-Kac law [Darling and Kac (1957)]. The typical fluctuations described by Eq. (7.83) is plotted by the dashed lines; see Figs. 7.16 and 7.17. Besides, for $\alpha = 0.5$ the typical result Eq. (7.83) implies that $f_{T^+/t}(x)$ blows up when $x \to 0$ and $x \to 1$.

Next we analyze the case of $T^+ \ll t$, i.e., $s \ll u$. Based on Eq. (7.81), we find the infinite density

$$f_{T^+}(t, T^+) \sim \mathcal{L}_{T^+}^{-1} \left[-\frac{1}{2} + \frac{1}{1 - \widehat{\phi}(u)} \right] \Phi(t), \qquad (7.85)$$

where $\Phi(t)$ is the survival probability defined by Eq. (1.35). Note that Eq. (7.85) is not normalized, which is not a problem since it is valid for $T^+ \ll t$.

We now investigate the infinite density Eq. (7.85) with two choices of $\phi(\tau)$. Similar to our previous examples the infinite density depends on the spectfics of $\phi(\tau)$ unlike the Lamperti law Eq. (7.82). Using the example of a Mittag-Leffler PDF $\phi(\tau)$, pluging Eq. (7.4) into Eq. (7.85) and then taking the inverse Laplace transform

$$f_{T^+}(t, T^+)/\Phi(t) \sim \frac{1}{2}\delta(T^+) + \frac{1}{\Gamma(\alpha)}(T^+)^{\alpha-1}; \qquad (7.86)$$

see Fig. 7.16. The first term on the right hand side is a delta function, it describes events where the process starts at state '$-$' and remains there for time t (the factor $1/2$ is due to the initial condition, the probability of $1/2$ to start in the state 'up' or 'down'). Furthermore, it is interesting to find that the typical result Eq. (7.82) is consistent with the theoretical result with Mittag-Leffler time statistics for all kinds of $0 < T^+ \ll t$, not including the delta function in Eq. (7.86).

Comparing Eq. (7.86) with typical fluctuations Eq. (7.84), we observe that for $\alpha = 1/2$ the occupation time with Mittag-Leffler waiting time produces large deviation statistics that are very similar to typical event statistics. But even in this very close scenario, we find an feature being exclusively revealed by the large deviation analysis. Namely, there is a discrete probability to find the occupation time being trapped in an initial state.

Now we derive a formal solution for the rare events. Using the relation $1/(1 - \widehat{\phi}(u)) = \sum_{n=0}^{\infty} \widehat{\phi}^n(u)$ and performing double inverse Laplace transforms, lead to

$$f_{T^+}(t, T^+) \sim \left(\frac{1}{2}\delta(T^+) + \sum_{n=1}^{\infty} \mathcal{L}_{T^+}^{-1} \left[\widehat{\phi}^n(u) \right] \right) \Phi(t). \qquad (7.87)$$

When $\phi(\tau)$ is one sided Lévy stable distribution, Eq. (7.87) reduces to

$$f_{T^+}(t, T^+) \sim \left(\frac{1}{2}\delta(T^+) + \sum_{n=1}^{\infty} \left(\frac{1}{n^{1/\alpha}} \ell_\alpha \left(\frac{T^+}{n^{1/\alpha}} \right) \right) \right) \Phi(t). \qquad (7.88)$$

It can be noticed that the behavior of $f_{T^+}(t, T^+)$ is determined by the shape of $\phi(\tau)$ for small T^+. Equation (7.88), or more precisely the limit $t \to \infty$ of $f_{T^+}(t, T^+)/\Phi(t)$, is the infinite density describing the occupation time statistics when $\phi(\tau)$ is the one sided Lévy distribution; see Fig. 7.16 for illustration.

We further consider a more general case. Equation (7.81) can be rewritten as

$$\widehat{f}_{T^+}(s, u) = \frac{2s - u}{2s(s + u)} + \frac{u(\widehat{\phi}(s + u) - \widehat{\phi}(s))}{2s(s + u)(1 - \widehat{\phi}(s + u)\widehat{\phi}(s))}.$$

Expanding the above equation and taking double inverse Laplace transforms, we have

$$\widehat{f}_{T^+}(s, u) = \frac{2s - u}{2s(s + u)} + \frac{u(\widehat{\phi}(s + u) - \widehat{\phi}(s))}{2s(s + u)} \sum_{n=0}^{\infty} \widehat{\phi}^n(s + u)\widehat{\phi}^n(s).$$

After some calculations, there exists

$$\widehat{f}_{T^+}(s, u) = \frac{2s - u}{2s(s + u)} + \frac{u}{2(s + u)} \frac{1 - \widehat{\phi}(s)}{s} + \frac{u}{2s} \frac{\widehat{\phi}(s + u) - 1}{s + u}$$

$$+ \left(\frac{1 - \widehat{\phi}(s)}{s} \frac{u}{s + u} + \frac{\widehat{\phi}(s + u) - 1}{s + u} \frac{u}{2s}\right) \sum_{n=1}^{\infty} \widehat{\phi}^n(s + u)\widehat{\phi}^n(s).$$

Performing double Laplace transforms, we have

$$f_{T^+}(t, T^+) = \delta(t - T^+) \int_0^{T^+} \phi(y)dy + \frac{1}{2}\phi(t - T^+) - \phi(T^+)$$

$$- \frac{1}{2}\delta(T^+) \int_0^t \phi(y)dy$$

$$+ \sum_{n=1}^{\infty} \left(\int_t^{\infty} \phi(y)dy *_t \mathcal{L}_t^{-1}[\exp(-sT^+)\widehat{\phi}^n(s)]\mathcal{L}_{T^+}^{-1}[\widehat{\phi}^n(u)]\right.$$

$$+ \int_0^{T^+} \mathcal{L}_\tau^{-1}[\widehat{\phi}^n(u)]d\tau \left(\mathcal{L}_t^{-1}\left[\exp(-sT^+)\left(\widehat{\phi}^{n+1}(s) - \widehat{\phi}^n(s)\right)\right]\right)$$

$$+ \frac{1}{2}T^+ \int_0^{t - T^+} \mathcal{L}_\tau^{-1}[\widehat{\phi}^n(s)]d\tau \int_{T^+}^{\infty} \phi(y)dy *_{T^+} \mathcal{L}_{T^+}^{-1}[\widehat{\phi}^n(u)]\right).$$

$$(7.89)$$

Doing some calculations, yields

$$
\begin{aligned}
f_{T^+}(t, T^+) &= \delta(t - T^+) \int_0^{T^+} \phi(y) dy + \frac{1}{2}\phi(t - T^+) - \phi(T^+) \\
&\quad - \frac{1}{2}\delta(T^+) \int_0^t \phi(y) dy \\
&\quad + \sum_{n=1}^{\infty} \left(\int_t^{\infty} \phi(y) dy *_t \mathcal{L}_t^{-1}[exp(-sT^+)\phi^n(s)] \mathcal{L}_{T^+}^{-1}[\hat{\phi}^n(u)] \right. \\
&\quad - \int_0^{T^+} \mathcal{L}_\tau^{-1}[\hat{\phi}^n(u)] d\tau \phi(t - T^+) \\
&\quad \left. + \frac{1}{2}T^+ \int_0^{t-T^+} \mathcal{L}_\tau^{-1}[\hat{\phi}^n(s)] d\tau \int_{T^+}^{\infty} \phi(y) dy *_{T^+} \mathcal{L}_{T^+}^{-1}[\hat{\phi}^n(u)] \right).
\end{aligned}
\tag{7.90}
$$

Especially, if $\phi(t)$ is the one sided Lévy stable density, Eq. (7.90) can be shown by

$$
\begin{aligned}
f_{T^+}(t, T^+) &= \delta(t - T^+) \int_0^{T^+} \ell_\alpha(y) dy + \frac{1}{2}\ell_\alpha(t - T^+) - \ell_\alpha(T^+) \\
&\quad - \frac{1}{2}\delta(T^+) \int_0^t \ell_\alpha(y) dy \\
&\quad + \sum_{n=1}^{\infty} \left(\frac{1}{n^{2/\alpha}} \int_t^{\infty} \ell_\alpha(y) dy *_t \ell_\alpha\left(\frac{t - T^+}{n^{1/\alpha}}\right) \ell_\alpha\left(\frac{T^+}{n^{1/\alpha}}\right) \right. \\
&\quad + \frac{1}{n^{1/\alpha}} \int_0^{T^+} \ell_\alpha\left(\frac{\tau}{n^{1/\alpha}}\right) d\tau \ell_\alpha(t - T^+) \\
&\quad \left. + \frac{T^+}{2n^{2/\alpha}} \int_0^{t-T^+} \ell_\alpha\left(\frac{\tau}{n^{1/\alpha}}\right) d\tau \int_{T^+}^{\infty} \ell_\alpha(y) dy *_{T^+} \ell_\alpha\left(\frac{T^+}{n^{1/\alpha}}\right) \right).
\end{aligned}
\tag{7.91}
$$

Now we investigate the total probability to find $0 < T^+ < T_1^+$, defined by $P(T_1^+) = \int_0^{T_1^+} f_{T^+}(t, T^+) dT^+$. To simplify the discussion, we just consider Mittag-Leffler time statistics. Using Eq. (7.86) and the asymptotic behaviors of $t^{\alpha-1} E_{\alpha,\alpha}(-t^\alpha)$ yields

$$
P(T_1^+) \sim \frac{1}{2}\Phi(t) + \frac{\sin(\pi\alpha)}{\pi\alpha} \frac{(T_1^+)^\alpha}{t^\alpha}
\tag{7.92}
$$

with $T_1^+ \ll t$. On the other hand, in the particular case $\alpha = 1/2$, utilizing the typical fluctuations Eq. (7.84) gives the arcsine distribution

$$P(T_1^+) \sim \frac{2}{\pi} \arcsin\left(\frac{T_1^+}{\sqrt{tT_1^+}}\right). \tag{7.93}$$

It can be noted that Eq. (7.93) reduces to $2\sqrt{T_1^+}/(\pi\sqrt{t})$ for $T_1^+ \ll t$. In this case, we see that Eqs. (7.92) and (7.93) are consistent with each other except for the first term of Eq. (7.92). It implies that, though Eq. (7.85) is not normaized, we can use it to calculate some observables.

7.7.2   *Occupation Time with* $1 < \alpha < 2$

Now we study the random variable $\epsilon = T^+ - t/2$, shifting the symmetry axis of $f_{T^+}(t, T^+)$ to zero. Similar to the derivation of Eq. (7.50), the double Laplace transforms of $f_\epsilon(t, \epsilon)$ is

$$\widehat{f}_\epsilon(s, u_\epsilon) = \left(s + \frac{u_\epsilon(\widehat{\phi}(s + \frac{u_\epsilon}{2}) - \widehat{\phi}(s - \frac{u_\epsilon}{2}))}{1 - \widehat{\phi}(s + \frac{u_\epsilon}{2})\widehat{\phi}(s - \frac{u_\epsilon}{2})}\right) \frac{1}{(s - \frac{u_\epsilon}{2})(s + \frac{u_\epsilon}{2})}. \tag{7.94}$$

Since the sign of ϵ is not fixed, i.e., it can be positive or negative, we replace u_ϵ with $-ik$ and move it into the Fourier space. For the typical case, i.e., $|\epsilon| \sim t^{1/\alpha}$, we find

$$\widetilde{f}_\epsilon(s, k) \sim \frac{1}{s - \frac{b_\alpha}{2\langle\tau\rangle}\left((s + \frac{ik}{2})^\alpha + (s - \frac{ik}{2})^\alpha\right)}. \tag{7.95}$$

Taking inverse Laplace and Fourier transforms yields [Schulz and Barkai (2015)]

$$f_\epsilon(t, \epsilon) \sim \begin{cases} \dfrac{C_{\text{occ}}}{t^{1/\alpha}} l_\alpha\left(\dfrac{C_{\text{occ}}}{t^{1/\alpha}}|\epsilon|\right), & \text{for } -t/2 < \epsilon < t/2; \\ 0, & \text{otherwise}, \end{cases} \tag{7.96}$$

where $C_{\text{occ}} = \langle\tau\rangle^{1/\alpha}/(2(b_\alpha|\cos(\pi\alpha/2)|)^{1/\alpha})$ and $l_\alpha(x)$ denotes the symmetric stable Lévy law with the index α, so the Fourier transform of $l_\alpha(x)$ is $\exp(-|k|^\alpha)$, which is a special case of $l_{\alpha,\beta}(x)$; see Sec. 7.8.

Since $0 < T^+ < t$, we find that $-t/2 < \epsilon < t/2$. It means that the order of ϵ can be as large as the observation time t. Hence to investigate the rare events, we consider s is of the order $|k|$. By inverting the Fourier and Laplace transforms, we find (see also [Schulz and Barkai (2015)])

$$f_\epsilon(t, \epsilon) \sim \frac{\alpha b_\alpha}{t^\alpha |\Gamma(1-\alpha)|} \times \begin{cases} \chi\left(\dfrac{2|\epsilon|}{t}\right), & -\dfrac{t}{2} < \epsilon < \dfrac{t}{2}; \\ 0, & \text{otherwise} \end{cases} \tag{7.97}$$

with

$$\chi(z) = \theta(0 < z \le 1)z^{-1-\alpha}\left(1 - \frac{\alpha-1}{\alpha}z\right).$$

We see that $t^\alpha f_\epsilon(t, \epsilon)$ does not depend on the exact shape of $\phi(\tau)$ besides the parameters α and b_α. Further the integral of Eq. (7.97) w.r.t. ϵ, in the limit $\epsilon \to 0$, is divergent. Thus, $f_\epsilon(t, \epsilon)$ is a non-normalized solution since its behavior, at $T^+ \to t/2$, is non-integrable. See the discussions and numerical examples in [Schulz and Barkai (2015)].

7.8 Some Properties of Stable Distribution

Now we discuss the series representation and the asymptotic behavior of Lévy stable distribution $l_{\alpha,\beta}(x)$ [Metzler and Klafter (2000, 2004)]. The corresponding PDF $l_{\alpha,\beta}(x)$ is given by the inverse Fourier transform

$$l_{\alpha,\beta}(x) = \frac{1}{2\pi}\int_{-\infty}^{\infty} \exp\left(-ikx - c|k|^\alpha\left(1 + i\beta\frac{z}{|z|}h(z,\alpha)\right)\right)dk, \quad (7.98)$$

where α, β, c are constants and

$$h(z,\alpha) = \begin{cases} \tan\left(\frac{\pi\alpha}{2}\right), & \alpha \ne 1; \\ \frac{\pi}{2}\log(z), & \alpha = 1. \end{cases}$$

Especially, for $\beta = 0$ and $c = 1$, Eq. (7.98) reduces to the symmetric stable distribution $l_\alpha(x)$. For simplification of analysis, let $\beta = 1$, $\alpha \ne 1$, and $c = -\cos(\pi\alpha/2)$,

$$l_{\alpha,1}(x) = \frac{1}{2\pi}\int_{-\infty}^{\infty} \exp(-ikx)\exp[(ik)^\alpha]dk.$$

Expanding the integrand in the right hand side as a Taylor series in x yields the convergent series

$$l_{\alpha,1}(x) = \sum_{n=0}^{\infty} \frac{(-1)^n}{\alpha\pi(2n)!}\Gamma\left(\frac{2n+1}{\alpha}\right)\cos(g(n,\alpha))x^{2n}$$
$$- \sum_{n=0}^{\infty} \frac{\text{sign}(x)(-1)^n\Gamma(\frac{2n+2}{\alpha})}{\pi\alpha(2n+1)!}\sin\left(g\left(n+\frac{1}{2},\alpha\right)\right)x^{2n+1}, \quad (7.99)$$

where $g(n,\alpha) = (2n+1)\pi(1/2 - 1/\alpha)$ and $\text{sign}(x) = x/|x|$ for $|x| > 0$ and zero otherwise. Especially, for $x \to 0$, Eq. (7.99) reduces to

$$l_{\alpha,1}(x) \sim \frac{1}{\alpha\pi}\Gamma\left(\frac{1}{\alpha}\right)\cos(g(0,\alpha)),$$

which is a constant and strictly less than $l_\alpha(0)$ for $\alpha < 2$. Furthermore, the asymptotic behavior of $l_{\alpha,1}(x)$ is

$$l_{\alpha,1}(x) \sim \sum_{n=1}^{\infty} \frac{\Gamma(1+\alpha n)\sin(\alpha n\pi)}{2\pi n! |x|^{1+\alpha n}}(-1 + \text{sign}(x)), \qquad (7.100)$$

which implies that $l_{\alpha,1}(x) \sim |x|^{-1-\alpha}/|\Gamma(-\alpha)|$, being the same as the left hand side of the tail of the symmetric Lévy stable distribution for $x \to -\infty$, and the tails of $l_{\alpha,1}(x)$ are asymmetric w.r.t. x.

7.9   Discussion

It is well known that when the averaged time interval between renewal events diverges, i.e., $0 < \alpha < 1$, the typical scale of the process is the measurement time and so observables of interest scale with t. Hence the rare fluctuations, and the far tails of the distributions of observables considered in this chapter, have corrections when the observable is of the order t^0. This leads to non-normalized states which describe these rare events. The opposite takes place when $1 < \alpha < 2$ namely when the mean sojourn time is finite but the variance is diverging. Here, we have a finite scale, but when observables like B, F, Z or T^+ become large, namely when they are of the order t, one naturally finds deviations from typical laws. Since the approximation in the far tail of the distribution must match the typical fluctuations which are described by heavy-tailed densities, we get by non-normalized states.

The uniform approximation provided in the text (for example Eq. (7.61) and the corresponding Fig. 7.12) bridges between the typical and rare fluctuations. It is obtained by matching the far tail distribution with its bulk fluctuations. Technically we find unifrom approximation by using exact theoretical results (see Eq. (7.27)), an approximation where we take $s \to 0$ (meaning $t \to \infty$) leaving the second variable u (corresponding for example to F) finite (see example Eq. (7.37)), and for special choice of the waiting time PDF we can get the solution in terms of infinite sums (for instance Eq. (7.69)). In principle the uniform approximation can be used to calculate quantities of the process like moments. However, it is much simpler to classify observables based on their integrability w.r.t. the non-normalized state, as is done in infinite ergodic theory. In the case of integrable observables, we may use the non-normalized state for the calculation of integrable expectations, somewhat similar to the calculation averages observables from normalized densities.

Importantly, the non-normalized states are not only a tool with which we obtain moments. As we have demonstrated both theoretically and numerically, they describe the perfectly normalized probability density of the observables, when the latter are properly scaled with time (see Fig. 7.6). It is rewarding that while the rare fluctuations are non-universal, in the sense that they depend on the details of the waiting time PDF, they can be obtained rather generally. Further, as we have shown for the backward and forward recurrence time, the density describing the typical fluctuations for $1 < \alpha < 2$ Eq. (7.36), describes the non-typical events for $0 < \alpha < 1$; all we need to do is replacing the finite mean waiting time with an effective time dependent one; see Eq. (7.24).

As mentioned in the introduction the distribution of the occupation time for Brownian motion and random walks is the arcsine law, and the same holds for the backward recurrence time (here $\alpha = 1/2$). While analyzing the rare events of these well known results, we see that the large deviations for these two observables behave differently (compare Fig. 7.10 with Fig. 7.17). For the backward recurrence time B we have deviations from arcsine law which differ for the case $B \propto t^0$ and $B \propto t$ (see Fig. 7.10). The same symmetry breaking is not found for the occupation time since by construction of the model the probability to be in the up ($+$) and down ($-$) state is the same (Fig. 7.17).

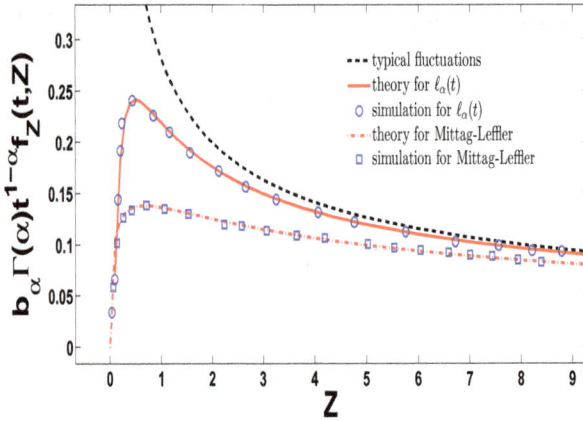

Fig. 7.13 Theory and simulations of the statistical behaviors of the rare events of the time interval straddling time t, with $\phi(\tau)$ Eqs. (1.29) and (7.3) for $t = 2000$ and $\alpha = 0.5$. The full and dash-dot lines are theory Eq. (7.65), showing the large deviations and the corresponding simulation results are presented by symbols obtained by averaging 2×10^7 trajectories.

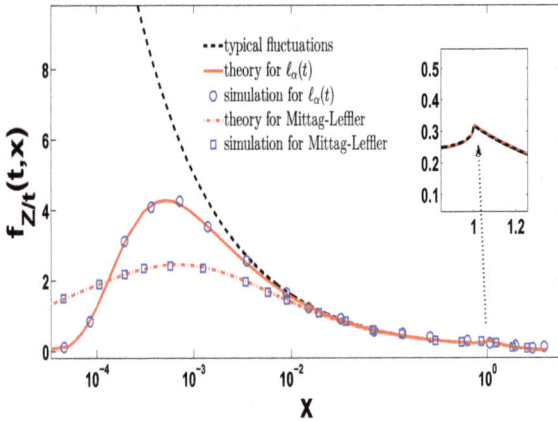

Fig. 7.14 The rescaled PDF of Z with the scale $x = Z/t$. The parameters are the same to Fig. 7.13. The solid and dash-dot lines are the theory results getting from Eq. (7.66). Furthermore, the small figure demonstrates that the first derivative of $f_Z(t, Z)$ is not continuous at $Z = t$.

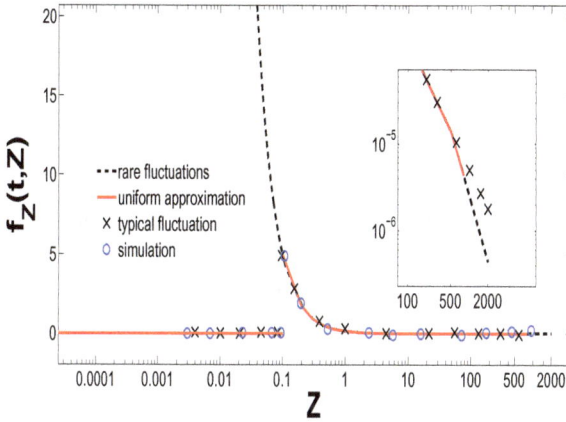

Fig. 7.15 The PDF of the straddling time Z versus Z for $\phi(\tau)$ Eq. (1.26). The rare fluctuations are given by Eq. (7.73) (the dashed line), depicting the behaviors when $Z \propto t$. The solid line is the uniform approximation Eq. (7.74). For the typical fluctuations, we use Eq. (7.72) which is shown by the symbols ($\times$). When $Z > t$, the rare fluctuations deviate from the typical fluctuations (see the inset). Here the observation time is 500 and the number of trajectories is 2×10^7.

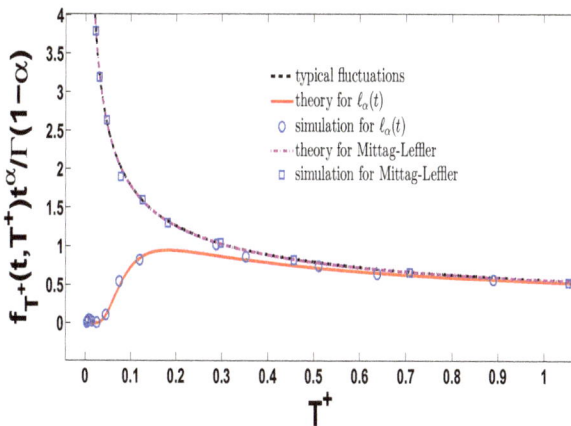

Fig. 7.16 The scaled PDF of the occupation time versus T^+ generated by the trajectories of particles with $\alpha = 0.5$ and $t = 2000$ for $T^+ \ll t$. The solid line and the dash-dot line correspond to the theoretical results given by Eqs. (7.88) and (7.86), respectively, depicting the large deviations with $T^+ \propto t^0$. The dashed line, given by Eq. (7.83), shows the typical fluctuations. Note that it overlaps with theoretical result of Mittag-Leffler waiting time, the top curve in the figure.

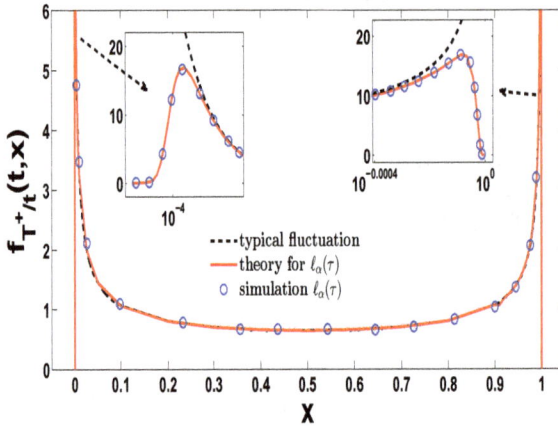

Fig. 7.17 The occupation time density $f_{T^+/t}(t, x)$ of a renewal process with $\phi(\tau)$ Eq. (1.30). The parameters are $\alpha = 0.5$ and $t = 2000$. The simulations, plotted by the symbols, are generated by averaging 2×10^7 trajectories and the curve is the theoretical result obtained from Eq. (7.88) and symmetry of $f_{T^+}(t, x)$. Note that the results for typical fluctuations Eq. (7.83), diverge on $x = 0$ and $x = 1$, while the large deviations theory predicts correctly finite value of the PDF.

Chapter 8

Governing Equation for Average First Passage Time and Transitions among Anomalous Diffusions

Actually, there are so many phenomena found in the nature indicating the transitions among different kinds of diffusions. One of the kinds is the CTRW model with tempered waiting time introduced in Sec. 3.3.1 whose MSD always transits from subdiffusion to normal diffusion. Therefore one question comes into mind naturally is how to model the other kinds of transitions, and what the governing equation of the PDF will be? In this chapter, we will further discuss these questions. Besides, the first passage time always plays an important role, and in this chapter we also find the governing equations for this statistical quantity. However, it will be very hard to directly calculate it through the framework of CTRW introduced in Chap. 2. Thus in Sec. 8.1, we begin to calculate the governing equations for this statistical quantity by utilizing the subordinated process introduced in Sec. 1.4.3. However, for the subdiffusion the average first passage time is infinite due to the diverging average waiting time, and the governing equation for this statistical quantity in this case is meaningless. Naturally, here we consider the tempered subordinated process and calculate its governing equation for the average first passage time.

Here in this chapter, we first consider the governing equations for statistical quantities. We mainly consider the average first exit time of tempered subordinated Langevin equation. Then we discuss the transitions among anomalous diffusions with different diffusion exponents. For the second part of this chapter, the corresponding equation of PDF is given. Besides, the ways of transitions among different kinds of diffusions (transitions among subdiffusion, normal diffusion and superdiffusion) are also illustrated. We also calculate the MSDs and the fractional moments. In the last part of this chapter we obtain the asymptotic result of first exit time of the particle moving in a harmonic external potential.

8.1 Governing Equation for Average First Passage Time

In this section, we mainly discuss the governing equation for the average first passage time of tempered subordinated Langevin equation. Actually according to the equations of the PDF of particle paths functional, also known as Feynman-Kac equations, we can obtain various governing equations for different kinds of statistical quantities by constructing different kinds of functions introduced in Sec. 3.2 and Sec. 3.3, such as the first exit time, the area under the particle's trajectory, etc [Carmi *et al.* (2010)]. In this section, we use another method to derive the governing equations for the first passage time. In fact, the average first passage time for the subdiffusion is infinite because of the diverging average waiting time, and it will lead to the discussion of this case making no sense. Therefore, in the following we mainly consider the tempered process.

Here we consider the n-dimensional stochastic process governed by the following Langevin equations:

$$\dot{X}(z) = F(X(z)) + \sqrt{2\varepsilon}\sigma(X(z))\xi(z),$$
$$\dot{T}(z) = \eta(z),$$
(8.1)

where $\xi(z)$ is Gaussian white noise and $\eta(z)$ is a tempered one sided Lévy-stable noise with tempering index λ and stability index $0 < \alpha < 1$. Besides, we assume these two noises are independent with each other. And the subordinated stochastic process is given by $Y(t) = X(S(t))$, where the process S is the inverse process of T, that is

$$S(t) = \inf_{z>0}\{z : T(z) > t\}.$$
(8.2)

And here we consider the characteristic function of process $T(z)$

$$\left\langle e^{-sT(z)} \right\rangle = e^{-z((s+\lambda)^\alpha - \lambda^\alpha)},$$

which represents the tempered waiting time distribution with diverging average. In the following we consider the subordinated stochastic process $Y(t)$. Here we take a brief look at the derivations of the governing equation of average first passage time $u(\mathbf{x})$, where $\mathbf{x}$ represents the start point of the process. We first consider the transition PDF of process $Y(t)$

$$p(\mathbf{x}, \mathbf{y}, t)d\mathbf{y} = P\{\mathbf{x}(t) \in \mathbf{y} + d\mathbf{y} | \mathbf{x}(0) = \mathbf{x}\}.$$

And $p(\mathbf{x}, \mathbf{y}, t)$ satisfies the tempered fractional forward Kolmogorov equation

$$\frac{\partial}{\partial t}p(\mathbf{x}, \mathbf{y}, t) = \frac{\partial}{\partial t}\int_0^t K(t - t', \lambda)L_\mathbf{y} p(\mathbf{x}, \mathbf{y}, t')dt',$$
(8.3)

where the memory kernel is given by its Laplace transform

$$\widehat{K}(s,\lambda) = \frac{1}{(s+\lambda)^\alpha - \lambda^\alpha} \tag{8.4}$$

and the Laplacian operator

$$L_{\mathbf{y}} = -\sum_{i=1}^{n} \frac{\partial}{\partial y^i} F^i(\mathbf{y}) + \varepsilon \sum_{i,j=1}^{n} \frac{\partial^2}{\partial y^i \partial y^j} a^{ij}(\mathbf{y}),$$

and the diffusion matrix $a(\mathbf{y})$ is given by $a(\mathbf{y}) = \frac{1}{2}\sigma(\mathbf{y})\sigma(\mathbf{y})^T$. The initial condition of Eq. (8.3) is $p(\mathbf{x},\mathbf{y},0) = \delta(\mathbf{y} - \mathbf{x})$, and the boundary condition is $p(\mathbf{x},\mathbf{y},t)|_{\mathbf{x}\in D, \mathbf{y}\in\partial D} = 0$ implying absorbing boundary condition. Besides, the transition PDF $p(\mathbf{x},\mathbf{y},t)$ also satisfies the following equation

$$\frac{\partial}{\partial t}p(\mathbf{x},\mathbf{y},t) = \frac{\partial}{\partial t}\int_0^t K(t-t',\lambda)L_{\mathbf{x}}^* p(\mathbf{x},\mathbf{y},t')dt', \tag{8.5}$$

where the operator

$$L_{\mathbf{x}}^* = \sum_{i=1}^{n} F^i(\mathbf{x})\frac{\partial}{\partial x^i} + \varepsilon \sum_{i,j=1}^{n} a^{ij}(\mathbf{x})\frac{\partial^2}{\partial x^i \partial y^i}. \tag{8.6}$$

According to the physical meaning of first exit time, denoted as τ, there exists the following equation

$$P\{\tau > t | \mathbf{x}(0) = \mathbf{x}\} = \int_D p(\mathbf{x},\mathbf{y},t)d\mathbf{y}, \tag{8.7}$$

where $p(\mathbf{x},\mathbf{y},t)$ is the solution of Eq. (8.3). Thus the average of τ satisfies

$$\begin{aligned} u(\mathbf{x}) &= E(\tau | \mathbf{x}(0) = \mathbf{x}) \\ &= \int_0^\infty t d_t[P\{\tau > t | \mathbf{x}(0) = \mathbf{x}\} - 1]. \end{aligned} \tag{8.8}$$

Besides, under the assumption of $P\{\tau < \infty\} = 1$ there exists

$$u(\mathbf{x}) = \int_0^\infty P\{\tau > t | \mathbf{x}(0) = \mathbf{x}\}dt. \tag{8.9}$$

Then utilizing Eq. (8.7), we obtain

$$u(\mathbf{x}) = \int_0^\infty \int_D p(\mathbf{x},\mathbf{y},t)d\mathbf{y}dt. \tag{8.10}$$

Define

$$P(\mathbf{x},\mathbf{y},t) = \int_0^t p(\mathbf{x},\mathbf{y},t')dt',$$

that is $\widehat{P}(\mathbf{x}, \mathbf{y}, s) = \frac{1}{s}\widehat{p}(\mathbf{x}, \mathbf{y}, s)$ after taking Laplace transform w.r.t. time t. According to the final value theorem of Laplace transform, there exists

$$\begin{aligned}
P(\mathbf{x}, \mathbf{y}) &:= P(\mathbf{x}, \mathbf{y}, t = \infty) \\
&= \lim_{s \to 0} s\widehat{P}(\mathbf{x}, \mathbf{y}, s) \\
&= \lim_{s \to 0} \widehat{p}(\mathbf{x}, \mathbf{y}, s).
\end{aligned} \tag{8.11}$$

From Eq. (8.3), we obtain the following equation in Laplace space

$$s\widehat{p}(\mathbf{x}, \mathbf{y}, s) = \delta(\mathbf{y} - \mathbf{x}) + s\widehat{K}(s, \lambda)L_{\mathbf{y}}p(\mathbf{x}, \mathbf{y}, s). \tag{8.12}$$

Then by substituting $\widehat{K}(s, \lambda)$ defined in Eq. (8.4) into the Eq. (8.12) and letting $s \to 0$, we have

$$L_{\mathbf{y}} \lim_{s \to 0} p(\mathbf{x}, \mathbf{y}, s) = -\alpha\lambda^{\alpha-1}\delta(\mathbf{y} - \mathbf{x}). \tag{8.13}$$

Basing on Eq. (8.11) and Eq. (8.13), we have

$$L_{\mathbf{y}}P(\mathbf{x}, \mathbf{y}) = -\alpha\lambda^{\alpha-1}\delta(\mathbf{y} - \mathbf{x}). \tag{8.14}$$

Similarly, we can also obtain the following equation from Eq. (8.5) with the same method above

$$L_{\mathbf{x}}^*P(\mathbf{x}, \mathbf{y}) = -\alpha\lambda^{\alpha-1}\delta(\mathbf{y} - \mathbf{x}). \tag{8.15}$$

Here we assume that the integrals can exchange the order, that is

$$u(\mathbf{x}) = \int_D \int_0^\infty p(\mathbf{x}, \mathbf{y}, t)dtd\mathbf{y} = \int_D P(\mathbf{x}, \mathbf{y})d\mathbf{y}. \tag{8.16}$$

Applying operator $L_{\mathbf{x}}^*$ on both sides of Eq. (8.16), there exists

$$\begin{aligned}
L_{\mathbf{x}}^*u(\mathbf{x}) &= \int_D L_{\mathbf{x}}^*P(\mathbf{x}, \mathbf{y})d\mathbf{y} \\
&= -\int_D \alpha\lambda^{\alpha-1}\delta(\mathbf{y} - \mathbf{x})d\mathbf{y}.
\end{aligned} \tag{8.17}$$

That is the governing equation of the average first exit time

$$L_{\mathbf{x}}^*u(\mathbf{x}) = -\alpha\lambda^{\alpha-1} \quad \text{for } \mathbf{x} \in D \tag{8.18}$$

with the boundary condition

$$u(\mathbf{x}) = 0 \quad \text{for } \mathbf{x} \in \partial D. \tag{8.19}$$

Example 8.1. Here we give two examples of the average of first exit time. We first consider the stochastic dynamics in one dimensional space. And take $a(x) = 1$, $F(x) = 0$, $\varepsilon = 1$ and $D = (-r, r)$ with $r > 0$. By utilizing the boundary condition Eq. (8.19) and solving Eq. (8.18), we arrive at

$$u(x) = a\lambda^{\alpha-1} \frac{r^2 - x^2}{2}.$$

The above result indicates that the average first exit time is not a monotone function w.r.t. α if $\lambda < 1$. And when $\alpha \in (0, 1)$, with the increase of λ, the heavy-tailed PDF of waiting time will be stronger suppressed and the waiting time between each step will be smaller. And it will make the average first exit time decrease, which is also what the result above indicates. Moreover, if $\lambda \to 0$, the average first exit time $u(x)$ is infinite, which is also expected because the average waiting time of subdiffusion is infinite.

Then we consider the particles move in two-dimensional space, and take $a^{ij}(\mathbf{x}) = \begin{pmatrix} 1 & 0 \\ 0 & 1 \end{pmatrix}$, $F(\mathbf{x}) = 0$, $\varepsilon = 1$ and the domain is $D = \{\mathbf{x} : |\mathbf{x}| < r\}$ representing a circle with $r > 0$. Then combining Eq. (8.18) and the corresponding boundary condition Eq. (8.19), we have

$$u(\mathbf{x}) = a\lambda^{\alpha-1} \frac{r^2 - |\mathbf{x}|^2}{4}.$$

Now we conclude the methods of obtaining the first passage time except for the methods introduced in Sec. 3.3.5 and this section, we will additionally introduce another method to obtain the first passage time directly from the Fokker-Planck equation with harmonic potential in Sec. 8.5.2.

8.2 Transition among Anomalous Diffusions: CTRW Description

From this section, we will turn to focus on the transition among anomalous diffusions under the framework of CTRW model, whereas the ordinary CTRW simply depicts a pure subdiffusion or a transition from subdiffusion to normal diffusion by utilizing the tempered waiting time. Now we begin to model the process that can transit freely among different kinds of diffusions. Here we should note that the significant feature of transition among different anomalous diffusions mainly represents the exponent of MSD changes for the one into the other one. In this section we construct the CTRW model to describe the transition among anomalous diffusion with different diffusion exponents. Now, we have known that the PDF $p(x, t)$ of the positions

x of the particles and time t has the following equation in Fourier-Laplace space introduced in Eq. (1.43) in Sec. 1.3.2.

$$\widetilde{p}(k, s) = \frac{1 - \widehat{\phi}(s)}{s} \frac{\widetilde{p}_0(k)}{1 - \widehat{\phi}(s)\widetilde{\lambda}(k)}, \tag{8.20}$$

where $\widehat{\phi}(s)$ and $\widetilde{\lambda}(k)$ represent the Laplace transform of waiting time distribution and Fourier transform of jump length distribution, respectively, and denote the initial condition of $p(x,t)$ and its corresponding Fourier transform as $p_0(x)$ and $\widetilde{p}_0(k)$, respectively. Here in this chapter, we consider the waiting time PDF with the following form in Laplace space

$$\widehat{\phi}(s) = \frac{1}{1 + (s\tau)^\mu [1 + (s\tau)^{-\rho}]^\gamma}, \tag{8.21}$$

where $0 < \rho < \mu < 1$, $0 < \gamma < 1$. Some concepts and conclusions of completely monotone function and Bernstein function are introduced first [Schilling *et al.* (2010); Berg and Forst (1975)] in order to help us ensure the above choice of waiting time distribution makes sense, that is the non-negativity of the $\phi(t)$ with Laplace transform shown in Eq. (8.21).

Definition 8.1. The function $g(x)$ is a completely monotone function if

$$(-1)^n g^{(n)}(x) \geq 0$$

for all $n \geq 0$ and $x > 0$.

It is obvious that the product of completely monotone functions is still a completely monotone function. An obvious example of completely monotone function is x^α, with $\alpha < 0$.

Definition 8.2. A function $f(x)$ is a Bernstein function if the following equation holds

$$(-1)^{n-1} f^{(n)}(x) \geq 0$$

for all $n \in N$ and $x > 0$.

An example of Bernstein function is x^α, where $\alpha \in (0, 1)$. Obviously the following properties of Bernstein function hold:

(1) The linear combination of Bernstein functions is still a Bernstein function.

(2) The composition function denoted as $f_1 \circ f_2$ of two Bernstein functions denoted as f_1 and f_2 is still a Bernstein function.

(3) If $f(x)$ is a Bernstein function then $g(x) := 1/f(x)$ is a completely monotone function.

In order to ensure waiting time PDF $\phi(t)$ making sense, its Laplace transform $\widehat{\phi}(s)$ needs to be completely monotone function [Schilling *et al.* (2010)]. According to the above relationship between completely monotone function and Bernstein function, we transfer our problem into proving $1 + (s\tau)^\mu [1 + (s\tau)^{-\rho}]^\gamma$, i.e., $(s\tau)^\mu [1 + (s\tau)^{-\rho}]^\gamma$ is a Bernstein function. Therefore, the function

$$(s\tau)^\mu \left[1 + (s\tau)^{-\rho}\right]^\gamma = \left[(s\tau)^{\mu/\gamma} + (s\tau)^{\mu/\gamma-\rho}\right]^\gamma$$

is a Bernstein function when both $s^{\mu/\gamma}$ and $s^{\mu/\gamma-\rho}$ are Bernstein functions, that is $0 < \mu/\gamma < 1$ and $0 < \mu/\gamma - \rho < 1$. Substituting the waiting time PDF in Eq. (8.21) and the jump length distribution of the Gaussian form, i.e., its Fourier transform with the form $\widetilde{\lambda}(k) = e^{-\sigma^2 k^2} \sim 1 - \sigma^2 k^2$, into Eq. (8.20), then we obtain

$$\widetilde{\widehat{p}}(k,s) = \frac{s^{\mu-1} \left[1 + (s\tau)^{-\rho}\right]^\gamma}{s^\mu \left[1 + (s\tau)^{-\rho}\right]^\gamma + \mathcal{K}_\mu k^2} \widetilde{p}_0(k), \qquad (8.22)$$

where $\mathcal{K}_\mu = \sigma^2/\tau^\mu$ is the generalized diffusion coefficient with the dimension $[\sigma^2/\tau^\mu] = \text{m}^2\text{s}^{-\mu}$. We note that the last approximation of Gaussian form jump length distribution in Fourier space only makes sense when k is small enough. After some rearrangements, we arrive at

$$s^\mu \left[1 + (s\tau)^{-\rho}\right]^\gamma \widetilde{\widehat{p}}(k,s) - s^{\mu-1} \left[1 + (s\tau)^{-\rho}\right]^\gamma \widetilde{p}_0(k) = -\mathcal{K}_\mu k^2 \widetilde{\widehat{p}}(k,s). \quad (8.23)$$

Here we introduce the regularized Prabhakar derivative $^C\mathcal{D}_{\rho,\omega,0+}^{\gamma,\mu}$ with $0 < \mu < 1$, defined as [Garra *et al.* (2014)]

$$^C\mathcal{D}_{\rho,\nu,0+}^{\gamma,\mu} f(t) = \left(\mathcal{E}_{\rho,1-\mu,\nu,0+}^{-\gamma} \frac{d}{dt} f\right)(t), \qquad (8.24)$$

where $\mu, \nu, \gamma, \rho \in C$, $\text{Re}(\mu) > 0$, $\text{Re}(\rho) > 0$, and

$$\left(\mathcal{E}_{\rho,\mu,\nu,0+}^\gamma f\right)(t) = \int_0^t (t-t')^{\mu-1} E_{\rho,\mu}^\gamma \left(\nu(t-t')^\rho\right) f(t') \, dt' \qquad (8.25)$$

is the Prabhakar integral [Prabhakar (1971)], and $E_{\rho,\mu}^\gamma(t)$ is the three parameter Mittag-Leffler function, defined as

$$E_{\rho,\mu}^\gamma(t) = \sum_{k=0}^\infty \frac{(\gamma)_k}{\Gamma(\rho k + \mu)} \frac{t^k}{k!}, \qquad (8.26)$$

where $(\gamma)_k = \Gamma(\gamma + k)/\Gamma(\gamma)$ is the Pochhammer symbol. For $\gamma = 0$ the Prabhakar integral reduces to the Riemann-Liouville fractional integral [Hilfer (2000)]

$$^{\text{RL}}I_{0+}^{\mu} f(t) = \frac{1}{\Gamma(\mu)} \int_0^t (t - t')^{\mu-1} f(t') \, dt'. \tag{8.27}$$

Besides the Laplace transform of Prabhakar derivative is given by [Garra *et al.* (2014)]

$$\mathcal{L}\left[{}^{\text{C}}\mathcal{D}_{\rho,\nu,0+}^{\gamma,\mu} f(t)\right](s) = s^{\mu} \left(1 - \nu s^{-\rho}\right)^{\gamma} \mathcal{L}\left[f(t)\right](s) - s^{\mu-1} \left(1 - \nu s^{-\rho}\right)^{\gamma} f(0+), \tag{8.28}$$

where $\text{Re}(s) > |\nu|^{1/\rho}$. The above result can be obtained from the Laplace transform formula of the Mittag-Leffler function [Prabhakar (1971)]

$$\mathcal{L}\left[t^{\mu-1} E_{\rho,\mu}^{\gamma}(\nu t^{\rho})\right](s) = \frac{s^{\rho\gamma-\mu}}{(s^{\rho} - \nu)^{\gamma}},$$

where $\text{Re}(s) > |\nu|^{1/\rho}$. Here we note that for $\gamma = 0$, Prabhakar derivative corresponds to Caputo derivative in Laplace space as expected [Mainardi (2010)]

$$\mathcal{L}\left[{}^{\text{C}}\mathcal{D}_{0+}^{\mu} f(t)\right](s) = s^{\mu}\mathcal{L}\left[f(t)\right](s) - s^{\mu-1} f(0+).$$

Then we reconsider the inverse Laplace and Fourier transforms of Eq. (8.23) w.r.t. s and k, respectively. Finally we arrive at the time fractional diffusion equation

$$^{\text{C}}\mathcal{D}_{\alpha,-\nu,0+}^{\gamma,\mu} p(x,t) = \mathcal{K}_{\mu} \frac{\partial^2}{\partial x^2} p(x,t), \tag{8.29}$$

where $\nu = \tau^{-\alpha}$, τ is time parameter with the dimension $[\tau] = \sec$, and $\mathcal{K}_{\mu}$ is the generalized diffusion coefficient with physical dimension $[\mathcal{K}_{\mu}] = \text{m}^2\text{sec}^{-\mu}$. Besides the initial condition is

$$p(x,0+) = p_0(x),$$

and the boundary condition is chosen to be absorbing, i.e., $p(x,t) = \frac{\partial}{\partial x} p(x,t) = 0$ when $x = \pm\infty$.

Next we make a further analysis of the transition of the waiting time distribution, which is given in Eq. (8.21) with the form of its Laplace transform. According to the series expansion approach [Podlubny (1999)], there exists

$$\phi(t) = \frac{1}{\tau} \sum_{n=0}^{\infty} (-1)^n \left(\frac{t}{\tau}\right)^{\mu n + \mu - 1} E_{\rho,\mu n+\mu}^{\gamma n+\gamma} \left(-\left[\frac{t}{\tau}\right]^{\rho}\right), \tag{8.30}$$

and the series of three parameter Mittag-Leffler functions shown in Eq. (8.30) is convergent [Sandev *et al.* (2011); Paneva-Konovska (2014)]. From the properties of three parameter Mittag-Leffler function, which is defined in Eq. (8.26), the asymptotic behavior of the waiting time distribution for the short time $t/\tau \ll 1$ is

$$\phi(t) \sim \frac{1}{\tau}\left(\frac{t}{\tau}\right)^{\mu-1} E_{\mu,\mu}\left(-\left[\frac{t}{\tau}\right]^{\mu}\right) \sim \frac{1}{\tau}\frac{(t/\tau)^{\mu-1}}{\Gamma(\mu)}, \qquad (8.31)$$

while for the long time limit $t/\tau \gg 1$, it behaves as

$$\phi(t) \sim \frac{1}{\tau}\left(\frac{t}{\tau}\right)^{\mu-\rho\gamma-1} E_{\mu-\rho\gamma,\mu-\rho\gamma}\left(-\left[\frac{t}{\tau}\right]^{\mu-\rho\gamma}\right) \sim \frac{\mu-\rho\gamma}{\tau}\frac{(t/\tau)^{-\mu+\rho\gamma-1}}{\Gamma(1-\mu+\rho\gamma)}. \qquad (8.32)$$

From the analytical results shown in Eq. (8.31) and Eq. (8.32), we conclude that the parameters ρ and γ have no influence on the waiting time distribution of the movement for the short time limit. However, when time t is sufficiently long, the influences of ρ and γ begin to emerge. Besides we can also conclude that the exponent of the waiting time distribution transits from $\mu-1$ to $-\mu+\rho\gamma-1$ as time moving on. Because of the transition of exponent of waiting time distribution, the movement of particle will also be influenced. The transition of the waiting time is illustrated in Fig. 8.2.

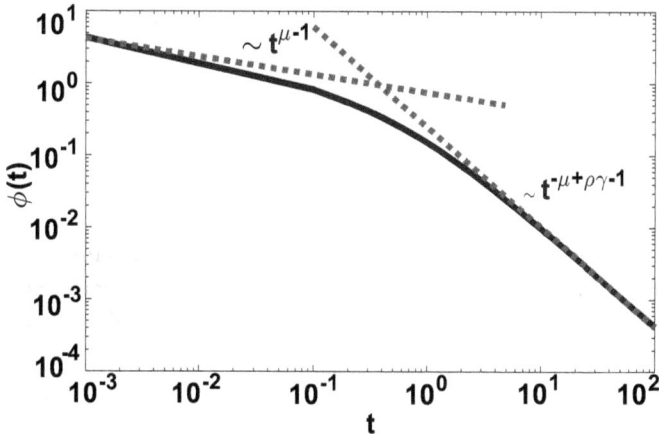

Fig. 8.1   The transition of the waiting time distribution Eq. (8.32). The parameters are chosen to be $\mu = 3/4$, $\rho = 7/16$, $\gamma = 5/6$ and $\tau = 1$.

Besides, we consider another waiting time distribution in Laplace space with the form of the following equation

$$\widehat{\phi}(s) = \frac{1}{1 + s\tau((s+b)\tau)^{\mu-1}\left[1 + ((s+b)\tau)^{-\rho}\right]^{\gamma}}, \tag{8.33}$$

where $b > 0$ with the physical dimension of $[b] = s^{-1}$ plays the role of tempering of waiting time distribution. By substituting the above waiting time distribution and the Gaussian jump length PDF into Eq. (8.20), we obtain the diffusion equation for this case

$$^{TC}\mathcal{D}_{\alpha,-\nu,0+}^{\gamma,\mu}P(x,t) = \mathcal{K}_{\mu}\frac{\partial^2}{\partial x^2}P(x,t), \tag{8.34}$$

where $^{TC}\mathcal{D}_{\alpha,-\nu,0+}^{\gamma,\mu}$ is the tempering regularized Prabhakar derivative introduced in [Sandev (2017)],

$$^{TC}\mathcal{D}_{\alpha,-\nu,0+}^{\gamma,\mu}f(t) = \left(^{T}\mathcal{E}_{\alpha,1-\mu,-\nu,0+}^{-\gamma}\frac{d}{dt}f\right)(t)$$

with

$$\left(^{T}\mathcal{E}_{\alpha,\mu,-\nu,0+}^{\gamma}f\right)(t) = \int_0^t e^{-b(t-t')}(t-t')^{\mu-1}E_{\alpha,\mu}^{\gamma}\left(-\nu[t-t']^{\alpha}\right)f(t')\,dt'.$$

Next we perform inverse Laplace transform of the waiting time PDF shown in Eq. (8.33) and obtain

$$\begin{aligned}
\phi(t) = \frac{1}{\tau}\sum_{n=0}^{\infty}\frac{(-1)^n}{\tau^{n+1}}\,{}^{RL}I_{0+}^{n+1}&\left(e^{-bt}\left(\frac{t}{\tau}\right)^{(\mu-1)(n+1)-1}\right.\\
&\left.\times E_{\rho,(\mu-1)(n+1)}^{\gamma n+\gamma}\left(-\left[\frac{t}{\tau}\right]^{\rho}\right)\right),
\end{aligned} \tag{8.35}$$

where $^{RL}I_{0+}^{\alpha}$ is the Riemann-Liouville integral defined in Eq. (8.27). From the above result Eq. (8.35) we can conclude that for short time because of the negligible influence of exponential tempering, the waiting time distribution asymptotically behaves the same as Eq. (8.30), i.e., Eq. (8.31), while for the long time the heavy-tail of waiting time distribution is strongly suppressed by the exponential tempering, finally it yields exponential waiting time PDF

$$\phi(t) = \frac{1}{\tau^*}\exp\left(-t/\tau^*\right), \tag{8.36}$$

where $\tau^* = \tau(b\tau)^{\mu-1}\left[1 + (b\tau)^{-\rho}\right]^{\gamma}$ with the physical dimension of $[\tau^*] = [\tau] = \sec$.

For the last part of this section, we further generalize our model. We consider the jump length distribution with the following form in Fourier space

$$\widetilde{\lambda}(k) = 1 - (\sigma|k|)^{\alpha_1} [1 + (\sigma|k|)^{-\rho_2}]^{\alpha_2},$$

where $1 < \alpha_1 < 2$ and $1 < \alpha_1 - \alpha_2\rho_2 < 2$. Substituting the above jump length distribution and waiting time PDF Eq. (8.21) into Eq. (8.20), the following result can be obtained

$$\widetilde{p}(k,s) = \frac{s^{-1}(s\tau)^\mu(1 + (s\tau)^{-\rho})^\gamma \widetilde{p}_0(k)}{(s\tau)^\mu(1 + (s\tau)^{-\rho})^\gamma + (\sigma|k|)^{\alpha_1}(1 + (\sigma|k|)^{-\rho_2})^{\alpha_2}}. \tag{8.37}$$

Here we note that for $\gamma = 0$ and $\alpha_2 = 0$, the jump length distribution of Lévy flight is the form of Lévy distribution, i.e., $\widetilde{\lambda}(k) = e^{-\sigma^{\alpha_1}|k|^{\alpha_1}} \sim 1 - \sigma^{\alpha_1}|k|^{\alpha_1}$ for small k. The purpose of such kind of jump length distribution is to obtain a transition from the 'slow' to the 'fast'. Besides, based on this jump length distribution, the transition can expand its area to all kind of diffusion. For the detailed discussion of Eq. (8.37), we will show in Sec. 8.4.

8.3 Non-Negativity of Solution: Subordinated Approach, and Stochastic Representation

In this section we construct a subordinated stochastic process, which is governed by Eq. (8.29). According to the discussion in Sec. 1.4.3, in order to find such stochastic representation, we must build the connection between the time t (known as physical time) and the Wiener process on another time scale denoted as u (known as operational time). Therefore, the PDF $p(x,t)$ of such kind of process $x(t)$ can be represented as

$$p(x,t) = \int_0^\infty w(x,u)h(u,t)\,du, \tag{8.38}$$

where

$$w(x,u) = \frac{1}{\sqrt{4\pi K_1 u}} \exp\left(-\frac{x^2}{4K_1 u}\right), \tag{8.39}$$

representing the PDF of a Wiener process, and here $h(u,t)$ represents the PDF of subordinating process. Then we can rewrite Eq. (8.22) in the form

$$\widetilde{p}(k,s) = \int_0^\infty e^{-uK_1 k^2} \widehat{h}(u,s)\,du, \tag{8.40}$$

where

$$\widehat{h}(u, s) = s^{\mu-1} \left[1 + (s\tau)^{-\rho}\right]^{\gamma} e^{-us^{\mu}\left[1+(s\tau)^{-\rho}\right]^{\gamma}}$$

$$= -\frac{\partial}{\partial u} \frac{1}{s} \widehat{L}(s, u) \tag{8.41}$$

and

$$\widehat{L}(s, u) = e^{-us^{\mu}\left[1+(s\tau)^{-\rho}\right]^{\gamma}}. \tag{8.42}$$

Taking inverse Fourier-Laplace transform of Eq. (8.40) leads to

$$p(x, t) = \int_0^{\infty} \frac{1}{\sqrt{4\pi\mathcal{K}_1 u}} e^{-\frac{x^2}{4\mathcal{K}_1 u}} h(u, t) \, du, \tag{8.43}$$

which indicates $h(u, t)$ is the PDF of subordinating process we want to find. It should be clarified that $p(x, t)$ needs to be non-negative if $h(u, t)$ is non-negative. That is also to say $\widehat{h}(u, s)$ is a completely monotone function w.r.t. s. In order to prove $\widehat{h}(u, s)$ is completely monotone, we only need to ensure both of $s^{\mu-1} \left[1 + (s\tau)^{-\rho}\right]^{\gamma}$ and $e^{-us^{\mu}\left[1+(s\tau)^{-\rho}\right]^{\gamma}}$ are completely monotone, and $e^{-us^{\mu}\left[1+(s\tau)^{-\rho}\right]^{\gamma}}$ is completely monotone if $s^{\mu} \left[1 + (s\tau)^{-\rho}\right]^{\gamma}$ is a Bernstein function. These conditions will be satisfied if $0 < \mu/\gamma$ and $0 < \mu/\gamma - \rho < 1$. Thus such construction of subordinated approach makes sense indeed.

In the final part of this section, we give a specific stochastic representation of the process $x(t)$, which is governed by the time fractional derivative differential equation Eq. (8.29) with Prabhakar derivative. Based on Eq. (8.40), Eq. (8.41) and Eq. (8.42), we build the stochastic representation of the following form

$$x(t) = \sqrt{2\mathcal{K}_{\mu}} B\left[\mathcal{S}(t)\right], \tag{8.44}$$

where $B(u)$ represents the Brownian motion, $\mathcal{S}(t)$ is the inverse Lévy stable subordinator and independent of $B(u)$ defined as

$$\mathcal{S}(t) = \inf \{u > 0 : \mathcal{T}(u) > t\},$$

where $\mathcal{T}(u)$ is a strictly increasing Lévy process defined by its characteristic function

$$\left\langle e^{-s\mathcal{T}(u)} \right\rangle = e^{-u\widehat{\Psi}(s)}$$

with the Lévy exponent $\widehat{\Psi}(s) = s^{\mu} \left[1 + (s\tau)^{-\rho}\right]^{\gamma}$. Still under the condition of $0 < \mu/\gamma < 1$ and $0 < \mu/\gamma - \rho < 1$, the function $\widehat{\Psi}(s) = s^{\mu} \left[1 + (s\tau)^{-\rho}\right]^{\gamma}$ is a Bernstein function and the stochastic representation we construct is well defined.

8.4 MSD, Fractional Moments, and Multi-Scale

Here in this section we first calculate MSD of the stochastic process with the PDF governed by the time fractional derivative equation Eq. (8.29). The initial condition in this section is chosen to be $p_0(x) = \delta(x)$ that is $\widetilde{p}_0(k) = 1$ in Fourier space. We still use the following equation to calculate the MSD according the Fourier-Laplace transform of PDF $p(x, t)$

$$\langle x^2(t) \rangle = \mathcal{L}^{-1} \left[-\frac{\partial^2}{\partial k^2} \widehat{\widetilde{p}}(k, s) \right] \Big|_{k=0}. \tag{8.45}$$

Then from Eq. (8.22), there exists

$$\langle x^2(t) \rangle = 2\mathcal{K}_\mu \mathcal{L}^{-1} \left[\frac{s^{-\mu-1}}{[1 + (s\tau)^{-\rho}]^\gamma} \right]$$
$$= 2\mathcal{K}_\mu t^\mu E^\gamma_{\rho, \mu+1} \left(-\left[\frac{t}{\tau} \right]^\rho \right). \tag{8.46}$$

According to the properties of three parameter Mittag-Leffler function, from Eq. (8.46) we obtain the short time asymptotic behavior

$$\langle x^2(t) \rangle \sim 2\mathcal{K}_\mu \frac{t^\mu}{\Gamma(\mu + 1)},$$

while for the long time, MSD asymptotically behaves as

$$\langle x^2(t) \rangle \sim 2\mathcal{K}_\mu \tau^\mu \frac{(t/\tau)^{\mu-\rho\gamma}}{\Gamma(\mu - \rho\gamma + 1)}.$$

From the above results, we observe the transition of the exponent from μ for the short time to $\mu - \rho\gamma$ for the long time, which means the diffusion decelerates as time moving on. Besides the role of parameter τ can also be concluded from the above equations, i.e., to control the speed of transition. Specifically, the larger τ is, the slower transition will be, and vice versa. In fact what the role parameter τ plays can also be obtained from the waiting time PDF in Laplace space shown in Eq. (8.21).

Next we mainly concern about the MSD of the particles whose PDF satisfies tempered time fractional diffusion equation shown in Eq. (8.34). With the same method above, there exists

$$\langle x^2(t) \rangle = 2\mathcal{K}_\mu {}^{RL}I^2_{0+} \left(e^{-bt} t^{\mu-2} E^\gamma_{\rho, \mu-1} \left(-\left[\frac{t}{\tau} \right]^\rho \right) \right), \tag{8.47}$$

where the operator ${}^{RL}I^2_{0+}$ represents the Riemann-Liouville integral. We can still analyse the asymptotic behavior of the MSD shown in Eq. (8.47). For the short time, the MSD behaves as $\langle x^2(t) \rangle \sim t^\mu$, while for the long time, it behaves asymptotically like a normal diffusion, that is $\langle x^2(t) \rangle \sim t$. This indicates the subdiffusion accelerates to a normal diffusion. However, such transition can only reach the normal diffusion at most. It can not break the barrier of normal diffusion and arrive at regime of superdiffusion. And the reason why normal diffusion appears for the long time is the exponential tempering strongly suppressed the heavy-tail of the waiting time PDF when time t is sufficiently long. And because of the tempering of the waiting time distribution, the tempered Prabhakar derivative is also obtained.

Then we are going to calculate the fractional order moments of the process whose PDF is given by Eq. (8.29). The definition of the fractional order moments of the particle is given by

$$\langle |x(t)|^q \rangle = \int_{-\infty}^{\infty} |x|^q p(x,t)\, dx.$$

Substituting $y = (s^{\mu/2}(1+(s\tau)^{-\rho})^{\gamma/2} |x|$, and according to the expression of PDF $p(x,t)$ in Laplace space, there exists

$$\langle |x(t)|^q \rangle = \Gamma(q+1)(K_\mu t^\mu)^{q/2}\, E_{\rho,\mu q/2+1}^{\gamma q/2}\left(-\left[\frac{t}{\tau}\right]^\rho\right). \tag{8.48}$$

Thus for the short time $t/\tau \ll 1$, we obtain the asymptotic behavior

$$\langle |x(t)|^q \rangle \sim \Gamma(q+1)\frac{(K_\mu t^\mu)^{q/2}}{\Gamma(1+\mu q/2)}\left[1-\frac{\gamma q}{2}\frac{\Gamma(1+\mu q/2)}{\Gamma(1+\rho+\mu q/2)}(t/\tau)^\rho\right]. \tag{8.49}$$

Here we note that for the short time $t/\tau \ll 1$, i.e., $s\tau \gg 1$, the leading term of the above equation is

$$\langle |x(t)|^q \rangle \sim \Gamma(q+1)\frac{(K_\mu t^\mu)^{q/2}}{\Gamma(1+\mu q/2)},$$

which coincides with the fractional Fokker-Planck equation with the time fractional order μ. This result is expected because the waiting time distribution shown in Eq. (8.21) behaves as

$$\widehat{\phi}(s) \sim \frac{1}{1+(s\tau)^\mu}$$

when $s\tau \gg 1$. On the contrary, for long time the asymptotic behavior is

$$\langle |x(t)|^q \rangle \sim \Gamma(q+1)(K_\mu \tau^\mu)^{q/2}\frac{(t/\tau)^{(\mu-\rho\gamma)q/2}}{\Gamma(1+(\mu-\rho\gamma)q/2)}$$
$$\times\left[1-\frac{\gamma q}{2}\frac{\Gamma(1+(\mu-\rho\gamma)q/2)}{\Gamma(1-\rho+(\mu-\rho\gamma)q/2)}(t/\tau)^{-\rho}\right]. \tag{8.50}$$

The leading term of Eq. (8.50) reads

$$\langle |x(t)|^q \rangle \sim \Gamma(q+1) \, (\mathcal{K}_\mu \tau^\mu)^{q/2} \, \frac{(t/\tau)^{(\mu-\rho\gamma)q/2}}{\Gamma(1+(\mu-\rho\gamma)q/2)}$$

and this result coincides with the one obtained from the fractional Fokker-Planck equation with the fractional order $\mu - \rho\gamma$. This result is also expected because for the long time limit the waiting time in Eq. (8.21) behaves asymptotically as $\hat{\phi}(s) \sim \frac{1}{1+(s\tau)^{\mu-\rho\gamma}}$. From the results above we can conclude that the fractional order moment behaves as the following form

$$\langle |x(t)|^q \rangle = C(q) t^{\nu(\mu,q,\gamma,t)}, \tag{8.51}$$

where the exponent $\nu(\mu, q, \gamma, t)$ represents it is concerned with μ, q, γ and even t and it is called the multi-scaling exponent. In [Mandelbrot (1999)], one can obtain the self-affine behavior $\langle |x(t)|^q \rangle = C(q) t^{qH}$, which is a special case of Eq. (8.51). And here $H > 0$ represents the Hurst exponent; for ordinary Brownian motion $H = 1/2$; for fractional Brownian motion, $0 < H < 1$. Here we can conclude the exponent of fractional order moment for the fractal or self-affine processes linearly depends on the fractional order q, $\nu \propto q$. And the process governed by the fractional diffusion equation with Caputo time fractional derivative belongs to such class. More general, if $\langle |x(t)|^q \rangle = C(q) t^{\nu(q)}$ and $\nu(q)$ is a given nonlinear function, we will call it a multi-fractal or multi-affine process [Mandelbrot (1999)].

In the above results, we take the jump length distribution simply as

$$\tilde{\lambda}(k) = 1 - \sigma^2 k^2 \tag{8.52}$$

in Fourier space and substitute it into Eq. (8.20), that is we consider the jump length distribution in Eq. (8.52) as the exact form of Fouier transform of some function. Then we consider the jump length distribution has the Gaussian form in Fourier space as shown in the following equation

$$\begin{aligned} \tilde{\lambda}(k) = e^{-\sigma^2 k^2} &= 1 - \sigma^2 k^2 + \frac{\sigma^4 k^4}{2} - \frac{\sigma^6 k^6}{3!} + \cdots \\ &= 1 - \frac{m_2 k^2}{2} + \frac{m_4 k^4}{4!} - \frac{m_6 k^6}{6!} + \cdots, \end{aligned} \tag{8.53}$$

where $m_2 = \Sigma^2 = 2\sigma^2$, $m_4 = 12\sigma^4$, $m_6 = 120\sigma^6$, i.e., $m_{2n} = 2\sigma^{2n}$, $n = 1, 2, \ldots$ are the finite moments. For the long time, the MSDs obtained from the CTRW model with jump length distribution shown in Eq. (8.52) and Eq. (8.53) and the same waiting time distribution are the same. However, if we calculate the higher order moment, the differences begin to emerge. Here we calculate the fourth moment of the process, that is

$$\langle x^4(t)\rangle = 6\,m_2 \left(\frac{t}{\tau}\right)^{2\mu} E_{\rho,2\mu+1}^{2\gamma}\left(-\left[\frac{t}{\tau}\right]^\rho\right) + m_4 \left(\frac{t}{\tau}\right)^\mu E_{\rho,\mu+1}^\gamma\left(-\left[\frac{t}{\tau}\right]^\rho\right).$$
(8.54)

From the results shown in Eq. (8.54), we can conclude that both m_2 and m_4, i.e., the second and the fourth moments of jump length respectively, will affect the fourth moment of the process. Then basing on Eq. (8.54) we calculate the asymptotic behavior of Eq. (8.54) by utilizing the large argument asymptotic expansion formula for three parameter Mittag-Leffler function. And the leading term is the same as Eq. (8.50) which is obtained from the jump length distribution Eq. (8.52). This also indicates the leading term of fourth moment obtained from the jump length Eq. (8.54) for the long time is only concerned with m_2, while the difference between the $\langle x^4(t)\rangle$ obtained from the CTRW model with jump length distributions shown in Eq. (8.52) and Eq. (8.53) appears for the short time limit. Then we use the properties of three parameter Mittag-Leffler functions to obtain the asymptotic behavior of the $\langle x^4(t)\rangle$ for the short time limit from the Eq. (8.53), i.e.,

$$\langle x^4(t)\rangle \sim \frac{m_4}{\tau^\mu}\frac{(t/\tau)^\mu}{\Gamma(1+\mu)}\left[1 - \gamma\frac{\Gamma(1+\mu)}{\Gamma(1+\mu+\rho)}(t/\tau)^\rho\right].$$
(8.55)

Comparing the result of Eq. (8.55) with the leading term of Eq. (8.49), we conclude that m_4 plays an important role in Eq. (8.55) while it does not appear in Eq. (8.49). These results are also discussed in [Barkai (2002)].

Then we consider the fractional moment of the stochastic process with PDF Eq. (8.37). First we consider the short time behavior, that is, s and $|k|$ are large enough. And we get the following equation

$$\widetilde{p}(k,s) \sim \frac{1}{s}\frac{(s\tau)^\mu}{(s\tau)^\mu + (\sigma|k|)^{\alpha_1}}\widetilde{p}_0(k).$$

By rearranging the above equation, we have

$$s^\mu\widetilde{p}(k,s) - s^{\mu-1}\widetilde{p}_0(k) = -\frac{\sigma^{\alpha_1}}{\tau^\mu}|k|^{\alpha_1}\widetilde{p}(k,s).$$

Then performing inverse Fourier-Laplace transform, we finally arrive at the space-time fractional diffusion equation with the form

$$^C\mathcal{D}_{0+}^\mu p(x,t) = \frac{\sigma^{\alpha_1}}{\tau^\mu}\frac{\partial^{\alpha_1}}{\partial|x|^{\alpha_1}}p(x,t),$$

where $\frac{\partial^\alpha}{\partial|x|^\alpha}$ represents the Riesz fractional derivative [Feller (1971)].

Definition 8.3. The Riesz fractional derivative of order α $(0 < \alpha \le 2)$ is given as a pseudo-differential operator with the Fourier symbol $-|k|^\alpha$, $k \in R$, i.e.,

$$\frac{\partial^\alpha}{\partial |x|^\alpha} f(x) = \mathcal{F}^{-1}\left[-|k|^\alpha \widetilde{F}(k)\right].$$

According to the properties of Lévy flight, if we still calculate the second order moment by using the method of Eq. (8.45), the conclusion will always be infinite. Then we calculate the fractional moment of the process as well [Metzler and Klafter (2000)],

$$\langle |x(t)|^q \rangle \sim \frac{4\pi}{\alpha_1} \frac{\Gamma(1+q)\Gamma(1+q/\alpha_1)\Gamma(-q/\alpha_1)}{\Gamma(1+q/2)\Gamma(-q/2)} \frac{\left(\frac{\sigma^{\alpha_1}}{\tau^\mu} t^\mu\right)^{q/\alpha_1}}{\Gamma(1+\mu q/\alpha_1)}, \tag{8.56}$$

where $0 < q < \alpha_1 < 2$. Taking limit $q \to 2$, we obtain $\langle |x(t)|^q \rangle^{2/q} \sim t^{\frac{2\mu}{\alpha_1}}$, and consider it as the MSD. Thus, there always exists an competition between the long waiting and long jump. Specifically, for $\frac{2\mu}{\alpha_1} < 1$ the subdiffusion will be observed; for $\frac{2\mu}{\alpha_1} = 1$ we can treat it as a normal diffusion. However we treat it as normal diffusion simply because its MSD is a linear function of time t. Many properties between these two processes are different; for the last case $\frac{2\mu}{\alpha_1} > 1$, the superdiffusion is obtained.

On the other hand, we consider the long time limit behavior of the process whose PDF in Fourier-Laplace space has the form of Eq. (8.37). Then by taking s and $|k|$ to be sufficiently small, there exists

$$\widetilde{\widehat{p}}(k, s) \sim \frac{1}{s} \frac{(s\tau)^{-\rho\gamma+\mu}}{(s\tau)^{-\rho\gamma+\mu} + (\sigma|k|)^{-\rho_2\alpha_2+\alpha_1}} \widehat{p}_0(k).$$

Thus for the long time limit, the following space-time fractional diffusion equation holds

$$^C\mathcal{D}_{0+}^{\mu-\rho\gamma} p(x,t) = \frac{\sigma^{\alpha_1-\rho_2\alpha_2}}{\tau^{\mu-\rho\gamma}} \frac{\partial^{\alpha_1-\rho_2\alpha_2}}{\partial |k|^{\alpha_1-\rho_2\alpha_2}} p(x,t).$$

Besides, its fractional moment can be obtained as well

$$\langle |x(t)|^q \rangle = \frac{4\pi}{\alpha_1 - \rho_2\alpha_2} \frac{\Gamma(1+q)\Gamma(1+q/(\alpha_1-\rho_2\alpha_2))\Gamma(-q/(\alpha_1-\rho_2\alpha_2))}{\Gamma(1+q/2)\Gamma(-q/2)}$$

$$\times \frac{\left(\frac{\sigma^{\alpha_1-\rho_2\alpha_2}}{\tau^{\mu-\rho\gamma}} t^{\mu-\rho\gamma}\right)^{q/(\alpha_1-\rho_2\alpha_2)}}{\Gamma\left(1+\frac{(\mu-\rho\gamma)q}{\alpha_1-\rho_2\alpha_2}\right)}, \tag{8.57}$$

where $0 < q < \alpha_1 - \rho_2\alpha_2 < 2$, and $\left\langle |x(t)|^q \right\rangle^{2/q} \sim t^{\frac{2(\mu-\rho\gamma)}{\alpha_1-\rho_2\alpha_2}}$. Thus we can also classify the corresponding process for the long time according to the exponent.

In order to make the following statement clearer we choose $\gamma = 0$ (Caputo time fractional derivative obtained), and the fractional moment (q-th moment) of the process transits from $t^{(q\mu)/\alpha_1}$ to $t^{(q\mu)/(\alpha_1-\rho_2\alpha_2)}$ as time t moves on, that is the process can accelerate. Thus we can control the parameters μ, ρ, γ, α_1, α_2 and ρ_2 to achieve an accelerating or decelerating transition. Besides, the process can freely transit among subdiffusion, normal diffusion and superdiffusion.

8.5 Fractional Fokker-Planck Equation with Prabhakar Derivative

After introducing the CTRW description of the process that can transit among different kinds of anomalous diffusions in Sec. 8.2, finding the stochastic representations in Sec. 8.3 and obtaining the MSDs and fractional moments of the process in Sec. 8.4, in this section we begin to derive the corresponding Fokker-Planck equation with Prabhakar derivative, which can be considered as a special fractional operator.

Here we consider the particles move in the external potential $V(x)$ with the waiting time Eq. (8.21) and Gaussian jump length distribution. Then the following fractional Fokker-Planck equation can be obtained by utilizing the same methods in Sec. 1.3,

$$
{}^{C}\mathcal{D}^{\gamma,\mu}_{\rho,-\nu,0+}p(x,t) = \left[\frac{\partial}{\partial x} \frac{V'(x)}{m\eta_\mu} + \mathcal{K}_\mu \frac{\partial^2}{\partial x^2} \right] p(x,t), \qquad (8.58)
$$

where $\nu = \tau^{-\rho}$, m is the mass of the particle, $\mathcal{K}_\mu$ is the generalized diffusion coefficient, and η_μ is the friction coefficient with physical dimension $[\eta_\mu] = \sec^{\mu-2}$. Here we note that, according to the definition of the Prabhakar derivative and $\frac{\partial}{\partial t}p(x,t) = 0$, the generalized Einstein-Stokes relation

$$
\mathcal{K}_\mu = \frac{k_B T}{m\eta_\mu}
$$

is obtained.

We first consider the case of the constant external force, i.e., $F(x) = -\frac{dV(x)}{dx} = F\Theta(t)$ ($V(x) = -Fx$), where $\Theta(t)$ is the Heaviside step function. By taking Fourier-Laplace transform of Eq. (8.58), one can obtain

$$
\widetilde{p}(k,s) = \frac{s^{\mu-1}\left[1 + (s\tau)^{-\rho}\right]^{\gamma}}{s^{\mu}\left[1 + (s\tau)^{-\rho}\right]^{\gamma} + i\frac{F}{m\eta_\mu}k + \mathcal{K}_\mu k^2}. \qquad (8.59)
$$

Then by inverse Fourier transform $k \to x$, the following equation holds

$$\widehat{p}(x, s) = \widehat{p}_0(x, s) \exp\left[-\frac{F}{2m\eta_\mu \mathcal{K}_\mu}x\right], \tag{8.60}$$

where

$$\widehat{p}_0(x, s) = \frac{\exp\left[-\sqrt{\frac{s^\mu [1+(s\tau)^{-\rho}]^\gamma}{\mathcal{K}_\mu} + \left(\frac{F}{2m\eta_\mu \mathcal{K}_\mu}\right)^2}\, |x|\right]}{\sqrt{\frac{s^\mu [1+(s\tau)^{-\rho}]^\gamma}{\mathcal{K}_\mu} + \left(\frac{F}{2m\eta_\mu \mathcal{K}_\mu}\right)^2}} \frac{s^{\mu-1} [1+(s\tau)^{-\rho}]^\gamma}{2\mathcal{K}_\mu}.$$

And the PDF of the case without external force is obtained as

$$\widehat{p}(x, s) = \frac{1}{2s}\sqrt{\frac{s^\mu [1+(s\tau)^{-\rho}]^\gamma}{\mathcal{K}_\mu}}\, e^{-\sqrt{\frac{s^\mu [1+(s\tau)^{-\rho}]^\gamma}{\mathcal{K}_\mu}}|x|}$$

simply by taking the external force $F = 0$.

Then one can calculate the moments of the process according to Eq. (8.59) with the method

$$\langle x^n(t)\rangle = \mathcal{L}^{-1}\left[i^n \frac{\partial^n}{\partial k^n}\widetilde{\widehat{p}}(k, s)\right]\Bigg|_{k=0}.$$

Then the first moment can be obtained as

$$\langle x(t)\rangle_F = \frac{F}{m\eta_\mu}\mathcal{L}^{-1}\left[\frac{s^{-\mu-1}}{[1+(s\tau)^{-\rho}]^\gamma}\right]$$
$$= \frac{F}{m\eta_\mu}t^\mu E^\gamma_{\rho,\mu+1}\left(-\left[\frac{t}{\tau}\right]^\rho\right); \tag{8.61}$$

while for the second moment, there exists

$$\langle x^2(t)\rangle_F = 2\mathcal{K}_\mu \mathcal{L}^{-1}\left[\frac{s^{-\mu-1}}{[1+(s\tau)^{-\rho}]^\gamma}\right] + 2\left(\frac{F}{m\eta_\mu}\right)^2 \mathcal{L}^{-1}\left[\frac{s^{-2\mu-1}}{[1+(s\tau)^{-\rho}]^{2\gamma}}\right]$$
$$= 2\mathcal{K}_\mu t^\mu E^\gamma_{\rho,\mu+1}\left(-\left[\frac{t}{\tau}\right]^\rho\right) + 2\left(\frac{F}{m\eta_\mu}\right)^2 t^{2\mu} E^{2\gamma}_{\rho,2\mu+1}\left(-\left[\frac{t}{\tau}\right]^\rho\right). \tag{8.62}$$

From Eq. (8.61) and Eq. (8.62), the second Einstein relation is obtained, i.e.,

$$\langle x(t)\rangle_F = \frac{F}{2k_B T}\langle x^2(t)\rangle_{F=0}.$$

8.5.1   *Relaxation of Modes*

Here we consider space and time are independent and assume the PDF can make the variable separation, i.e., $p(x,t) = X(x)T(t)$. Then according to Eq. (8.58), we have

$$^C\mathcal{D}^{\gamma,\mu}_{\rho,-\nu,0+}T(t) = -\lambda T(t), \tag{8.63}$$

$$\left[\frac{\partial}{\partial x}\frac{V'(x)}{m\eta_\mu} + \mathcal{K}_\mu\frac{\partial^2}{\partial x^2}\right]X(x) = -\lambda X(x), \tag{8.64}$$

where λ is a separation constant. Then the solution of Eq. (8.58) is given as $p(x,t) = \sum_n X_n(x)T_n(t)$, where $X_n(x)$, $T_n(n)$ are the eigenfunctions corresponding to the eigenvalue λ_n. Then by taking Laplace transform of Eq. (8.63) w.r.t. time t and doing some rearrangements, taking inverse Laplace transform, we finally get the following relaxation law corresponding to the eigenvalue λ_n

$$T_n(t) = T_n(0)\sum_{j=0}^{\infty}(-\lambda_n)^j t^{\mu j}E^{\gamma j}_{\rho,\mu j+1}\left(-\left[\frac{t}{\tau}\right]^\rho\right), \tag{8.65}$$

where $T_n(0) = \langle p_0(x), X_n(x)\rangle$ and $\langle\cdot,\cdot\rangle$ represents the inner product. Taking long time limit $t/\tau \gg 1$ on Eq. (8.65), the power law decay can be obtained

$$T_n(t) \sim \frac{T_n(0)}{\lambda_n\tau^\mu}\frac{(t/\tau)^{-(\mu-\rho\gamma)}}{\Gamma\left(1-(\mu-\rho\gamma)\right)}. \tag{8.66}$$

Here we note that if we take $\gamma = 0$, then the conclusion of monofractional diffusion equation will be recovered

$$T_n(t) = T_n(0)\sum_{j=0}^{\infty}\frac{(-\lambda_n)^j t^{\mu j}}{\Gamma(\mu j+1)} = T_n(0)E_\mu\left(-\lambda_n t^\mu\right).$$

8.5.2   *Harmonic External Potential*

Here we take the harmonic external potential $V(x) = \frac{1}{2}m\omega^2 x^2$, where ω is a frequency, which is one of the most important potentials in physics. And basing on spatial eigenequation Eq. (8.64), the solution will be given in terms of Hermite polynomials $H_n(z)$ [Erdélyi *et al.* (1955)]

$$X_n(x) = C_nH_n\left(\sqrt{\frac{m\omega^2}{2k_BT}}x\right)\exp\left(-\frac{m\omega^2}{2k_BT}x^2\right), \tag{8.67}$$

where the eigenvalue spectrum of the corresponding Sturm-Liouville problem is given by $\lambda_n = n\frac{\omega^2}{\eta_\mu}$ for $n = 0, 1, 2, ...$, and C_n here is the normalisation constant. According to the normalisation condition $\langle X_n(x), X_n(x) \rangle = 1$, the solution can be obtained

$$p(x,t) = \left(\frac{m\omega^2}{2\pi k_B T}\right)^{\frac{1}{2}} \sum_n \frac{1}{2^n n!} T_n(0) H_n\left(\sqrt{\frac{m\omega^2}{2k_B T}}x\right)$$

$$\times \exp\left(-\frac{m\omega^2}{2k_B T}x^2\right) \sum_{j=0}^{\infty} \left(-\frac{n\omega^2}{\eta_\mu}\right)^j t^{\mu j} E^{\gamma j}_{\rho,\mu j+1}\left(-\left[\frac{t}{\tau}\right]^\rho\right).$$

$$(8.68)$$

For the corresponding results of monofractional diffusion equation, one can find in [Metzler *et al.* (1999a)]. For $n = 0$ in Eq. (8.68), the Gaussian stationary solution

$$p(x,t) = \sqrt{\frac{m\omega^2}{2\pi k_B T}} \exp\left(-\frac{m\omega^2}{2k_B T}x^2\right).$$

Then we calculate the approximate behavior of the first passage time for this case. According to [Dybiec and Sokolov (2015)], the following relation between the first passage time density $f(t)$ and the survival probability $S(t)$ which is defined as

$$S(t) = \int_\Omega W(x,t|x_0,0)\,dx$$

is satisfied

$$F(t) := \int_0^t f(u)\,du = 1 - S(t).$$

That is $f(t) = -\frac{d}{dt}S(t)$. First we consider the sum over j in Eq. (8.68) for the long time yields

$$T_n(t) \sim T_n(0)\frac{\eta_\mu}{n\omega^2}\frac{\tau^{-\rho\gamma}t^{-(\mu-\rho\gamma)}}{\Gamma(1-\mu+\rho\gamma)}.$$

Then

$$p(x,t|x_0,0) \sim \sqrt{\frac{m\omega^2}{2\pi k_B T}} \exp\left(-\frac{m\omega^2}{2k_B T}x^2\right) + \left(\frac{m\omega^2}{2\pi k_B T}\right)^{\frac{1}{2}} \sum_{n=1}^{\infty} \frac{T_n(0)}{2^n n!}$$

$$\times H_n\left(\sqrt{\frac{m\omega^2}{2k_B T}}x\right) \exp\left(-\frac{m\omega^2}{2k_B T}x^2\right) \frac{\eta_\mu}{n\omega^2}\frac{\tau^{-\rho\gamma}t^{-(\mu-\rho\gamma)}}{\Gamma(1-\mu+\rho\gamma)}.$$

Here we consider the one-dimensional process and the domain Ω is an interval $[-L, L]$. Then

$$S(t) \sim \left(\frac{m\omega^2}{2\pi k_B T}\right)^{\frac{1}{2}} \int_{-L}^{L} \exp\left(-\frac{m\omega^2}{2k_B T} x^2\right) dx + \left(\frac{m\omega^2}{2\pi k_B T}\right)^{\frac{1}{2}} \sum_{n=1}^{\infty} \frac{T_n(0)}{2^n n!}$$

$$\times \left[\int_{-L}^{L} H_n\left(\sqrt{\frac{m\omega^2}{2k_B T}} x\right) \exp\left(-\frac{m\omega^2}{2k_B T} x^2\right) dx\right] \frac{\eta_\mu \tau^{-\mu}}{n\omega^2} \frac{(t/\tau)^{-(\mu-\rho\gamma)}}{\Gamma(1-\mu+\rho\gamma)}$$

$$= \mathrm{erf}\left(\sqrt{\frac{m\omega^2}{2k_B T}} L\right) + \left(\frac{m\omega^2}{2\pi k_B T}\right)^{\frac{1}{2}} \sum_{n=1}^{\infty} \frac{T_n(0)}{2^n n!} \left[\int_{-L}^{L} H_n\left(\sqrt{\frac{m\omega^2}{2k_B T}} x\right)\right.$$

$$\left. \times \exp\left(-\frac{m\omega^2}{2k_B T} x^2\right) dx\right] \frac{\eta_\mu \tau^{-\mu}}{n\omega^2} \frac{(t/\tau)^{-(\mu-\rho\gamma)}}{\Gamma(1-\mu+\rho\gamma)},$$

$$(8.69)$$

where

$$\mathrm{erf}(x) := \frac{1}{\sqrt{\pi}} \int_{-x}^{x} e^{-t^2}\, dt = \frac{2}{\sqrt{\pi}} \int_{0}^{x} e^{-t^2}\, dt$$

is the error function [Erdélyi *et al.* (1955)]. Next we take a specific initial condition $p_0(x) = \delta(x)$. Then $T_n(0) = H_n(0)$. Besides, we assume that $\frac{m\omega^2}{2k_B T} = 1$.

According to the numerical calculations (we take the first one hundred terms instead of the infinite series), we observe that for $L \geqslant 2$ the second term of the right hand side of Eq. (8.69) can be approximately treated as zero. For this case the survival probability asymptotically behaves as $S(t) \sim \mathrm{erf}(L)$. Then we can also obtain the density of the first passage time $f(t) \sim \delta(t)$ for $L \geqslant 2$. This result is expected, because if the particle moves in the harmonic external potential, then the potential will constrain the particle in a small domain and almost keep the particles from moving outside.

On the other hand, we consider the $0 < L < 2$. Also according to the numerical calculations, we calculate the first 10 terms and the first 200 terms respectively and observe between the difference between them is so little that we consider the first 10 terms instead of the infinite series.

Then we have the following approximation

$$S(t) \approx \mathrm{erf}(L) + \frac{\exp(-L^2)L}{115200\sqrt{\pi}}(104745 - 61270L^2 + 18572L^4 - 2328L^6 + 96L^8)$$

$$\times \frac{\eta_\mu}{\omega^2} \frac{\tau^{-\rho\gamma}t^{-(\mu-\rho\gamma)}}{\Gamma(1-\mu+\rho\gamma)}.$$

Sum

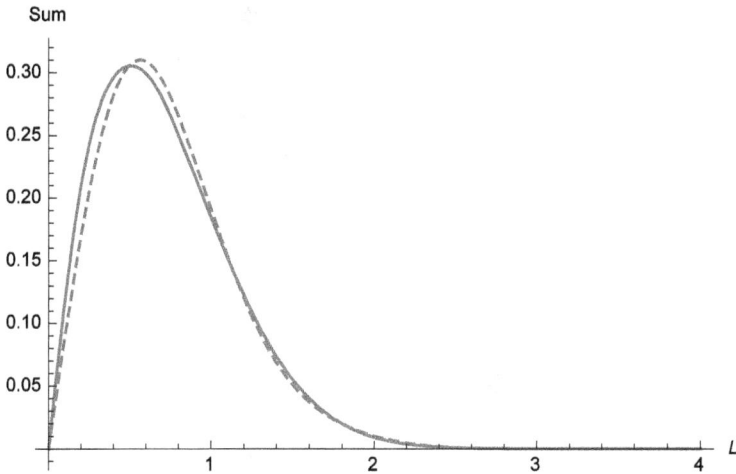

Fig. 8.2   Comparison between the first 10 terms (dashed line) and the first 100 terms (solid line) of the infinite series in Eq. (8.69).

Then the distribution of the first passage time approximately behaves as

$$f(t) \approx \frac{\exp(-L^2)L}{115200\sqrt{\pi}}(104745 - 61270L^2 + 18572L^4 - 2328L^6 + 96L^8)$$

$$\times \frac{\eta_\mu}{\tau^{\rho\gamma}\omega^2}\frac{\mu - \rho\gamma}{\Gamma(1 - \mu + \gamma\rho)}t^{-1+\rho\gamma-\mu}$$

for $0 < L < 2$.

In the last part of this section, we derive the differential equations for the first and the second moments of the process moving in the harmonic potential. For the first moment, we can easily obtain the following time fractional equation with Prabhakar derivative

$$^C\mathcal{D}^{\gamma,\mu}_{\rho,-\nu,0+}\langle x(t)\rangle + \frac{\omega^2}{\eta_\mu}\langle x(t)\rangle = 0 \qquad (8.70)$$

with the initial condition $x_0 = \int_{-\infty}^{\infty} xW_0(x)\,dx$. After Laplace transform w.r.t. time t, rearranging the equation and taking inverse Laplace transform, we finally obtain the following relaxation form of the first moment

$$\langle x(t)\rangle = \sum_{j=0}^{\infty}\left(-\frac{\omega^2}{\eta_\mu}\right)^j t^{\mu j} E^{\gamma j}_{\rho,\mu j+1}\left(-\left[\frac{t}{\tau}\right]^\rho\right).$$

Taking $\gamma = 0$, the result of monofractional diffusion can be recovered. That is

$$\langle x(t)\rangle = x_0 E_\mu\left(-\frac{\omega^2}{\eta_\mu}t^\mu\right),$$

which has the power law scaling form $\langle x(t) \rangle \sim \frac{x_0 \eta_\mu}{\omega^2} \frac{t^{-\mu}}{\Gamma(1-\mu)}$ for the long time.

On the other hand we can also obtain the second order moment and obtain the following equation

$$^C\!D^{\gamma,\mu}_{\rho,-\nu,0+}\left\langle x^2(t) \right\rangle + 2\frac{\omega^2}{\eta_\mu}\left\langle x^2(t) \right\rangle = 2\mathcal{K}_\mu.$$

With the same skill of Laplace transform, we obtain the following form

$$\left\langle x^2(t) \right\rangle = x_0^2 \mathcal{L}^{-1}\left[\frac{s^{\mu-1}\left[1+(s\tau)^{-\rho}\right]^\gamma}{s^\mu\left[1+(s\tau)^{-\rho}\right]^\gamma + 2\frac{\omega^2}{\eta_\mu}}\right] + \mathcal{L}^{-1}\left[\frac{2\mathcal{K}_\mu}{s^\mu\left[1+(s\tau)^{-\rho}\right]^\gamma + 2\frac{\omega^2}{\eta_\mu}}\right],$$

equivalently for the long time

$$\left\langle x^2(t) \right\rangle = x_{\text{th}}^2 + \left(x_0^2 - x_{\text{th}}^2\right)\sum_{j=0}^{\infty}\left(-2\frac{\omega^2}{\eta_\mu}\right)^j t^{\mu j} E^{\gamma j}_{\rho,\mu j+1}\left(-\left[\frac{t}{\tau}\right]^\rho\right),$$

where $x_0 = x(0)$ represents the initial value of the position, and $x_{\text{th}}^2 = \frac{k_B T}{m\omega^2}$ is the stationary (thermal) value. Different from the Brownian motion of which the second moment approaches the stationary value exponentially, the second moment here approaches the stationary value with Mittag-Leffler relaxation, i.e.,

$$\left\langle x^2(t) \right\rangle \sim x_{\text{th}}^2 + \left(x_0^2 - x_{\text{th}}^2\right) E_{\mu-\rho\gamma}\left(-\frac{2\omega^2}{\eta_\mu \tau^{-\mu}}\left[\frac{t}{\tau}\right]^{\mu-\rho\gamma}\right).$$

And for the long time, the power law scaling also appears, that is

$$\left\langle x^2(t) \right\rangle \sim x_{\text{th}}^2 + \left(x_0^2 - x_{\text{th}}^2\right)\frac{\eta_\mu \tau^{-\mu}}{2\omega^2}\frac{(t/\tau)^{-\mu+\rho\gamma}}{\Gamma(1-\mu+\rho\gamma)}.$$

8.6   A Brief Introduction of Three Parameter Mittag-Leffler Functions

Here in this section, we make a brief introduction to the three parameter Mittag-Leffler function in order to complete the theory of this chapter. We begin this section with the definition [Prabhakar (1971)].

Definition 8.4. The three parameter Mittag-Leffler function is defined as

$$E^\delta_{\alpha,\beta}(z) = \sum_{k=0}^{\infty}\frac{(\delta)_k}{\Gamma(\alpha k+\beta)}\frac{z^k}{k!}, \qquad (8.71)$$

where $(\delta)_k$ represents the Pochhammer symbol. If we take $\beta = \delta = 1$ or $\delta = 1$ then the one parameter Mittag-Leffler function $E_\alpha(z)$ and two parameter Mittag-Leffler function $E_{\alpha,\beta}(z)$ will be recovered respectively. For the three parameter Mittag-Leffler function, the following formula also exists

$$E_{\alpha,\beta}^\delta(-z) = \frac{z^{-\delta}}{\Gamma(\delta)} \sum_{n=0}^\infty \frac{\Gamma(\delta+n)}{\Gamma(\beta - \alpha(\delta+n))} \frac{(-z)^{-n}}{n!} \tag{8.72}$$

for $|z| > 1$. From the definition of Mittag-Leffler function shown in Eq. (8.71), we can obtain the asymptotic behavior of $E_{\alpha,\beta}^\delta(-t^\alpha)$ for short time t

$$E_{\alpha,\beta}^\delta(-t^\alpha) \sim \frac{1}{\Gamma(\beta)} - \delta \frac{t^\alpha}{\Gamma(\alpha+\beta)}$$
$$\sim \frac{1}{\Gamma(\beta)} \exp\left(-\delta \frac{\Gamma(\beta)}{\Gamma(\alpha+\beta)} t^\alpha\right).$$

On the other hand, for the long time limit, according to Eq. (8.72) we obtain the power law behaviour of the three parameter Mittag-Leffler function as

$$E_{\alpha,\beta}^\delta(-t^\alpha) \sim \frac{t^{-\alpha\delta}}{\Gamma(\beta - \alpha\delta)}.$$

Another useful property is the Laplace transform of the three parameter Mittag-Leffler function and it is given by [Prabhakar (1971)]

$$\mathcal{L}\left[t^{\beta-1} E_{\alpha,\beta}^\delta\right](s) = \frac{s^{\alpha\delta-\beta}}{(s^\alpha \mp a)^\delta}$$

for $\mathrm{Re}(s) > |a|^{\frac{1}{\alpha}}$.

Bibliography

Aaronson, J. (1997). *An Introduction to Infinite Ergodic Theory*, Vol. 50 (American Mathematical Society, Providence).

Abad, E., Yuste, S. B., and Lindenberg, K. (2010). Reaction-subdiffusion and reaction-superdiffusion equations for evanescent particles performing continuous-time random walks, *Phys. Rev. E* **81**, p. 031115.

Abramowitz, M. and Stegun, I. A. (1984). *Handbook of Mathematical Functions with Formulas, Graphs, and Mathematical Tables* (John Wiley & Sons, New York).

Aghion, E., Kessler, D. A., and Barkai, E. (2017). Large fluctuations for spatial diffusion of cold atoms, *Phys. Rev. Lett.* **118**, p. 260601.

Aghion, E., Kessler, D. A., and Barkai, E. (2019). From non-normalizable Boltzmann-Gibbs statistics to infinite-ergodic theory, *Phys. Rev. Lett.* **122**, p. 010601.

Akimoto, T. (2012). Distributional response to biases in deterministic superdiffusion, *Phys. Rev. Lett.* **108**, p. 164101.

Akimoto, T. and Yamamoto, E. (2016). Distributional behaviors of time-averaged observables in the Langevin equation with fluctuating diffusivity: Normal diffusion but anomalous fluctuations, *Phys. Rev. E* **93**, p. 062109.

Allegrini, P., Aquino, G., Grigolini, P., Palatella, L., Rosa, A., and West, B. J. (2005). Correlation function and generalized master equation of arbitrary age, *Phys. Rev. E* **71**, p. 066109.

Altshuler, B. L., Aronov, A. G., and Khmelnitsky, D. E. (1982). Effects of electron-electron collisions with small energy transfers on quantum localisation, *J. Phys. C* **15**, pp. 7367–7386.

Applebaum, D. (2009). *Lévy Processes and Stochastic Calculus, Cambridge Studies in Advanced Mathematics*, Vol. 116, 2nd edn. (Cambridge University Press, Cambridge).

Baeumer, B. and Meerschaert, M. M. (2010). Tempered stable Lévy motion and transient super-diffusion, *J. Comput. Appl. Math.* **233**, pp. 2438–2448.

Baldassarri, A., Bouchaud, J. P., Dornic, I., and Godrèche, C. (1999). Statistics of persistent events: An exactly soluble model, *Phys. Rev. E* **59**, pp. R20–R23.

Bar-Haim, A. and Klafter, J. (1998). On mean residence and first passage times in finite one-dimensional systems, *J. Chem. Phys.* **109**, pp. 5187–5193.

Barkai, E. (2001). Fractional Fokker-Planck equation, solution, and application, *Phys. Rev. E* **63**, p. 046118.

Barkai, E. (2002). CTRW pathways to the fractional diffusion equation, *Chem. Phys.* **284**, pp. 13–27.

Barkai, E. (2006). Residence time statistics for normal and fractional diffusion in a force field, *J. Stat. Phys.* **123**, pp. 883–907.

Barkai, E. (2007). Strong correlations between fluctuations and response in aging transport, *Phys. Rev. E* **75**, p. 060104.

Barkai, E., Aghion, E., and Kessler, D. A. (2014). From the area under the Bessel excursion to anomalous diffusion of cold atoms, *Phys. Rev. X* **4**, p. 021036.

Barkai, E. and Cheng, Y. C. (2003). Aging continuous time random walks, *J. Chem. Phys.* **118**, pp. 6167–6178.

Baule, A. and Friedrich, R. (2006). Investigation of a generalized Obukhov model for turbulence, *Phys. Lett. A* **350**, pp. 167–173.

Bel, G. and Barkai, E. (2005). Weak ergodicity breaking in the continuous-time random walk, *Phys. Rev. Lett.* **94**, p. 240602.

Bel, G. and Barkai, E. (2006). Weak ergodicity breaking with deterministic dynamics, *EPL* **74**, pp. 15–21.

Berg, C. and Forst, G. (1975). *Potential Theory on Locally Compact Abelian Groups* (Springer, Berlin).

Bertin, E. M. and Bouchaud, J.-P. (2003). Linear and nonlinear response in the aging regime of the one-dimensional trap model, *Phys. Rev. E* **67**, p. 065105.

Bertoin, J., Fujita, T., Roynette, B., and Yor, M. (2006). On a particular class of self-decomposable random variables: the durations of Bessel excursions straddling independent exponential times, *Probab. Math. Statist.* **26**, pp. 315–366.

Bianco, S., Grigolini, P., and Paradisi, P. (2005). Fluorescence intermittency in blinking quantum dots: Renewal or slow modulation? *J. Chem. Phys.* **123**, p. 174704.

Boettcher, S., Robe, D. M., and Sibani, P. (2018). Aging is a log-Poisson process, not a renewal process, *Phys. Rev. E* **98**, p. 020602.

Bouchaud, J.-P. and Georges, A. (1990). Anomalous diffusion in disordered media: Statistical mechanisms, models and physical applications, *Phys. Rep.* **195**, pp. 127–293.

Brokmann, X., Hermier, J.-P., Messin, G., Desbiolles, P., Bouchaud, J.-P., and Dahan, M. (2003). Statistical aging and nonergodicity in the fluorescence of single nanocrystals, *Phys. Rev. Lett.* **90**, p. 120601.

Bruno, R., Sorriso-Valvo, L., Carbone, V., and Bavassano, B. (2004). A possible truncated-Lévy-flight statistics recovered from interplanetary solar-wind velocity and magnetic-field fluctuations, *EPL* **66**, pp. 146–152.

Buonocore, A., Nobile, A. G., and Ricciardi, L. M. (1987). A new integral equation for the evaluation of first-passage-time probability densities, *Adv. Appl. Prob.* **19**, pp. 784–800.

Burov, S. and Barkai, E. (2012). Weak subordination breaking for the quenched trap model, *Phys. Rev. E* **86**, p. 041137.

Cairoli, A. and Baule, A. (2015). Anomalous processes with general waiting times: Functionals and multipoint structure, *Phys. Rev. Lett.* **115**, p. 110601.

Cairoli, A. and Baule, A. (2017). Feynman-Kac equation for anomalous processes with space- and time-dependent forces, *J. Phys. A: Math. Theor.* **50**, p. 164002.

Carmi, S. and Barkai, E. (2011). Fractional Feynman-Kac equation for weak ergodicity breaking, *Phys. Rev. E* **84**, p. 061104.

Carmi, S., Turgeman, L., and Barkai, E. (2010). On distributions of functionals of anomalous diffusion paths, *J. Stat. Phys.* **141**, pp. 1071–1092.

Cartea, A. and del Castillo-Negrete, D. (2007). Fluid limit of the continuous-time random walk with general Lévy jump distribution functions, *Phys. Rev. E* **76**, p. 041105.

Chambers, J. M., Mallows, C. L., and Stuck, B. W. (1976). A method for simulating stable random variables, *J. Amer. Statist. Assoc.* **71**, pp. 340–344.

Chechkin, A. V., Gorenflo, R., and Sokolov, I. M. (2005). Fractional diffusion in inhomogeneous media, *J. Phys. A: Math. Gen.* **38**, pp. L679–L684.

Chen, M. H. and Deng, W. H. (2015). Discretized fractional substantial calculus, *ESAIM Math. Model. Numer. Anal.* **49**, pp. 373–394.

Cherstvy, A. G. and Metzler, R. (2013). Population splitting, trapping, and non-ergodicity in heterogeneous diffusion processes, *Phys. Chem. Chem. Phys.* **15**, pp. 20220–20235.

Chung, K. L. (1976). Excursions in Brownian motion, *Ark. Mat.* **14**, pp. 155–177.

Cifarelli, D. M. and Regazzini, E. (1975). Contributi intorno ad un test per lhomogeneita tra du campioni. *Giornale Degli Economiste* **34**, pp. 233–249.

Compte, A. (1996). Stochastic foundations of fractional dynamics. *Phys. Rev. E* **53**, pp. 4191–4193.

Darling, D. A. and Kac, M. (1957). On occupation times for Markoff processes, *Trans. Amer. Math. Soc.* **84**, pp. 444–458.

del Castillo-Negrete, D. (2009). Truncation effects in superdiffusive front propagation with Lévy flights, *Phys. Rev. E* **79**, p. 031120.

Deng, W. H., Chen, M. H., and Barkai, E. (2015). Numerical algorithms for the forward and backward fractional Feynman–Kac equations, *J. Sci. Comput.* **62**, pp. 718–746.

Deng, W. H., Li, B. Y., Tian, W. Y., and Zhang, P. W. (2018). Boundary problems for the fractional and tempered fractional operators, *Multiscale Model. Simul.* **16**, pp. 125–149.

Deng, W. H., Wang, W. L., Tian, X. C., and Wu, Y. J. (2016). Effects of the tempered aging and the corresponding Fokker–Planck equation, *J. Stat. Phys.* **164**, pp. 377–398.

Deng, W. H., Wu, X. C., and Wang, W. L. (2017). Mean exit time and escape probability for the anomalous processes with the tempered power-law waiting times, *EPL* **117**, p. 10009.

Deng, W. H. and Zhang, Z. J. (2019). *High Accuracy Algorithm for the Differential Equations Governing Anomalous Diffusion: Algorithm and Models for Anomalous Diffusion* (World Scientific, Singapore).

Denisov, S. I., Horsthemke, W., and Hänggi, P. (2009). Generalized Fokker-Planck equation: Derivation and exact solutions, *Eur. Phys. J. B* **68**, pp. 567–575.

Diethelm, K., Ford, N. J., Freed, A. D., and Luchko, Y. (2005). Algorithms for the fractional calculus: a selection of numerical methods, *Comput. Methods Appl. Mech. Engrg.* **194**, pp. 743–773.

Dybiec, B. and Sokolov, I. M. (2015). Estimation of the smallest eigenvalue in fractional escape problems: Semi-analytics and fits, *Comput. Phys. Comm.* **187**, pp. 29–37.

Dyke, P. (2014). *An Introduction to Laplace Transforms and Fourier Series*, Springer Undergraduate Mathematics Series (Springer, London).

Dynkin, E. B. (1961). *Selected Translations in Mathematical Statistics and Probability* (American Mathematical Society, Providence, Providence, RI).

Edery, Y., Berg, S., and Weitz, D. (2018). Surfactant variations in porous media localize capillary instabilities during haines jumps, *Phys. Rev. Lett.* **120**, p. 028005.

Erdélyi, A., Magnus, W., Oberhettinger, F., and Tricomi, F. G. (1954). *Tables of Integral Transforms* (McGraw-Hill, New York).

Erdélyi, A., Magnus, W., Oberhettinger, F., and Tricomi, F. G. (1955). *Higher Transcedential Functions* (McGraw-Hill, New York).

Fedotov, S. (2010). Non-Markovian random walks and nonlinear reactions: Subdiffusion and propagating fronts, *Phys. Rev. E* **81**, p. 011117.

Fedotov, S., Al-Shamsi, H., Ivanov, A., and Zubarev, A. (2010). Anomalous transport and nonlinear reactions in spiny dendrites, *Phys. Rev. E* **82**, p. 041103.

Feller, W. (1971). *An Introduction to Probability Theory and Its Applications*, Vol. II (John Wiley & Sons, New York).

Fischer Black, M. S. (1973). The pricing of options and corporate liabilities, *J. Political Econ.* **81**, pp. 637–654.

Fogedby, H. C. (1994). Langevin equations for continuous time Lévy flights, *Phys. Rev. E* **50**, pp. 1657–1660.

Friedrich, R., Jenko, F., Baule, A., and Eule, S. (2006). Anomalous diffusion of inertial, weakly damped particles, *Phys. Rev. Lett.* **96**, p. 230601.

Froemberg, D. and Barkai, E. (2013). Time-averaged Einstein relation and fluctuating diffusivities for the Lévy walk, *Phys. Rev. E* **87**, p. 030104.

Froemberg, D., Schmidt-Martens, H. H., Sokolov, I. M., and Sagués, F. (2011). Asymptotic front behavior in an $A + B \to 2A$ reaction under subdiffusion, *Phys. Rev. E* **83**, p. 031101.

Gajda, J. and Magdziarz, M. (2010). Fractional Fokker-Planck equation with tempered α-stable waiting times: Langevin picture and computer simulation, *Phys. Rev. E* **82**, p. 011117.

Garra, R., Gorenflo, R., Polito, F., and Živorad Tomovski (2014). Hilfer-Prabhakar derivatives and some applications, *Appl. Math. Comput.* **242**, pp. 576–589.

Getoor, R. K. and Sharpe, M. J. (1979). Excursions of Brownian motion and Bessel processes, *Z. Wahrsch. Verw. Gebiete* **47**, pp. 83–106.

Godrèche, C. and Luck, J. M. (2001). Statistics of the occupation time of renewal processes, *J. Stat. Phys.* **104**, pp. 489–524.

Gorenflo, R., Mainardi, F., and Vivoli, A. (2007). Continuous-time random walk and parametric subordination in fractional diffusion, *Chaos Solitons Fractals* **34**, pp. 87–103.

Gradshteyn, I. S. and Ryzhik, I. M. (1980). *Table of Integrals, Series, and Products* (Academic Press, USA).

Grebenkov, D. S. (2007). NMR survey of reflected Brownian motion, *Rev. Mod. Phys.* **79**, pp. 1077–1137.

Heinsalu, E., Patriarca, M., Goychuk, I., and Hänggi, P. (2007). Use and abuse of a fractional Fokker-Planck dynamics for time-dependent driving, *Phys. Rev. Lett.* **99**, p. 120602.

Henry, B. I., Langlands, T. A. M., and Wearne, S. L. (2006). Anomalous diffusion with linear reaction dynamics: From continuous time random walks to fractional reaction-diffusion equations, *Phys. Rev. E* **74**, p. 031116.

Hilfer, R. (2000). *Applications of Fractional Calculus in Physics* (World Scientific, Singapore).

Hou, R. and Deng, W. H. (2018). Feynman-Kac equations for reaction and diffusion processes, *J. Phys. A* **51**, p. 155001.

Iomin, A. and Méndez, V. (2013). Reaction-subdiffusion front propagation in a comblike model of spiny dendrites, *Phys. Rev. E* **88**, p. 012706.

Iomin, A. and Sokolov, I. M. (2012). Application of hyperbolic scaling for calculation of reaction-subdiffusion front propagation, *Phys. Rev. E* **86**, p. 022101.

Itô, K. (1950). Stochastic differential equations in a differentiable manifold, *Nagoya Math. J.* **1**, pp. 35–47.

Jespersen, S., Metzler, R., and Fogedby, H. C. (1999). Lévy flights in external force fields: Langevin and fractional Fokker-Planck equations and their solutions, *Phys. Rev. E* **59**, pp. 2736–2745.

Jose, K. K., Uma, P., Lekshmi, V. S., and Haubold, H. J. (2010). Generalized Mittag-Leffler distributions and processes for applications in astrophysics and time series modeling, in *Proceedings of the Third UN/ESA/NASA Workshop on the International Heliophysical Year 2007 and Basic Space Science*, Astrophys. Space Sci. Proc., pp. 79–92.

Kac, M. (1949). On distributions of certain Wiener functionals, *Trans. Amer. Math. Soc.* **65**, pp. 1–13.

Kac, M. (1951). On some connections between probability theory and differential and integral equations, in *Proceedings of the Second Berkeley Symposium on Mathematical Statistics and Probability* (University of California Press, Berkeley), pp. 189–215.

Kenkre, V. M., Montroll, E. W., and Shlesinger, M. F. (1973). Generalized master equations for continuous-time random walks, *J. Stat. Phys.* **9**, pp. 45–50.

Klafter, J. and Sokolov, I. M. (2011). *First Steps in Random Walks: From Tools to Applications* (Oxford University Press, Oxford).

Klafter, J. and Zumofen, G. (1994). Probability distributions for continuous-time random walks with long tails, *J. Phys. Chem.* **98**, pp. 7366–7370.

Korabel, N. and Barkai, E. (2009). Pesin-type identity for intermittent dynamics with a zero Lyaponov exponent, *Phys. Rev. Lett.* **102**, p. 050601.

Kozubowski, T. J. (2001). Fractional moment estimation of Linnik and Mittag-Leffler parameters, *Math. Comput. Modelling* **34**, pp. 1023–1035.

Krüsemann, H., Godec, A., and Metzler, R. (2014). First-passage statistics for aging diffusion in systems with annealed and quenched disorder, *Phys. Rev. E* **89**, p. 040101.

Krüsemann, H., Schwarzl, R., and Metzler, R. (2016). Ageing Scher–Montroll transport, *Transp. Porous Media* **115**, pp. 327–344.

Kubo, R. (1966). The fluctuation-dissipation theorem, *Rep. Progr. Phys.* **29**, pp. 255–284.

Kutner, R. and Masoliver, J. (2017). The continuous time random walk, still trendy: fifty-year history, state of art and outlook, *Eur. Phys. J. B* **90**, p. 50.

Lamperti, J. (1958). An occupation time theorem for a class of stochastic processes, *Trans. Amer. Math. Soc.* **88**, pp. 380–387.

Langlands, T., Henry, B., and Wearne, S. (2008). Anomalous subdiffusion with multispecies linear reaction dynamics, *Phys. Rev. E* **77**, p. 021111.

Lévy, P. (1940). Sur certains processus stochastiques homogènes, *Compositio Math.* **7**, pp. 283–339.

Luchinin, A. G. and Dolin, L. S. (2014). Application of complex-modulated waves of photon density for instrumental vision in turbid media, *Dokl. Phys.* **59**, pp. 170–172.

Magdziarz, M. and Weron, A. (2007). Numerical approach to the fractional Klein-Kramers equation, *Phys. Rev. E* **76**, p. 066708.

Magdziarz, M., Weron, A., and Klafter, J. (2006). Anomalous diffusion schemes underlying the cole cole relaxation: The role of the inverse-time α-stable subordinator, *Physica A* **367**, pp. 1–6.

Magdziarz, M., Weron, A., and Klafter, J. (2008). Equivalence of the fractional Fokker-Planck and subordinated Langevin equations: the case of a time-dependent force, *Phys. Rev. Lett.* **101**, p. 210601.

Mainardi, F. (2010). *Fractional Calculus and Waves in Linear Viscoelesticity: An Introduction to Mathematical Models* (Imperial College Press, London).

Mainardi, F., Gorenflo, R., and Scalas, E. (2004). A fractional generalization of the Poisson processes, *Vietnam J. Math.* **32**, Special Issue, pp. 53–64.

Mainardi, F., Gorenflo, R., and Vivoli, A. (2007). Beyond the Poisson renewal process: a tutorial survey, *J. Comput. Appl. Math.* **205**, pp. 725–735.

Majumdar, S. N. (1999). Persistence in nonequilibrium systems, *Current Sci.* **77**, pp. 370–375.

Majumdar, S. N. (2005). Brownian functionals in physics and computer science, *Current Sci.* **89**, pp. 2076–2092.

Majumdar, S. N. and Bray, A. J. (2002). Large-deviation functions for nonlinear functionals of a Gaussian stationary Markov process, *Phys. Rev. E* **65**, p. 051112.

Majumdar, S. N. and Comtet, A. (2002). Local and occupation time of a particle diffusing in a random medium, *Phys. Rev. Lett.* **89**, p. 060601.

Mandelbrot, B. B. (1999). *Multifractals and 1/f Noise: Wild Self-Affinity in*

Physics (Springer, Berlin).

Mantegna, R. N. and Stanley, H. E. (1994). Stochastic process with ultraslow convergence to a Gaussian: The truncated Lévy flight, *Phys. Rev. Lett.* **73**, pp. 2946–2949.

Margolin, G., Protasenko, V., Kuno, M., and Barkai, E. (2005). *Power-Law Blinking Quantum Dots: Stochastic and Physical Models*, chap. 4 (Wiley, Blackwell), pp. 327–356.

Mathai, A. M. and Saxena, R. K. (1978). *The H-Function with Applications in Statistics and other Disciplines* (Halsted Press, New York).

Meerschaert, M. M., Sabzikar, F., Phanikumar, M. S., and Zeleke, A. (2014). Tempered fractional time series model for turbulence in geophysical flows, *J. Stat. Mech. Theory Exp.* **2014**, p. 09023.

Meerschaert, M. M. and Sikorskii, A. (2012). *Stochastic Models for Fractional Calculus, De Gruyter Studies in Mathematics*, Vol. 43 (Walter de Gruyter & Co., Berlin).

Meerschaert, M. M., Zhang, Y., and Baeumer, B. (2008). Tempered anomalous diffusion in heterogeneous systems, *Geophys. Res. Lett.* **35**, p. 17403.

Mendez, V., Fedotov, S., and Horsthemke, W. (2010). *Reaction-Transport Systems: Mesoscopic Foundations, Fronts, and Spatial Instabilities* (Springer Science & Business Media, Berlin).

Metzler, R., Barkai, E., and Klafter, J. (1999a). Anomalous diffusion and relaxation close to thermal equilibrium: A fractional Fokker-Planck equation approach, *Phys. Rev. Lett.* **82**, pp. 3563–3567.

Metzler, R., Barkai, E., and Klafter, J. (1999b). Deriving fractional Fokker-Planck equations from a generalised master equation, *EPL* **46**, pp. 431–436.

Metzler, R., Jeon, J.-H., Cherstvy, A. G., and Barkai, E. (2014). Anomalous diffusion models and their properties: non-stationarity, non-ergodicity, and ageing at the centenary of single particle tracking, *Phys. Chem. Chem. Phys.* **16**, pp. 24128–24164.

Metzler, R. and Klafter, J. (2000). The random walk's guide to anomalous diffusion: A fractional dynamics approach, *Phys. Rep.* **339**, pp. 1–77.

Metzler, R. and Klafter, J. (2004). The restaurant at the end of the random walk: recent developments in the description of anomalous transport by fractional dynamics, *J. Phys. A: Math Theor.* **37**, pp. 161–208.

Miyaguchi, T., Akimoto, T., and Yamamoto, E. (2016). Langevin equation with fluctuating diffusivity: A two-state model, *Phys. Rev. E* **94**, p. 012109.

Monthus, C. and Bouchaud, J.-P. (1996). Models of traps and glass phenomenology, *J. Phys. A: Math Theor.* **29**, pp. 3847–3869.

Montroll, E. W. and Weiss, G. H. (1965). Random walks on lattices. II, *J. Math. Phys.* **6**, pp. 167–181.

Mörters, P. and Peres, Y. (2010). *Brownian Motion*, Vol. 30 (Cambridge University Press, Cambridge).

Niemann, M., Barkai, E., and Kantz, H. (2016). Renewal theory for a system with internal states, *Math. Model. Nat. Phenom.* **11**, pp. 191–239.

Nyberg, M., Lizana, L., and Ambjörnsson, T. (2018). Zero-crossing statistics for non-Markovian time series, *Phys. Rev. E* **97**, p. 032114.

Oldham, K. B. and Spanier, J. (1974). *The Fractional Calculus* (Academic Press, California).

Paneva-Konovska, J. (2014). Convergence of series in three parametric Mittag-Leffler functions, *Math. Slovaca* **64**, pp. 73–84.

Papoulis, A. (1984). *Probability, Random Variables, and Stochastic Processes*, 2nd edn., McGraw-Hill Series in Electrical Engineering. Communications and Information Theory (McGraw-Hill Book Co., New York).

Perret, A., Comtet, A., Majumdar, S. N., and Schehr, G. (2015). On certain functionals of the maximum of Brownian motion and their applications, *J. Stat. Phys.* **161**, pp. 1112–1154.

Piryatinska, A., Saichev, A. I., and Woyczynski, W. A. (2005). Models of anomalous diffusion: the subdiffusive case, *Physica A* **349**, pp. 375–420.

Pitman, J. (1999). *The Distribution of Local Times of a Brownian Bridge* (Springer, Berlin).

Podlubny, I. (1999). *Fractional Differential Equations* (Academic Press, San Diego).

Polyanin, A. D. and Manzhirov, A. V. (2007). *Handbook of Mathematics for Engineers and Scientists* (Chapman & Hall/CRC, Boca Raton).

Polyanin, A. D., Zaitsev, V. F., and Moussiaux, A. (2002). *Handbook of First-order Partial Differential Equations* (Taylor & Francis, London).

Prabhakar, T. R. (1971). A singular integral equation with a generalized Mittag Leffler function in the kernel, *Yokohama Math. J.* **19**, pp. 7–15.

Rebenshtok, A. and Barkai, E. (2008). Weakly non-ergodic statistical physics, *J. Stat. Phys.* **133**, pp. 565–586.

Rebenshtok, A., Denisov, S., Hänggi, P., and Barkai, E. (2014a). Infinite densities for Lévy walks, *Phys. Rev. E* **90**, p. 062135.

Rebenshtok, A., Denisov, S., Hänggi, P., and Barkai, E. (2014b). Non-normalizable densities in strong anomalous diffusion: Beyond the central limit theorem, *Phys. Rev. Lett.* **112**, p. 110601.

Redner, S. (2001). *A Guide to First-Passage Processes* (Cambridge University Press, Cambridge).

Risken, H. (1989). *The Fokker-Planck Equation* (Springer-Verlag, Berlin).

Robert, C. P. and Casella, G. (2004). *Monte Carlo Statistical Methods* (Springer-Verlag, New York).

Sadhu, T., Delorme, M., and Wiese, K. J. (2018). Generalized arcsine laws for fractional Brownian motion, *Phys. Rev. Lett.* **120**, p. 040603.

Saichev, A. I. and Zaslavsky, G. M. (1997). Fractional Kinetic equations: solutions and applications, *Chaos* **7**, pp. 753–764.

Samko, S. G., Kilbas, A. A., and Marichev, O. I. (1993). *Fractional Integrals and Derivatives - Theory and Applications* (Gordon and Breach, New York).

Sandev, T. (2017). Generalized Langevin equation and the Prabhakar derivative, *Mathematics* **5**, p. 66.

Sandev, T., Živorad Tomovski, and Dubbeldam, J. L. (2011). Generalized Langevin equation with a three parameter Mittag-Leffler noise, *Physica A* **390**, 21, pp. 3627–3636.

Schehr, G. and Le Doussal, P. (2010). Extreme value statistics from the real space renormalization group: Brownian motion, bessel processes and continuous time random walks, *J. Stat. Mech: Theory Exp.* **2010**, p. P01009.

Scher, H. and Lax, M. (1972). Continuous time random walk model of hopping transport: Application to impurity conduction, *J. Non-Cryst. Solids* **8**, pp. 497–504.

Schilling, R., Song, R., and Vondracek, Z. (2010). *Bernstein Functions* (De Gruyter, Berlin).

Schmidt, M., Sagués, F., and Sokolov, I. (2007). Mesoscopic description of reactions for anomalous diffusion: a case study, *J. Phys.: Condens. Matter* **19**, p. 065118.

Schneider, W. R. and Wyss, W. (1989). Fractional diffusion and wave equations, *J. Math. Phys.* **30**, pp. 134–144.

Schulz, J. H. P. and Barkai, E. (2015). Fluctuations around equilibrium laws in ergodic continuous-time random walks, *Phys. Rev. E* **91**, p. 062129.

Schulz, J. H. P., Barkai, E., and Metzler, R. (2013). Aging effects and population splitting in single-particle trajectory averages, *Phys. Rev. Lett.* **110**, p. 020602.

Schulz, J. H. P., Barkai, E., and Metzler, R. (2014). Aging renewal theory and application to random walks, *Phys. Rev. X* **4**, p. 011028.

Seaborn, J. B. (1991). *Hypergeometric Functions and Their Applications*, Vol. 8 (Springer-Verlag, New York).

Shemer, Z. and Barkai, E. (2009). Einstein relation and effective temperature for systems with quenched disorder, *Phys. Rev. E* **80**, p. 031108.

Shepp, L. A. (1982). On the integral of the absolute value of the pinned Wiener process, *Ann. Probab.* **10**, pp. 234–239.

Sokolov, I., Chechkin, A., and Klafter, J. (2004). Fractional diffusion equation for a power-law-truncated Lévy process, *Physica A* **336**, pp. 245–251.

Sokolov, I. M., Schmidt, M., and Sagués, F. (2006). Reaction-subdiffusion equations, *Phys. Rev. E* **73**, p. 031102.

Stanislavsky, A., Weron, K., and Weron, A. (2008). Diffusion and relaxation controlled by tempered α-stable processes, *Phys. Rev. E* **78**, p. 051106.

Takeuchi, K. A. and Akimoto, T. (2016). Characteristic sign renewals of Kardar–Parisi–Zhang fluctuations, *J. Stat. Phys.* **164**, pp. 1167–1182.

Thaler, M. (2002). A limit theorem for sojourns near indifferent fixed points of one-dimensional maps, *Ergodic Theory Dynam. Systems* **22**, pp. 1289–1312.

Thaler, M. and Zweimüller, R. (2006). Distributional limit theorems in infinite ergodic theory, *Probab. Theory Related Fields* **135**, pp. 15–52.

Touchette, H. (2009). The large deviation approach to statistical mechanics, *Phys. Rep.* **478**, pp. 1–69.

Tuckwell, H. C. and Wan, F. Y. (1984). First-passage time of Markov process to moving barriers, *J. Appl. Prob.* **21**, pp. 695–709.

Tunaley, J. K. E. (1974). Asymptotic solutions of the continuous-time random walk model of diffusion, *J. Stat. Phys.* **11**, pp. 397–408.

Turgeman, L., Carmi, S., and Barkai, E. (2009). Fractional Feynman-Kac equation for non-Brownian functionals, *Phys. Rev. Lett.* **103**, p. 190201.

Wang, W. L. and Deng, W. H. (2018). Aging Feynman-Kac equation, *J. Phys. A*

51, p. 015001.

Wang, W. L., Schulz, J. H. P., Deng, W. H., and Barkai, E. (2018a). Renewal theory with fat-tailed distributed sojourn times: Typical versus rare, *Phys. Rev. E* **98**, p. 042139.

Wang, X. D., Chen, Y., and Deng, W. H. (2018b). Feynman-Kac equation revisited, *Phys. Rev. E* **98**, p. 052114.

Weiss, G. H. (1994). *Aspects and Applications of the Random Walk*, Random Materials and Processes (North-Holland Publishing Co., Amsterdam).

Weron, A., Burnecki, K., Akin, E. J., Solé, L., Balcerek, M., Tamkun, M. M., and Krapf, D. (2017). Ergodicity breaking on the neuronal surface emerges from random switching between diffusive states, *Sci. Rep.* **7**, p. 5404.

Whitelam, S. (2018). Large deviations in the presence of cooperativity and slow dynamics, *Phys. Rev. E* **97**, p. 062109.

Wu, X. C., Deng, W. H., and Barkai, E. (2016). Tempered fractional Feynman-Kac equation: theory and examples, *Phys. Rev. E* **93**, p. 032151.

Xu, P. B. and Deng, W. H. (2018). Lévy walk with multiple internal states, *J. Stat. Phys.* **173**, pp. 1598–1613.

Yadav, A., Fedotov, S., Méndez, V., and Horsthemke, W. (2007). Propagating fronts in reaction-transport systems with memory, *Phys. Lett. A* **371**, pp. 374–378.

Yor, M. (2001). *Exponential Functionals of Brownian Motion and Disordered Systems* (Springer Berlin Heidelberg, Berlin), pp. 182–203.

Zaburdaev, V., Denisov, S., and Klafter, J. (2015). Lévy walks, *Rev. Mod. Phys.* **87**, pp. 483–530.

Index